Evaluating Climate Change Action for Sustainable Development

Juha I. Uitto • Jyotsna Puri
Rob D. van den Berg

Editors

Evaluating Climate Change Action for Sustainable Development

Editors
Juha I. Uitto
Independent Evaluation Office
Global Environment Facility
Washington, DC, USA

Jyotsna Puri
International Initiative for Impact
   Evaluation (3ie)
New Delhi, India

Rob D. van den Berg
King's College London
   and International Development
   Evaluation Association (IDEAS)
Leidschendam, The Netherlands

ISBN 978-3-319-43701-9   ISBN 978-3-319-43702-6   (eBook)
DOI 10.1007/978-3-319-43702-6

Library of Congress Control Number: 2016957874

Foreword

In 2015, the world achieved several crucial milestones for safeguarding the global commons. Most prominently were the adoption of the Sustainable Development Goals and the Paris Agreement on Climate Change. These agreements clearly recognize that the health of the global commons, like land, forests, oceans, and climate, is vital for our future development prospects – a recognition that was embedded in the very creation of the Global Environment Facility (GEF) 25 years ago. It is very timely to reflect on what we have achieved, what is still to be done, and how we can scale up our efforts in helping countries along the way to implementing these agreements.

Thanks to all the support we received from donor countries, partners, and other stakeholders, the GEF was able to invest over US$14 billion in grants and mobilized over US$74 billion in additional financing for over 4,000 projects in developing countries. Still, the key drivers of environmental degradation continue to intensify with a growing, and more affluent, global population and rapid urbanization, driving increased demand for food, fiber and materials. The associated pressures on forests, land, and oceans are increasingly being exacerbated by climate change, thereby threatening biodiversity and Earth's life support systems. If we are to succeed at the scale the problem deserves, we need to change key economic systems – how we produce food; our cities; how we live and move around; and our energy system, how we power our vehicles, industries, and homes. In a nutshell, we have our work cut out for ourselves as never before.

Fortunately, we are not flying in the dark. The GEF stands today on top of a quarter-century of experience dealing with the global environment and the stewardship of the global commons. This body of work offers a tremendous opportunity to learn – both from our successes and our failures. This is what makes the continuous, uninterrupted evaluation and assessments by GEF's Independent

Evaluation Office so important. If we are to successfully implement the landmark 2015 global agreements to put the world on a low-emission, climate-resilient, and sustainable trajectory it is an imperative to embrace evidence-based learning and monitoring and evaluation as an integral aspect of implementation, so that our approaches and priorities can be refined and optimized even when we are already on the road.

A few lessons are already emerging. First, we need to rapidly break down the sectoral walls that isolate the environment from economics – at the international and national scales – so as to start mainstreaming environmental considerations in the wider decision-making process. Second, we must bring the various lines of funding the GEF provides countries much closer together, making the best use possible of their interlinkages and addressing the systemic nature of the threats. Finally, we must move from just addressing the symptoms of environmental degradation – given we are running out of fingers to stick in the progressively more porous proverbial dyke – and start tackling the key drivers of environmental degradation.

Climate change is arguably one of the most complex of the global environmental phenomena, with its root causes ingrained across almost all sectors and industries. Agriculture, for instance, accounts for a significant share of global greenhouse gas emissions, including through methane emissions from livestock, nitrous oxide from fertilizer use, and land use change. Cattle, palm oil, and rice together contribute approximately half of all food production-related greenhouse gas emissions, requiring solution approaches that take into account the linkages between these individual emission sources. Learning from our past agriculture projects will help us tackle these emission sources at the systemic level.

Against this background, we also have to address the fact we are already locked in a path towards a warmer world irrespective of what we may do today. The Paris Agreement has embraced the utmost priority of promoting greater resilience in the ways we conduct our business and daily lives. This will be particularly important for the poorest and most vulnerable countries. We are confident we can also build our future work on the significant portfolio of climate adaptation work that has accumulated important lessons over the past decade – meaning promoting and replicating approaches that are most effective in helping communities adapt to droughts, sea level rise, and changing seasonal weather patterns.

This book could not come at a better time. We need to make choices that direct our limited resources to their best use on the ground – so as to most effectively help countries fulfill their commitments to achieving the goals and objectives of the 2030 Agenda for Sustainable Development. Evaluations and lessons learned from past projects and programs must be considered by all stakeholders involved, to enable informed decision-making to do justice to the urgency of the problems at hand. The efforts of the Independent Evaluation Office of the GEF to produce useful, concrete. and practical lessons, exemplified in this book and through the success of the wider Climate-Eval network, will help. I hope that this fresh body of knowledge captured hereinafter will reach an audience beyond the GEF and by doing so undoubtedly become useful to everyone in the broader environment and development community.

CEO and Chairperson   Naoko Ishii
Global Environment Facility
Washington, DC, USA

Foreword

Climate change is among the most difficult challenges facing the world. Its global nature, intergenerational impact, and the massive risks and uncertainty associated with it combine to create an unparalleled need for global collective action. It is also significant in that addressing it will go a very long way toward addressing other environmental problems – air pollution, water risks, soil degradation, and the loss of forests, natural habitats, and biodiversity. At the same time, the world continues to grapple with eradicating poverty and inequalities and spurring economic growth. Going full circle, it is primarily the poor who suffer from climate change and environmental degradation.

Monitoring and evaluating the efforts to address these concerns are particularly important. The stakes are high, and we have an incredibly short window to do things right. And in a time when public financing is decreasing and investment decisions are made without considering overall global environmental and development costs, evaluation is essential for us to understand how we can best make use of these limited resources. Evaluators are in a position to present evidence about how we can make a difference in promoting development that is both environmentally sound and equitable.

Using relatively small resources, what can be done to help decision-makers achieve the revolution needed to address our challenges? Those who have the privilege to manage precious resources have an obligation to help decision-makers get to the right place. Using evidence, evaluators can speak truth to power. It is not only about how money is being used and whether it has a decent return. It is about measuring the results of our action, knowing why and how things are working, or not working. It is about learning.

The book *Evaluating Climate Change Action for Sustainable Development* is a collaborative venture. The chapters provide an interdisciplinary perspective and document emerging and innovative evaluation knowledge and practice of climate change and its links to sustainable development. Such knowledge based on solid analysis of experiences in the field is indispensable as we move forward. This book is a welcome addition to the literature.

We are still a long way from achieving a resilient and low-carbon economy. We have a role and responsibility to help find solutions to our common global challenges. We, as individuals, have an impact much bigger than we realize. With passion and grit, let us all think bigger and out of the box to make the world a better place, not only for this generation but for the future as well.

President and CEO   Andrew Steer
World Resources Institute
Washington, DC, USA

Preface

Climate change is one of the preeminent challenges facing the world today. The consequences of climate change manifest themselves in multiple ways, including increased variability and intensity of extreme weather events and sea level rise. We are already seeing the impacts of climate change, and the first ones to feel them tend to be poor people and poor countries that are most vulnerable and have the least capacities to cope with them. The search for solutions to mitigate climate change and to adapt to its consequences is urgent. Rigorous evaluation of policies, programs, and projects can help the international community to identify technical solutions, economic strategies, and social innovations that improve our ability to deal with climate change. This is the focus of the present book.

The book has its genesis in the Climate-Eval Community of Practice hosted by the Independent Evaluation Office of the Global Environment Facility. Its overarching goal is to establish standards and norms, support capacity development, and share good practices in evaluations of climate change and development. In November 2014, Climate-Eval with its partners organized the Second International Conference on Evaluating Climate Change and Development in Washington, D.C. The aim of this event was to promote an interdisciplinary exchange of ideas and methods to evaluate climate change and sustainable development. This 3-day event brought together some 300 leading experts and policymakers in the field and included sessions on climate change mitigation, adaptation, and policy, as well as special sessions such as panel discussions and roundtables. Topics discussed ranged from theory of change approaches to evaluation and institutional capacity; to disaster risk reduction, resilience, and tracking adaptation; to monitoring and evaluation of ecosystem-based and natural resource management interventions and climate change funds.

Evaluating Climate Change Action for Sustainable Development builds upon a selection of the most relevant and practical papers and presentations given at the 2014 conference. Following the conference, the editors and the authors worked closely together to develop the presentations into a coherent set of articles organized around the three main themes of climate change evaluation: policy, mitigation, and adaptation. This book aims to provide an authoritative interdisciplinary perspective of innovative and emerging evaluation knowledge and practice around climate change and development. It focuses on lessons learned and gained from evaluating climate change projects, programs, and policies as they link to sustainable development, from the perspectives of international organizations, NGOs, multilateral and bilateral aid agencies, and the academia. Authors share methodologies and approaches used to better understand problems and assess interventions, strategies, and policies. They also share challenges encountered, what was done to solve these, and lessons learned from evaluations. Collectively, the authors illustrate the importance of evaluation in providing evidence to guide policy change and informed decision-making.

This book is written for policymakers, program and project proponents, practitioners, academics, and other informed audiences concerned with climate change, sustainable development, and evaluation.

Washington, DC, USA   Juha I. Uitto
New Delhi, India   Jyotsna Puri
Leidschendam, The Netherlands   Rob D. van den Berg

Acknowledgments

The editors wish to express their appreciation to the organizations and individuals who contributed valuable perspectives, ideas, and support throughout the course of this book's development. The book has its origins in the Climate-Eval Community of Practice hosted by the Independent Evaluation Office (IEO) of the Global Environment Facility (GEF) who organized the Second International Conference on Evaluating Climate Change and Development, in November 2014 in Washington, D.C. The Climate-Eval CoP and the conference were generously supported by the governments of Germany, Sweden, and Switzerland. Their contributions are noted with gratitude. We would also like to thank the Climate-Eval partners, the IEO of GEF, the African Evaluation Association (AfrEA), the International Program Evaluation Network (IPEN), and SEA Change, an Asian community of practice for the monitoring and evaluation of climate change interventions.

A number of experts and professional colleagues provided their support to the book by reviewing the manuscript and individual chapters at critical times: Dennis Bours, Carlo Carugi, Megan Kennedy-Chouane, Howard Stewart, Anna Viggh, and Molly Watts. The chapter contributors also kindly cross-reviewed each other's manuscripts during the preparation of the book. We would like to thank all of them for their valuable contributions. Juan Portillo and his team in the IEO of GEF provided important administrative support throughout the process.

We would specifically like to thank Lee Cando-Noordhuizen who acted as the assistant editor, keeping track of all details and ensuring that the book project was proceeding on schedule. Her professionalism and attention to detail were indispensable.

The Editors

Contents

Part II   Climate Change Mitigation

Part III   Climate Change Adaptation

Contributors

Tougiani Abasse National Agricultural Research Institute of Niger, Niamey, Niger

Aryanie Amellina Climate and Energy Area, Institute for Global Environmental Studies, Hayama, Japan

Babou André Bationo Environment and Agricultural Research Institute, Ouagadougou, Burkina Faso

Emilia Bretan World Bank, São Paulo, Brazil

Saaka Buah Savanna Agricultural Research Institute, Wa, Ghana

Lee Cando-Noordhuizen Independent Evaluation Office, Global Environment Facility, Washington, DC, USA

Michael Carbon Evaluation Office, United Nations Environment Programme, Nairobi, Kenya

Joanne Chong Institute for Sustainable Futures, University of Technology Sydney, Sydney, Australia

Monika Egger Kissling Evaluation and Corporate Controlling Division, Swiss Agency for Development and Cooperation, Berne, Switzerland

Nathan L. Engle Climate Policy Team, World Bank, Washington, DC, USA

Wiebke Förch Deutsche Gesellschaft für Internationale Zusammenarbeit (GIZ) GmbH, Gaborone, Botswana

Yann François GERES - Group for the Environment, Renewable Energy and Solidarity, Aubagne, France

Marina Gavaldão GERES - Group for the Environment, Renewable Energy and Solidarity, Aubagne, France

Anna Gero Institute for Sustainable Futures, University of Technology Sydney, Sydney, Australia

Jasmine Hyman School of Forestry & Environmental Studies, Yale University, New Haven, CT, USA

Irene Karani LTS Africa, Nairobi, Kenya

Nyachomba Kariuki LTS Africa, Nairobi, Kenya

Timo Leiter Climate Change and Climate Policy Group, Deutsche Gesellschaft für Internationale Zusammenarbeit (GIZ) GmbH, Eschborn, Germany

Debora Ley Environmental Change Institute, School of Geography and the Environment, University of Oxford, Oxford, UK

Takaaki Miyaguchi Ritsumeikan University, Kyoto, Japan

Neeraj Kumar Negi Global Environment Facility, Independent Evaluation Office, Washington, DC, USA

Jyotsna Puri International Initiative for Impact Evaluation (3ie), New Delhi, India

Issa Sawadogo Environment and Agricultural Research Institute, Ouagadougou, Burkina Faso

Tonya Schuetz Independent Consultant, Munich, Germany

Jacques Somda Planning, Monitoring, Evaluation and Learning, International Union for Conservation of Nature, Ouagadougou, Burkina Faso

Philip Thornton CGIAR Research Program on Climate Change, Agriculture and Food Security (CCAFS), International Livestock Research Institute (ILRI), Nairobi, Kenya

Pia Treichel Plan International, Melbourne, Australia

Juha I. Uitto Independent Evaluation Office, Global Environment Facility, Washington, DC, USA

Rob D. van den Berg King's College London and International Development Evaluation Association (IDEAS), Leidschendam, The Netherlands

Ioannis Vasileiou World Bank, Washington, DC, USA

Roman Windisch Quality and Resources Division WEQA, Swiss State Secretariat for Economic Affairs SECO, Berne, Switzerland

Aaron Zazueta Independent Consultant, New York, USA

Robert Zougmoré Africa Program, CGIAR Research Program on Climate Change, Agriculture and Food Security (CCAFS), Bamako, Mali

List of Acronyms

AAP	Africa Adaptation Programme
ADB	Asian Development Bank
AFD	Agence Française de Développement
APR	Ambiguous property rights
BAAC	Bank of Agriculture and Agricultural Credit
BMUB	German Federal Ministry for the Environment, Nature Conservation, Building and Nuclear Safety
BMZ	German Federal Ministry for Economic Cooperation and Development
CAF	County Adaptation Fund
CBD	Convention for Biological Diversity
CCAFS	CGIAR Research Program on Climate Change, Agriculture and Food Security
CC-CBA	Child-Centred Community-Based Adaptation
CCSP	Climate Change Subprogramme
CDM	Clean Development Mechanism
CDM	DOE CDM Designated Operational Entity
CDRF	Capacity Development Results Framework
CGIAR	Consultative Group for International Agricultural Research
CIF	Climate Investment Funds
CMO	Context + Mechanism = Outcome
COP	Conference of the Parties
CPR	Common property resources
CPWF	Challenge Program on Water and Food
CRGE	Climate-Resilient Green Economy
CRM	Climate risk management
CRP	CGIAR Research Programs
CSO	Civil society organizations
CTF	Clean Technology Fund
CWP	Ceramic water purifiers

DFID	Department for International Development
DOLA	Department of Local Administration
DPP	Drought preparedness plan
DRE	Decentralized Renewable Energy
DRE	Decentralized renewable energy
DRR	Disaster risk reduction
EWS	Early warning system
FAO	Food and Agriculture Organization
FAO	Food and Agriculture Organization
FCPF	Forest Carbon Partnership Facility
FGD	Focus group discussions
FSF	Fast-start financing
FSP	Full-sized projects
GACC	Global Alliance for Clean Cookstoves
GEF	Global Environment Facility
GEF	IEO Global Environment Facility Independent Evaluation Office
GERES	Group for the Environment, Renewable Energy and Solidarity, Groupe Energies Renouvelables, Environnement et Solidarités
GIZ	Deutsche Gesellschaft für Internationale Zusammenarbeit GmbH
GNI	Gross national income
HDR	Human Development Report
HMNDP	High-Level Meeting on National Drought Policy
ICIDP	Isiolo County Integrated Development Plan
IDB	Inter-American Development Bank
IDRC	International Development Research Center
IEA	International Energy Agency
IFAD	International Fund for Agriculture and Development
IO	Intermediate outcomes
IPCC	Intergovernmental Panel on Climate change
JCM	Joint Crediting Mechanism
JCM TPE	Third party entity
JICA	Japan International Cooperation Agency
JP	Joint Programme
LDC	Least development country
LFA	Logframe approach
LG	Local government
LI	Livelihood Index
MDB	Multilateral development bank
MDG	Millennium Development Goals
MDG-F	Millennium Development Goal Achievement Fund
MEL	Monitoring, evaluation and learning
METIJ	Ministry of Economy, Trade, and Industry Japan
MOEJ	Ministry of the Environment Japan
MRV	Measurement, Reporting and Verification

MSP	Medium-sized projects
MTS	Medium-term strategy
NAP	National Adaptation Plan
NCPC	National Cleaner Production Centres
NDC	Nationally Determined Contributions
NDMA	National Drought Management Authority
NDP	National drought policy
NESDB	National Economic and Social Development Board
NICFI	Norway's International Climate and Forest Initiative
NLS	New Laos Stove
NPR	No secure property rights
ODA	official development assistance
OECD-DAC	Organization for Economic Cooperation and Development-Development Assistance Committee
OH	Outcome harvesting
OM	Outcome mapping
PAR	Pressure and release
PAR	Participatory action research
PCVA	participatory, climate change vulnerability and capacity assessments
PIN	Project Idea Note
PIPA	Participatory impact pathways analysis
PM&E	Participatory monitoring and evaluation
PPCR	Pilot Program for Climate Resilience
PSH	Passive solar houses
QCA	Qualitative comparative analysis
R4D	Research for development
RBM	Results-based management
RCT	Randomized control trials
REDD+	Reducing Emissions from Deforestation and Forest Degradation
RFD	Royal Forest Department
SDC	Swiss Agency for Development and Cooperation
SDG	Sustainable Development Goals
SDIP	Sustainable Development Implementation Plan
SDIR	Sustainable Development Implementation Report
SEADEP	Socioeconomic Assessment of Domestic Energy Practices
SECO	State Secretariat for Economic Affairs
SGP	Small Grants Programme
SME	Small- and medium-sized enterprises
SNV	Netherlands Development Organisation
STAP	Scientific and Technical Advisory Panel
STAR	System for Transparent Allocation of Resources
TAMD	Tracking Adaptation and Measuring Development
TC	Technical cooperation

TOC	Theory of Change
TPB	Theory of planned behavior
UNDP	United Nations Development Programme
UNEP	United Nations Environment Programme
UNFCCC	United Nations Framework Convention on Climate Change
UNIDO	United Nations Industrial Development Organization
USAID	United States Agency for International Development
VCO	Voluntary carbon offset
WB	World Bank
WBI	World Bank Institute
WHO	World Health Organization

Chapter 1
Evaluating Climate Change Action for Sustainable Development: Introduction

Juha I. Uitto, Jyotsna Puri, and Rob D. van den Berg

Abstract This chapter considers evaluation as essential for learning and for reflecting on whether actions to address the complex challenges pertaining to climate change are on track to producing the desired outcomes. The Paris Agreement of 2015 was an important milestone on the road towards a zero-carbon, resilient, prosperous and fair future. However, while the world has agreed on the need to tackle climate change for sustainable development, it is critical to provide evidence-based analysis of past experiences and ongoing innovations to shed light on how we might enhance the effectiveness and efficiency of actions at various levels. Thorough and credible evaluations help us identify what works, for whom, when and where and under what circumstances in order to mitigate climate change, achieve win-win situations for the society, the economy and the environment, reduce risk and increase resilience in the face of changing climate conditions. This chapter serves as an introduction to the book on Evaluating Climate Change Action for Sustainable Development that sets the scene on the current state of climate change evaluation and brings together experiences on evaluating climate change policy, mitigation and adaptation.

Keywords Evaluation • Climate change • Global environment • Mitigation • Adaptation

Climate change has emerged as one of the preeminent challenges facing humankind in the twenty first century. The Intergovernmental Panel on Climate Change states

J.I. Uitto (✉)
Independent Evaluation Office, Global Environment Facility, Washington, DC, USA
e-mail: juitto@thegef.org

J. Puri
International Initiative for Impact Evaluation (3ie), New Delhi, India
e-mail: jpuri@3ieimpact.org

R.D. van den Berg
King's College London and International Development Evaluation Association (IDEAS), Leidschendam, The Netherlands
e-mail: rdwinterberg@gmail.com

© The Authors(s) 2017   1
J.I. Uitto et al. (eds.), *Evaluating Climate Change Action for Sustainable Development*, DOI 10.1007/978-3-319-43702-6_1

unequivocally that there has been an unprecedented warming of the global climate system since the 1950s and that this warming has been influenced by human actions (IPCC 2015). The anthropogenic emissions of greenhouse gases have increased constantly since the pre-industrial level, driven largely by economic and population growth, and are now at the highest historical peak. The impacts of the climate change will affect – and are already affecting – all people and parts of the world often in negative and sometimes unexpected ways. Urgent and concerted action is required to address climate challenges through mitigation efforts as well as through improving the ways in which societies and the global economic system adapt to the effects of climate change. Actions have been initiated on multiple fronts. What is needed is evidence-informed understanding of the effectiveness and efficiency of such actions. Therefore, robust evaluation is a must. That is what this book focuses on.

The year 2015 was a historic turning point for global action on climate change. The Paris Agreement under the UN Framework Convention on Climate Change (UNFCCC) was adopted by the Conference of Parties in its 21st Session.[1] The Paris Agreement is a binding commitment intended to set the world on a path towards a zero-carbon, resilient, prosperous and fair future. In 2015, the plan of action called the 2030 Agenda for Sustainable Development and its associated Sustainable Development Goals (SDGs) was also adopted by member States of the United Nations.[2] The seventeen SDGs are universal and share a common global vision of progress towards a safe, just and sustainable space for all human beings. They reflect the moral principles that no one and no country should be left behind and that everyone and every country should share a common responsibility for delivering the global vision. Specifically goal 13 calls for urgent action to combat climate change and its impacts, recognizing the key linkages of climate change to development and human wellbeing. Goal 13 also refers to the UNFCCC as the global forum to tackle climate change. Also in 2015, the Third UN Conference on Disaster Reduction in the Japanese city of Sendai adopted the Sendai Framework for Disaster Reduction 2015–2030 which also identifies climate change as one of the drivers of increased disaster risk.[3] All these political commitments at the global level demonstrate the urgent concern of the international community and individual governments for climate change and its direct impacts on sustainable development.

Impacts of changing climate express themselves in a multitude of ways.[4] Already now melting snow and ice, and changing precipitation patterns are altering hydrological systems affecting water resources quantity, quality and continuity, as

---

[1]Adoption of the Paris Agreement. Conference of the Parties Twenty-first session, Paris, 30 November to 11 December 2015. United Nations Framework Convention on Climate Change. FCCC/CP/2015/L.9/Rev.1. (https://unfccc.int/resource/docs/2015/cop21/eng/l09r01. pdf, downloaded 8 April 2016).

[2]https://sustainabledevelopment.un.org/sdgs (downloaded 8 April 2016).

[3]http://www.preventionweb.net/files/43291_sendaiframeworkfordrren.pdf (downloaded 8 April 2016).

[4]Unless otherwise indicated, this section is based on IPCC 2015.

streams of water from the glaciers and rainfall patterns become more erratic. Terrestrial, marine and freshwater species have started to alter their geographical range and migration patterns, and their abundance has started to be affected. IPCC projections indicate that climate change will in the future undermine food production through changed weather patterns and ecosystem impacts. Notably, production of three main crops that sustain humanity – wheat, rice and maize – is projected to be negatively affected. Similarly, fisheries productivity will likely be challenged, adding to the problems caused by overfishing. A large proportion of both animal and plant species will face extinction thus exacerbating the loss of biological diversity. Human health may also be affected negatively, as a warmer climate will facilitate the spread of vector borne and tropical diseases to higher latitudes. Extreme weather and climate events are on the rise. These include increased frequency and intensity of storms, as well as climatic variability. While rainfall will increase in some areas, others will face more frequent and prolonged droughts.

Climate risk and vulnerability vary considerably between different regions and groups. Coastal areas are generally the most vulnerable due to storms and sea level rise and associated saline intrusions to coastal ecosystems and aquifers. More and more people are concentrated in coastal areas: it is estimated that more than 40 % of the world's people live within 100 km from the coast and over the past decade more than 60 % of disaster losses have occurred in coastal areas (DasGupta and Shaw 2016). Despite these losses, concentration of the world population on coasts continues. The worst affected are low-lying coastal countries, which are exposed to rising seas and increasing storms. Small island developing states and poor countries, such as Bangladesh, face challenges of survival, but also rich countries like the Netherlands must invest increasing resources to deal with coastal hazards. A large proportion of major cities are located in the coastal areas and, thus, exposed to climate related hazards. Cities, such as London, New York and Tokyo are all coastal, but so are megacities in the poorer regions of the world: Lagos, Kolkata, Dhaka, Jakarta and others. Their ability to cope with and adapt to climate related disasters and rising sea levels is much lower. Similarly, mountain and highland areas experience the risk of climate change acutely. They are the water towers of the world and home to some of the poorest people in the world (FAO 2015). They are doubly vulnerable in terms of global freshwater availability and local food security. Adaptive capacity and vulnerability have deep social, economic and political determinants (Pelling 2011). Risk is defined as a function of hazard exposure and vulnerability to it (Wisner et al. 2004). Apart from direct physical factors, vulnerability has a strong social dimension: people with fewer economic means and political power have less ability to cope with and recover from disasters. In addition, they are often confined to living in the most hazardous places, such as informal settlements on denuded slopes (Surjan et al. 2016).

It is incumbent upon us to deal with climate change in a comprehensive manner. There is a need to address the root causes of climate change to mitigate it. IPCC links anthropogenic greenhouse gas emissions to key drivers that include: population size, economic activity, lifestyle, energy use, land use patterns, technology and climate policy. All are directly related to virtually all aspects of human activity and aspirations. As societies get richer, their energy use and emissions tend to increase. There is therefore a compelling need for decoupling economic growth from increases in energy use and emissions (Mulder and Groot, 2004). Development and spread of energy efficient technologies and renewable energy will play a key role in the process (Yang and Yu 2015; Edenhofer et al. 2011). The adoption of low-carbon transport systems is also high on the agenda (Dalkmann and Huizenga 2010). While such mitigation measures are needed, their impacts will be long-term and dependent on widespread societal adoption. Even in best scenarios, it will take many decades before they take effect. According to IPCC (2015), emissions scenarios that keep warming below 2°C over the twenty-first century relative to pre-industrial levels will involve 40–70 % reductions in global anthropogenic emissions by 2050 and near-zero emission levels by 2100. Although this is consistent with the Paris Agreement targets, the current voluntary mitigation efforts by signatory countries fall well short of this.

It is consequently necessary to invest in adaptation to climate change and to enhance societal resilience to climate change impacts. Adaptation refers to reducing the adverse effects of climate change on human and natural systems. At the 2010 UNFCCC conference in Cancun, Mexico, the parties adopted the Cancun Adaptation Framework[5] affirming that adaptation must be addressed with the same level of priority as mitigation. They further agreed that adaptation is a challenge faced by all parties, and that enhanced action and international cooperation is urgently required to enable and support the implementation of adaptation actions aimed at reducing vulnerability and building resilience in developing countries (para 11). IPCC (2015) recognizes that adaptation options exist in all sectors but their context and potential differ between sectors and regions. Furthermore, adaptation and mitigation responses are underpinned by common factors, including effectiveness of institutions and governance, innovation and investments in environmentally sound technologies and infrastructure, sustainable livelihoods and behavioural and lifestyle choices (SPM 4.1).

Several international financial and technical facilities have been set up to help countries address climate change challenges. The Global Environment Facility[6] (GEF) has already been in existence for a quarter century as the financial mechanism to the UNFCCC. It finances projects in developing countries that focus on mitigation efforts. The GEF recognizes the multidisciplinary nature of mitigation. While greenhouse gas emission reductions through promotion of sustainable transport, energy efficiency and renewable energy are important, emissions reductions from sectors, such as land use and forestry are also important, as is protecting global carbon sinks like the oceans. The World Bank manages the Climate Investment Funds[7] that operate through four key programmes that help developing countries pilot low-emissions and climate resilient development: the Clean Technology Fund, Forest Investment Programme, Pilot Programme for Climate Resilience, and Scaling up Renewable Energy Programme. The Adaptation Fund[8] helps developing

---

[5]http://unfccc.int/resource/docs/2010/cop16/eng/07a01.pdf#page=4 (downloaded 8 April 2016).
[6]https://www.thegef.org/gef/
[7]https://www-cif.climateinvestmentfunds.org/
[8]https://www.adaptation-fund.org/

countries to build resilience and adapt to climate change through financing projects and programmes that focus on vulnerable communities. The Green Climate Fund[9] was set up in anticipation of the Paris Agreement to mobilize funding and invest in low-emission and climate-resilient development. It intends to address both mitigation and adaptation equally. Importantly all these funds and facilities involve concrete strategies and action by all governments, the private sector, civil society, as well as individual citizens. But clearly global climate action is not limited to these funds.

The Paris Agreement fully recognizes what also emerged from the UN Conference on Financing for Development[10]: public funds will not be sufficient to tackle climate change, not to prevent it nor to adapt to its effects. In this regard public-private partnerships and private initiatives are envisaged to play a key role. Governments are invited to ensure an enabling environment and level playing field for initiatives of the private sector and civil society. These have emerged over the past decade and include social and environmental impact investing, corporate responsibility to contribute to sustainable development, market oriented social initiatives and so on. The G8 has published several documents regarding the promise of impact investing[11] and the World Business Council for Sustainable Development has been instrumental in developing the inclusion of natural resources accounting in business practices.[12] Civil society initiatives range from fair trade to climate-smart agricultural practices (the range of initiatives is staggering), as is demonstrated in the Social Enterprises World Forum.[13]

Climate change is a complex issue encompassing physical, technological, institutional, economic, social and political spheres. For a sustainable future for all of us, it is essential that we identify the best and most suitable measures and make the right choices for mitigation and adaptation to climate change. This is where evaluation comes in.

1.1   Critical Role of Evaluation

Evaluation is essential for learning and for reflecting on whether we are taking the right actions for the right things for current and future generations. Evaluation of climate change policies, mitigation and adaptation actions helps us assess progress on the complex challenges we are facing. Evaluation also helps us identify what

---

[9]http://www.greenclimate.fund/home

[10]See http://www.un.org/esa/ffd/ffd3/ – the UN Conference for Financing for Development took place in Addis Ababa, 13–16 July 2015.

[11]See http://www.socialimpactinvestment.org/ for the work sponsored by the G8 on impact investing.

[12]See http://www.wbcsd.org for the World Business Council for Sustainable Development.

[13]See http://sewf2015.org/about-sewf/ for the Social Enterprises World Forum.

works, under what circumstances and for whom. Evidence-based analysis of past experiences and ongoing innovations is likely to shed light on how we might enhance the effectiveness and efficiency of actions at various levels and achieve win-win situations and multiple benefits, such as reducing risk and increasing resilience.

Evaluation makes a judgement of the value or worth of the evaluand – the subject of evaluation – be it a policy, strategy, programme, project or any other type of intervention. It can take several forms. It can be formative, looking into the ways an intervention is implemented in order to identify ways in which the intervention and its performance could be improved. It can be summative to determine the extent to which the intervention has achieved its anticipated desired results. An evaluation can also be prospective, assessing the likely outcomes of proposed interventions a priori (Morra Imas and Rist 2009). A category of summative evaluations is impact evaluation that looks into whether the programme or intervention has contributed in a measurable way to a larger longer term goal (such as transforming national policy or the market towards a more climate friendly directions) than just the direct outputs and outcome of the intervention itself. An impact evaluation can use a range of approaches and methodologies that are rigorous (Stern et al. 2012). It has been argued that it is important to distinguish between the 'direct' and 'final or ultimate' impact of interventions (van den Berg 2013). As we address issues critical to climate change, we must ensure that interventions make a difference and help to significantly increase mitigation or adaptation or both while also ensuring sustainable development, or be able to identify and measure trade-offs.

In this context, a special challenge is posed by the private sector and civil society initiatives on impact investing, corporate responsibility and sustainable development, as well as civic initiatives and social enterprise. The role of evaluation in these relatively new areas of work is not yet established, which is why they have been identified as the 'New Frontiers for Evaluation', an initiative of the Centre for Development Impact in the UK.[14] A Wilton Park conference in July 2015 discussed the potential role of evaluation in various initiatives, calling for a gap analysis of what has been evaluated and where methods and capacity need to be developed.[15] Much of this is relevant for climate change to inform investments in green technologies and transitions towards sustainable resource use in business practices. We also need to take stock of what we know to start, operationalize and manage climate smart enterprises. Evaluation can play a very important role in measuring effectiveness, cost-effectiveness and longer term impact.

Evaluating climate change can be challenging primarily because climate change is a global good (Puri and Dhody 2016). Other challenges include the fact that climate change programmes are frequently multi-sector, multi-objective complex programmes that aim to affect not just environment but also poverty, livelihoods,

---

[14] See http://www.cdimpact.org/projects/new-frontiers-evaluation

[15] See https://www.wiltonpark.org.uk/wp-content/uploads/WP1411-Report.pdf

health, income and food security. Additionally climate change programmes aim to affect not just immediate outcomes but outcomes over generations. Last but not least, is the absence of data and capacity in this area – most evaluators are trained in more traditional sectors and hence think about evaluations in traditional ways. Indeed Picciotto (2007) identifies climate change as a significant challenge for development evaluation. Evaluating climate change action is a relatively new frontier for the field that has only emerged in the first decade of the 2000s (van den Berg and Feinstein, 2009). We maintain that evaluation in the field of environment and sustainable development has the opportunity to leapfrog, while borrowing from other disciplines, but also innovating to generate high quality and relevant evidence to inform national and international efforts directed at environment and sustainable development (Rowe 2012; Uitto 2014). This book has its origins in the Climate-Eval community of practice started and hosted by the Independent Evaluation Office of the GEF. The book brings together state-of-the-art contributions of evaluations pertaining to climate change policy, mitigation and adaptation.

1.2 Book Structure

The book contains 18 chapters in which leading authors examine innovative and emerging evaluation knowledge and practice of climate change and its link to sustainable development. The authors discuss methodologies and approaches to better understand, learn from and assess interventions, strategies and policies. The contributions also discuss evaluation challenges encountered and lessons learned to better understand and tackle difficult areas of evaluation.

Chapter 2 or overview chapter by Rob D. van den Berg and Lee Cando-Noordhuizen, 'Action on climate change: What does it mean and where does it lead to?' discusses the micro-macro paradox of climate change action. There is evidence that climate action works and achieves direct impact – yet climate change seems unstoppable. An analysis of multiple comprehensive evaluations indicates that technology and knowledge are available to fight climate change. However, economic development and subsidies harmful to the climate still outweigh remedial climate action with at least a factor of one hundred. Current successes of programmes and projects will not impact global trends unless unsustainable subsidies and actions are stopped.

Chapter 3 written by Rob D. van den Berg, 'Mainstreaming impact evidence in climate change and sustainable development' examines the demand for impact evidence and concludes that this demand goes beyond the experimental evidence that is produced during the lifetime of an intervention. Van den Berg argues for impact considerations to be mainstreamed throughout interventions, programmes and policies and for evaluations to gather evidence where available, rather than focusing the search for impact and its measurements on one or two causal mechanisms that are chosen for verification through experimentation.

Chapter 4 by Tonya Schuetz, Wiebke Förch, Philip Thornton and Ioannis Vasileiou from the CGIAR Research Program on Climate Change, Agriculture and Food Security (CCAFS) describes the design of an impact pathway-based Monitoring, Evaluation and Learning (MEL) system that combines classic indicators of process in research with innovative indicators of change. The chapter highlights the importance of engaging users of research in the development of impact pathways and continuously throughout the life of the program. Results show that partnerships with diverse actors such as the private sector and policy makers are key to achieving change. The chapter concludes that research alone is insufficient to bring about change. However, research does generate knowledge that stakeholders can put to use to generate development outcomes.

Chapter 5 by Monika Egger Kissling and Roman Windisch, 'Lessons from taking stock on 12 years of Swiss international cooperation on climate change' highlights the challenges encountered and lessons learned from this assessment where a bilateral donor puts climate change lens on a longstanding development cooperation portfolio. The chapter discusses the need (1) for evaluators to put more effort in identifying best methodological practices amidst a large volume of information, diverse portfolio and absence of reliable data; (2) for practitioners to invest more in strategic project design and monitoring to provide accurate data; and (3) for policy makers to be cognizant of the value that evaluation brings, as it is an important tool that contributes to accountability.

Chapter 6 by Michael Carbon discusses the approach, process and lessons from the evaluation of UNEP's Climate Change Sub-programme. It shows the importance of developing an appropriate analytical framework that is well-suited for the scope and complexity of the object of evaluation, and how the Theory of Change approach helped make a credible assessment of UNEP's contribution towards impact, sustainability and upscaling.

Chapter 7 written by Aryanie Amellina focuses on an assessment of the initial phases of the Joint Crediting Mechanism (JCM) in Indonesia. It highlights JCM governance and ease of use of methodologies related to measurement, reporting and verification (MRV). The author concludes with recommendations to strengthen methods to determine reference emissions and for clarifying ways to allocate credit among countries to define a pathway to a tradeable credit mechanism.

Chapter 8 by Jyotsna Puri, 'Using mixed methods to assessing trade-offs between agricultural decisions and deforestation', demonstrates the importance of using qualitative and quantitative methods to assess and measure win-win development policies that also help mitigate climate change. The author's study explores the poverty and environment nexus using historical data on land rights and panel data on land use in Thailand. The chapter concludes that it is important to measure the differential effects of policies on different crops, agricultural intensity and agricultural frontier. In the case examined by the author, she advises that policies that encourage cultivation may not be detrimental to forest cover after all.

Chapter 9 written by Aaron Zazueta and Neeraj Negi presents the methodological approach adopted in the evaluation of climate change mitigation projects supported by the Global Environment Facility in four emerging markets, namely

China, India, Mexico and Russia. The authors demonstrate the use of the Theory of Change approach to carry out a comparative analysis across projects seeking to bring about changes across diverse markets or market segments in different countries. Zazueta and Negi highlight how the evaluation focused on understanding the extent and forms by which GEF projects are contributing to long-term market changes, leading to reduction in GHG emissions, and on assessing the added value of GEF support in the context of multiple factors affecting market change.

Chapter 10 written by Yann François and Marina Gavaldão explore how climate change mitigation projects can reduce greenhouse gas emissions, potentially have adaptation benefits, and achieve sustainable development objectives. 'Integrating avoided emissions in climate change evaluation policies for LDCs' provides an example of socio-economic benefits gained if accounting for avoided emissions are incorporated in projects, in this case, passive solar housing technology.

Chapter 11 by Debora Ley, 'Sustainable development, climate change, and renewable energy in rural Central America', demonstrates the potential and multiple benefits of decentralized renewable energy. The author also demonstrates how specific drivers can facilitate or hinder projects in achieving multiple objectives using on the ground, qualitative methods.

Chapter 12 by Jasmine Hyman, 'Unpacking the black box of technology distribution, development potential and carbon markets benefits' explores whether and how carbon markets can support a pro-poor development agenda. The author introduces a 'Livelihood Index' to understand the employment impact of a carbon intervention. Their study finds that variations in the distribution framework means that development outcomes may compete rather than complement one another. Methods used include value chain analysis and a qualitative analysis to understand how carbon finance recipients access the mechanism, perceive the project and conceptualise its benefits.

Chapter 13 by Takaaki Miyaguchi and Juha Uitto presents the methodology of a meta-analysis of ex-post evaluations of climate change adaptation (CCA) programmes in nine countries using a realist approach. The authors conclude that adopting a realist approach to evaluating complex development projects is a useful way of providing relevant explanations, instead of judgments, about what type of intervention may work for whom, how and under what circumstances for future programming.

Chapter 14 written by Jacques Somda, Robert Zougmoré, Tougiani Abasse, Babou André Bationo, Saaka Buah and Issa Sawadogo, 'Adaptation processes in agriculture and food security: Insights from evaluating behavioural changes in West Africa' focuses on the evaluation of adaptive capacities of community-level human systems related to agriculture and food security. The study highlights findings regarding approaches and domains to monitor and evaluate behavioural changes from CGIAR's research program on climate change, agriculture and food security (CCAFS). Results suggest that application of behavioural change theories can facilitate the development of climate change adaptation indicators that are complementary to indicators of development outcomes. The authors conclude that collecting stories on behavioural changes can contribute to biophysical adaptation and monitoring and evaluation.

Chapter 15, written by Irene Karani and Nyachomba Kariuki, 'Using participatory approaches in measuring resilience and development in Isiolo County, Kenya' highlights the use of participatory approaches through a Tracking Adaptation and Measuring Development (TAMD) Framework to measure resilience in Kenya. The authors outline the process of developing subjective indicators and demonstrate the advantage of empowering the local community in collection of baseline, monitoring and early outcome data as they develop Theories of Change. The article concludes by sharing lessons and policy implications.

Chapter 16 by Joanne Chong, Anna Gero and Pia Treichel, 'Evaluating climate change adaptation in practice: a child-centred, community-based project in the Philippines' documents a research and evaluation approach applied in a child-centred and community-based CCA project implemented across four provinces in the Philippines. The authors emphasise the success of the methodology due to its participatory foundations – local voices and perspectives matter in understanding the impact of the project.

Chapter 17 written by Emilia Bretan and Nathan L. Engle focuses on real time milestones and outcomes from Brazil's Drought Preparedness and Climate Resilience Programme (Drought NLTA). Evidence gathered through the participatory monitoring and evaluation (PM&E) approach showed that the programme was able to convene key-regional and federal level multi-sector stakeholders, resulting in a bottom-up and regionally-led collaboration. Through engagement and commitment of the partners, the programme illustrates good practice for coordination and continuous sharing of knowledge and data between service providers, secretariats, municipalities and other stakeholders from distinct sectors, states, and governmental levels.

Chapter 18 by Timo Leiter presents a decision-support tool developed by the German International Cooperation (GIZ GmbH), the Adaptation M&E Navigator. The author explains the rationale, structure and how this tool can help policy- and decision-makers select a suitable M&E approach by providing a list of specific M&E paradigms and matching them with relevant approaches.

Evaluation plays an ever crucial role in learning: why are things happening or not happening? Are we doing the right thing or not? Why and why not? Are there better ways? The evaluation profession has become more adept at introducing scientific tools and the link between science and evaluation is becoming stronger. Evaluation is helping bridge the science-policy divide.

The contributions included in this book demonstrate a good understanding not only of assumptions and outcomes, but also of context as they attempt to explain how and for whom interventions may work. Methodologies used are varied and may sometimes be sophisticated. However, they all answer operational and practical questions.

We are in a world with changing boundaries. Our boundaries have changed in terms of what we want from our programs and strategies, what we want from evaluations and what types of tools we have access to. We are now witnessing the surge and availability of big and open data and a variety of innovative techniques that will also enable this sector to leapfrog and push the frontiers of learning and evaluation. It is our hope that this book will contribute to this push.

References

Berg, R.D. van den, & Feinstein, O. (Eds.). (2009). *Evaluating climate change and development* (World Bank series on development, Vol. 8). New Brunswick/London: Transaction Publishers.

Berg, R. D. van den (2013). Evaluation in the context of global public goods. In R. Rist, M. H. Boily, & F. Martin (Eds.), *Devellopment evaluation in turbulent times: Dealing with crises that endanger our future* (pp. 33–50). Washington, DC: The World Bank.

Dalkmann, H., & Huizenga, C. (2010). *Advancing sustainable low-carbon transport through the GEF* (Scientific and Technical Advisory Panel (STAP) of the Global Environment Facility (GEF)). Nairobi: UNEP.

DasGupta, R., & Shaw, R. (2016). Sustainable development and coastal disasters: Linking policies to practices. In J. I. Uitto & R. Shaw (Eds.), *Sustainable development and disaster risk reduction* (pp. 161–172). Tokyo: Springer.

FAO. (2015). *Mapping the vulnerability of mountain peoples to food insecurity.* In R. Romeo, A. Vita, R. Testolin. & T. Hofer, (Eds.). Rome: Food and Agriculture Organization of the United Nations.

Edenhofer, O., Pichs-Madruga, R., Sokona, Y., Seyboth, K., Matschoss, P., Kadner, S., Zwickel, T., Eickemeier, P., Hansen, G., Schlömer, S., & Von Stechow, C. (Eds). (2011). *IPCC special report on renewable energy sources and climate change mitigation. Intergovernmental Panel on Climate change (IPCC).* Cambridge/New York: Cambridge University Press.

IPCC (2015). *Climate change 2014: Synthesis report. Intergovernmental Panel on Climate Change.* http://ar5-syr.ipcc.ch/ipcc/ipcc/resources/pdf/IPCC_SynthesisReport.pdf. Accessed Apr 2016.

Morra Imas, L. G., & Rist, R. C. (2009). *The road to results: Designing and conducting effective development evaluations.* Washington, DC: The World Bank.

Mulder, P., & de Groot, H. L. F. (2004). *Decoupling economic growth and energy use* (Tinbergen Institute Discussion Paper). Amsterdam/Rotterdam.

Pelling, M. (2011). *Adaptation to climate change: From resilience to transformation.* Oxon: Routledge.

Picciotto, R. (2007). The New environment for development evaluation. *American Journal of Evaluation, 28*(4), 509–521.

Puri, J., & Dhody, B. (2016). Missing the forests for the trees? Assessing the use of impact evaluations in forestry programmes. In J. I. Uitto & R. Shaw (Eds.), *Sustainable development and disaster risk reduction* (pp. 227–245). Tokyo: Springer.

Rowe, A. (2012). Evaluation of natural resource interventions. *American Journal of Evaluation, 33*(3), 384–394.

Stern, E., Stame, N., Mayne, J., Forss, K., Davies, R., & Befani, B. (2012). *Broadening the range of designs and methods for impact evaluations* (Working paper 38). UK Department for International Development (DFID).

Surjan, A., Kudo, S., & Uitto, J. I. (2016). Risk and vulnerability. In J. I. Uitto & R. Shaw (Eds.), *Sustainable development and disaster risk reduction* (pp. 37–55). Tokyo: Springer.

Uitto, J. I. (2014). Evaluating environment and development: Lessons from international cooperation. *Evaluation, 20*(1), 44–57.

Wisner, B., Blaikie, P., Cannon, T., & Davis, I. (2004). *At risk: Natural hazards, people's vulnerability and disasters* (2nd ed.). London: Routledge.

Yang, M., & Yu, X. (2015). *Energy efficiency: Benefits for environment and society.* London: Springer.

Chapter 2
Action on Climate Change: What Does It Mean and Where Does It Lead To?

Rob D. van den Berg and Lee Cando-Noordhuizen

Abstract In 2014, the second conference on evaluating climate change and development offered the opportunity to take stock of evaluative evidence of the challenges, failures and success of climate change action. In 2011 one of the authors raised the possibility of a micro-macro paradox of climate change action (van den Berg, Evaluation 17:405, 2011): in his view evaluations of climate change action provided evidence that climate action works and achieves direct impact – yet climate change seems unstoppable. Several major, comprehensive evaluations were presented at the 2014 conference and provided an overview of actions taken and their successes and failures, as well as obstacles on the way to global impact. This chapter presents an overview of issues, evidence and the way forward for evaluators tackling climate action and sustainable development. The evidence provides support for the micro-macro paradox of 2011 and indicates that the global community has the technology and knowledge on how to stop climate change. However, actions that promote climate change still outweigh remedial climate action with at least a factor of 100. Thus current successes of programs and projects will not impact global trends, unless at the same time the non-sustainable subsidies and actions are stopped.

Keywords Climate action • Climate change • Evaluation • Micro-macro paradox • Systems change

R.D. van den Berg (✉)
King's College London and International Development Evaluation Association (IDEAS),
Leidschendam, The Netherlands
e-mail: rdwinterberg@gmail.com

L. Cando-Noordhuizen
Independent Evaluation Office, Global Environment Facility, Washington, DC, USA
e-mail: leecando.eval@gmail.com

© The Author(s) 2017   13
J.I. Uitto et al. (eds.), *Evaluating Climate Change Action for Sustainable Development*, DOI 10.1007/978-3-319-43702-6_2

2.1   Introducing the Micro-Macro Paradox: Success at the Micro-level Does Not Lead to Success at the Macro-level?

In development economics the question whether aid contributed to economic growth was hotly debated after Mosley (1987, 139ff) identified this as the "micro-macro" paradox. He could not find any statistically significant correlation between development aid and the economic growth rate of recipient countries, taking into account other factors that cause growth. Mosley defended aid nonetheless, as benefits at the micro level were often shown to be substantial and essential. Nevertheless, economic growth was supposed to be the engine of future development that would make aid unnecessary, and if aid would not contribute to economic development, it could turn out to be ineffective in the longer run and not have meaning beyond just the benefits of a specific and localized project or intervention. Even if a project has significant short term outcomes, but it did not contribute to economic growth, it could be argued that the *sustainability* of its benefits are questionable.

A second milestone in this discussion was reached in 1998 with the publication of the World Bank report on "Assessing aid: what works, what doesn't and why?" (Dollar and Prichett 1998), which focused on the role of aid in reducing poverty, and rekindled the micro-macro paradox discussion, as it better identified when aid could potentially contribute to economic growth: when countries had good policies, good governance and management and well-functioning institutions. The ensuing debate in development economics revived the micro-macro paradox, until in (2010) Arndt, Jones and Tarp aimed to close the arguments by demonstrating a positive and statistically significant causal effect of aid on growth in poor countries over the long run. "There is no micro-macro paradox", they conclude (p. 27). What is interesting in their analysis is that they attribute their success in demonstrating evidence for growth to "methodological advances in the programme evaluation literature", which have "improved the profession's capacity to identify causal effects in economic phenomena" (p. 26).

Evaluation methodology thus has helped to solve the micro-macro paradox in development economics, according to Arndt, Jones and Tarp. If we accept that, let us explore whether evaluation methodology is also able to help us in solving the paradox of successful climate change interventions, versus a devastating trend of global warming and associated climate variability that does not appear to be influenced by climate change interventions.

The opportunity for a broad perspective on this issue presented itself at the 2nd International Conference on Evaluating Climate Change and Development, where several comprehensive evaluations of Climate Change aid were presented. They offered an opportunity for a meta-analysis of the results of some of the largest public sector efforts to address climate change in developing countries. Of special interest is whether these evaluations offer any hope regarding the micro-macro paradox of climate interventions that make a local difference but do not seem to impact at the global level.

Seven comprehensive evaluations will be assessed and the evidence they present of the discrepancy between micro and macro impact will be judged. The first four were presented at the 2nd International Conference on Evaluating Climate Change and Development; the last three were added as they emerged in the same year and complement the picture:

1. The Fifth Overall Performance Study of the Global Environment Facility, undertaken by the Independent Evaluation Office of the GEF;
2. The Independent Evaluation of the Climate Investment Funds, undertaken by ICF International on behalf of the five independent evaluation departments of the multilateral development banks;
3. The evaluation of climate change support in the Inter-American Development Bank, conducted by the Office of Evaluation and Oversight of the IDB;
4. The evaluation of the effectiveness of Swiss International Cooperation in climate change, conducted by a consortium led by Gaia Consulting Oy for the Swiss Agency for Development and the State Secretariat for Economic Affairs;
5. The real-time evaluation of initiatives of the Asian Development Bank to support access to climate finance, implemented by the Independent Evaluation Department of the ADB;
6. The real-time evaluation of Norway's International Climate and Forest Initiative (NICFI), undertaken by a consortium led by LTS International;
7. The external evaluation of the UN-REDD programme (Reducing Emissions from Deforestation and Forest Degradation in Developing Countries).

Furthermore, we will also include the older but still highly relevant climate change evaluations of the Independent Evaluation Group of the World Bank (World Bank/IEG 2009, 2010, 2012), as well as the Fourth Overall Performance Study of the Global Environment Facility (GEF/EO 2010), the precursor to OPS5.

2.2   The Micro-Macro Paradox: Successful Climate Action But No Global Impact?

The micro-macro paradox of successful environmental interventions was raised by one of the authors in a keynote address at the Second Global Assembly of the International Development Evaluation Association in Amman, Jordan, on 14 April 2011 (van den Berg 2011). The argument is that a sizeable proportion of interventions were demonstrated to have direct and long-term impact in the sense of achieving lasting success in for example reducing greenhouse gas emissions from a specific source, but they have made no impact on global environmental trends, that have continued their downward slide. This is the case for climate change, for the historical loss of biodiversity that is now increasingly seen as a human caused mass extinction to be compared with the extinction of dinosaurs 65 million years ago, and for the increasing pollution of our environment with chemical substances, which endanger human health and the health of our habitat.[1]

A first indication of the paradox and its solution emerged in the comprehensive evaluations of the Global Environment Facility (GEF). The GEF was established as (interim) financial instrument of the main environmental conventions resulting from the 1992 Earth Summit, on climate change (UNFCCC), biodiversity (CBD) and some of the various conventions on chemicals (most notably Stockholm). For more than two decades it had been the core organization for support to developing countries and countries with economies in transition, raising a considerable amount of funding itself, and as a co-funding agency, an even larger amount from other sources. The GEF is replenished every 4 years by its donors. One of the important documents of this replenishment is an independent comprehensive evaluation of the performance of the institution up to that time. In the fourth Overall Performance Study (OPS4) of the GEF some elements of the micro-macro paradox were first explored (GEF/EO 2010). Interventions financed by the GEF had started in 1992 and the 2010 Overall Performance Study was the first to be able to report on the longer term impact of these interventions. OPS4 concluded that the processes set in motion by GEF co-funded projects were progressing toward longer term impact, provided follow-up actions were taken by countries and stakeholders. Nevertheless, global environmental trends continued to "spiral downward" (conclusion 1, p. 15). A first indication of why this was the case was provided in a calculation of the purchasing power of GEF funds over time: the fourth replenishment of the GEF, while nominally higher than the first replenishment, represented 83 % of the value of the first replenishment, while at the same time funding needs had increased dramatically (p. 16–18).

The Fifth Overall Performance Study of the GEF (GEF/IEO 2014) provides more details to the same arguments. It concludes again that environmental trends "continue to decline" (p. 10), whereas the "intervention logic of the GEF is catalytic and successful in achieving impact over time" (p. 13). Like OPS4, the evaluation focuses on funding levels to explain the paradox between evidence of impact and declining global trends. This time the context is broadened and includes public funding that leads to environmental decline. At the time of OPS5, annual commitments of the GEF had reached the level of US\$ 1 billion. Overall public funding for environmental support to developing countries had reached the level of US\$ 10 billion annually. However, funding needs for action on global environmental issues "are conservatively assessed as at least US\$ 100 billion annually" (p. 17). Thus a funding gap emerges that in itself provides an explanation of the paradox.

---

[1]Rijk, van Duursen and van den Berg (2016). Health cost that may be associated with Endocrine Disrupting Chemicals: an inventory, evaluation and way forward to assess the potential socio-economic impact of EDC-associated health effects in the EU. University of Utrecht. They calculate the cost in 2028 in the EU from €46 to 288 billion per year, if no action is taken. This is just one example of a particular type of chemical substance; new chemical substances are introduced in food and packaging every year.

However, "global public funding of at least US$ 1 trillion annually is available for
(...) unsustainable environmental practices, such as subsidies for fossil fuels"
(p. 17). In other words, the GEF is ten times out-funded by others, whereas ten
times more than overall public funding for global environmental public goods is
required, and ten times more than that is actually spent through public funding on
subsidies that destroy our environment. The paradox is thus revealed as a *veridical*
paradox[2]: a seeming conflict between impact of GEF versus impact on global
environmental trends, that is resolved if competing funding channels are taking
into account.[3]

The questions we pose in this chapter are whether the other comprehensive
evaluations provide further evidence for this; whether the historical path that the
GEF has followed to arrive at this situation has been matched by others, or whether
they have been able to tackle the barriers to impact; and whether their specific
routes offer insights into what may be done to increase the chances of success at
systems level – i.e. whether they provide insights into what works, when and where,
for whom and under what circumstances to achieve success in humanity's efforts to
address the potentially disastrous consequences of climate change over time.

2.3 From Early Results to the Slow Materialization of Impact

In the years leading up to OPS5 the project portfolio of interventions supported by
the GEF has matured over time, since its inception in 1992, to enable a judgment on
the effectiveness and impact of these interventions. The First and Second Overall
Performance Studies of the GEF were not able to provide comprehensive assessments of the results and impact of the GEF, due to the fact that many interventions
had just been completed or were still on-going at the time of the evaluations[4] (this
paragraph based on ICF 2005, 21–22). The Third Overall Performance Study was
expected to be the first to report on results and impact, and it had to disappoint its
readers on impact. It was able to report on results, as these were mostly at the
outcome level. On longer term impact the OPS3 team was confronted with "general
unavailability of impact-level results data" (ICF 2005, 21). Several reasons were
identified why these data were unavailable: lack of an overall results measurement
framework including baselines, indicators and targets; lack of efforts at the project

---

[2]As defined by the logician W.V. Quine (1966) in *The Ways of Paradox and other essays*.
New York: Random House: a veridical paradox is a statement that seems to contradict itself but
may nonetheless be true.

[3]For an update on energy subsidies alone, see IMF Working Paper 15/105 *How large are Global
Energy Subsidies?* by David Coady, Louis Sears and Baoping Shang, that estimates subsidies and
related costs to be higher than $5 trillion in 2015.

[4]This paragraph based on ICF, 2005, 21–22.

level to generate data; lack of systematic efforts to conduct "end-of-project" evaluations and perhaps most importantly: the time horizon. Whereas GEF projects on average take no longer than 5 years, environmental change may take decades before it becomes measurable (ICF 2005, 22). However, OPS3 noted with muted optimism that monitoring and evaluation had become more important in the GEF and there was evidence of growing harmonization of goals and processes across the GEF (ICF 2005, 12). We will see these themes return in other organizations and their evaluations.

The Climate Investment Funds (CIF), initiated in 2008, were set up to overcome two obstacles that the GEF had to face: slow procedures and fragmentation of funding. The GEF had to spread its contributions over a large group of countries (more than 150) and not just in climate change, but in other priority areas such as biodiversity, international waters and persistent organic pollutants. The slow implementation of GEF interventions, also led to time delays in achieving impact, while time is of the essence in the fight against climate change. The CIF would focus on a relatively small number of countries, to enable it to provide higher levels of funding, "potentially allowing greater impact" (ICF 2014, viii) and it would apply a "light touch" approach to ensure quick decision making – relying on the multilateral development banks to provide the technical expertise to design, review and implement projects. However, up to May 2014 only a small proportion – about 9 % – of the approved funding had been disbursed to action on the ground (ICF, vii). The evaluation notes in 2014 that "most CIF projects are still on the drawing board or in early execution" (ICF, viii) and thus the effort to speed up procedures in comparison to the GEF largely failed. Failure to overcome the second barrier of insufficient funding to achieve longer term impact cannot yet be ascertained: the question cannot yet be answered.

Yet "transformative impact is a major goal of the CIF, and a justifiable one" (ICF, x). The evaluation notes that CIF resources, even though more focused and considerably higher than the GEF's in its partner countries, "are small relative to global needs", so they need to be focused on countries and on activities where they will be able to support transformative change. However, the evaluation also notes that many of the CIFs activities lack a convincing theory of change that provides a clear picture of how broader adoption would be achieved. On the positive side the evaluation commends the CIF for its learning and piloting objectives, and notes the "vast potential" for providing knowledge on how countries can respond to the challenge of climate change (ICF, xii).

The evaluations of climate change efforts of the World Bank Group go back in time from 2009 (when the first study was published) to 2012 (when the third report was published on the IEG website). They refer to a much broader and older portfolio of activities that the Bank implemented, many of which were undertaken with co-funding from the GEF. The longer term impact on several areas of work could be evaluated. However, the primary focus of many interventions was often on aspects such as support for energy policies, deforestation, low carbon technologies, and adaptation, and differed in how they related to climate change. The emerging picture is thus less straight-forward than the GEF assessment. Nevertheless, the three World Bank evaluations provide indirect support for the paradox and some hopeful signs of where the paradox may be solved.

First and foremost, the evaluations identify energy efficiency as a crucial pathway towards climate action that potentially funds itself.[5] Well guided efforts toward energy efficiency tend to have economic returns that dwarf those of most other development projects, while at the same time resulting in lower greenhouse gas emissions. Especially the second evaluation (World Bank 2010, p. 32) identifies several promising avenues: efficient lighting that offers very high economic returns and significant emission reductions; reducing losses in the transmission and distribution of energy; large-scale efforts in energy efficiency may reduce the need for power plants (World Bank 2010, p. xv). The 2010 evaluation was one of the first to provide evaluative evidence that energy subsidies are "expensive, damage the climate and benefit the rich" (World Bank 2010, p. 119).

These findings in the World Bank/IEG evaluations (most notably the second evaluation) were further supported by evaluative evidence from the Asian Development Bank, the European Bank for Reconstruction and Development and the GEF. In a briefing note the Evaluation Cooperation Group of the multilateral banks noted strong evidence from independent evaluations that[6]:

- Energy efficiency investments are highly cost-effective;
- Fossil fuel subsidies discourage energy efficiency;
- The financial sector can be persuaded to provide energy efficiency loans;
- Genuine demonstration projects can transform markets;
- Biases against energy efficiency projects can be overcome.

However, presenting this evidence to the climate change negotiators could to some extent be characterized as "preaching to the converted" and the evidence for these points still needs to sway governments to reduce fossil fuel subsidies and promote energy efficiency.

The Inter-American Development Bank's 2014 evaluation of its climate change strategy notes that the IDB has seen its largest contribution to greenhouse gas emission reductions from its support for renewable energy investments (mainly hydropower – IDB 2014, p. 34), rather than energy efficiency in which the Bank has not been as active. The 2014 evaluation aligns the IDB with the earlier ECG briefing note in suggesting that "improvements in energy efficiency have perhaps the greatest potential impact in reducing GHG emissions at the lowest costs", for which energy subsidies "remain a key barrier" (IDB 2014, p. x). A second sector that turned out to be highly relevant for climate change was transportation: bus

---

[5]IEG [2016]. *Four myths about climate change*. Webtext accompanying the publication of the three Climate Change and the World Bank Group reports. http://ieg.worldbankgroup.org/topic/climate-change, accessed May 9 2016.

[6]ECG (2011). *Overcoming barriers to energy efficiency: new evidence from independent evaluation*. S.l., Evaluation Cooperation Group. [Briefing note, November 23, 2011.] This note was presented to the 17th Conference of the Parties (COP17) of the UN Framework Convention for Climate Change, held from 28 November to 9 December 2011 in Durban, South Africa.

rapid transport systems and road projects, where especially the former have contributed to greenhouse gas emissions.

However, a lack of clear classification of climate change related activities and investments, as well as the lack of a transparent measurement system for greenhouse gas emission reductions, caused considerable difficulties in identifying which projects and loans of the Bank were of relevance for its climate change strategy and how much they contributed. The IDB climate change strategy was approved in 2011 to bring activities together that are relevant for climate change and to approach them in a more systematic way, to enable mainstreaming and upscaling. While the activities themselves have considerable history in the IDB, their relationship to climate change goals has been relatively recent. The benefit of an older portfolio is that it enables a look at finished projects, even if this means extra efforts to reconstruct what its specific contribution to climate change mitigation was. Compared to the CIF evaluation, the IDB evaluation is able to provide evaluative evidence on effectiveness, though longer term impact remains elusive due to measurement problems (IDB 2014, p. x). The Independent Evaluation Department of the Asian Development Bank in its real-time evaluation of the ADB's initiatives to support access to climate finance also noted the difficulty of assessing the climate impact of activities that may have other primary objectives and the lack of a consistent framework for measuring greenhouse gas emission reductions (ADB/IED 2014, p. xi).

The evaluation of the Swiss International Cooperation in Climate Change from 2000 to 2012 develops the same argument for the portfolio of interventions it looked at. The focus on climate change is relatively new in Swiss cooperation, and many of the older projects were formulated and implemented from development and poverty alleviation perspectives. As a result, no consistent data sets are available to measure the impact of especially the earlier interventions on climate change (Gaia Consulting 2014, p. 9). Yet the portfolio scores high on effectiveness, showing "moderate to strong" effectiveness in reducing greenhouse gas emissions and increasing people's abilities to cope with the impacts of climate change (Ibidem, p. 5). Due to the methodological challenges in evaluating a portfolio that emerged from other objectives, but is now seen as central to climate change efforts, the evaluation is reduced to noting that there are "numerous examples of successful emission reductions" but no overall picture emerging.

Norway's support to Reducing Emissions from Deforestation and Forest Degradation (REDD+) through Norway's International Climate and Forest Initiative (NICFI) has been evaluated in 2013 in a summative evaluation, looking at what had been achieved so far. This support has similarities with the CIFs in that funding is aimed to achieve impact through focusing on a few countries, so that the amounts of funding become catalytic. The evaluation concludes that the portfolio is "providing a substantial, direct contribution towards the conservation of natural forests" (LTS International et al. 2014, p. xxii), and that it is "likely" that this will lead to higher level and long term impact, as the supported activities contribute to "sustainable development" (LTS, p. xxiv). Yet this would be dependent on future funding, which is uncertain – it is this lack of certainty that the evaluation proclaims to be the "single greatest risk" to the sustainability of the REDD+ initiatives. Important to note is that greater coherence and consistency has been achieved in measuring greenhouse gas emission reductions – the deciding factor no doubt was that this support was set up as climate change support from the beginning (see LTS, p. xxx).

The picture emerging from the Norwegian evaluation is complemented by the independent evaluation of the UN-REDD+ programme, undertaken in 2014. This evaluation concludes that the programme has been moderately successful in delivering outputs, whereas its overall (programme) effectiveness is rated as moderately unsatisfactory (Frechette 2014, p. iv). Its efficiency is rated as unsatisfactory: the three UN partners in UN-REDD+ continue to have their separate procedures, which leads to inefficiency in the management of the programme (Frechette, p. 30).

We may draw the following conclusions from this overview of the findings of the seven comprehensive evaluations, which are presented in Table 2.1 First of all, three conditions at the portfolio level emerge for an evaluation to be able to provide evidence of direct impact and of impact at the global level:

1. Only funding agencies that have steadily built a **coherent portfolio** focused on climate change can expect evaluative evidence on the impact of this portfolio; portfolio's that are gathered from interventions with other aims as primary objective tend to show a lack of data related to climate change, different interpretations of what should be done and a wider range of activities to achieve outputs.
2. The portfolio needs to be coherent and **mature** to find solid evidence of direct impact; this is the case for the GEF only. The UN-REDD+ evaluation managed to gather evidence on the "likelihood" of impact and sustainability.
3. Only the GEF and UN-REDD+ have a consistent set of instructions for **measuring greenhouse gas emission reductions**. These instructions are still under development and will no doubt further improve over time; but they make it possible to aggregate GHG reductions at the portfolio level. The IDB, ADB and the Swiss Cooperation evaluations faced difficulties for using GHG reduction data because of the lack of coherence in the portfolio, with interventions now counted as important for climate change which were not set up for this purpose originally. Even though their portfolios are mature, they do not lend themselves to providing evidence at the impact level, as the lack of comparable data leads to problems of aggregation that cannot be overcome, at least not until the portfolios have matured further and measurement norms and standards are agreed.

The first important element of the micro-macro paradox is evident in the judgments on **efficiency** and **effectiveness**. Where these were rated, efficiency was deemed to be low or unsatisfactory. Where effectiveness was rated, evidence pointed in the direction of moderately satisfactory to fully satisfactory outputs. On the **direct impact level**, of amounts of GHG emission reductions in the new situation, only the GEF provided evidence at the portfolio level, but other evaluations certainly provided evidence at the intervention level, such as the IDB, ADB, and Swiss Cooperation. The only discrepancy in findings emerged between the NICFI and UN-REDD+ evaluations, where the Norwegian evaluation found a

Table 2.1 Evaluative evidence on progress towards global impact in seven comprehensive evaluations

Comparison of conclusions of 2014 comprehensive evaluations on climate change

Evaluation	Coherent portfolio	Mature portfolio	GHG measurement	Efficiency	Effectiveness	Direct impact at programme level	Global level impact
GEF – OPS5	Yes	Yes	Yes	Low	High	Evidence of higher level impact, yet trends continue to decline	Insufficient funding versus subsidies for non-sustainable use of natural resources
CIF	Yes	No	Data not always comparable	Low	Not evaluated	Not evaluated	Resources are small compared to global needs; focus on transformation yet many activities lack a clear pathway towards impact; learning potential yet unfulfilled
IDB/OVE	No	Yes	Data not always comparable	Not evaluated	High	Problems of aggregation	Not evaluated
ADB/IED	No	Yes	Data not always comparable	Not rated	Not rated	Problems of aggregation	Not evaluated
Swiss Cooperation	No	Yes	Data not always comparable	Not evaluated	High	Problems of aggregation	Not evaluated
NICFI	Yes		Yes	Low efficiency of UN-REDD+	High	Likelihood of higher level and longer term impact	Insufficient funding may be a barrier to reach longer term sustainability of forests as natural resource
UN-REDD+	Yes	No	Yes	Low	Moderately satisfactory	Moderately unsatisfactory	Not evaluated

"likelihood of higher level and longer term impact", but the UN-REDD+ evalua-tion rated the same as "moderately unsatisfactory". Although the NICFI and UN-REDD+ evaluations overlap to a large extent (NICFI being the biggest donor to REDD+ initiatives), the difference may be due to a willingness or reluctance to look into the future. On **global level impact** the evaluations that were willing to enter into a somewhat reflective and speculative mode – i.e. the CIF and NICFI evaluations – came to similar conclusions as the GEF, that funding in these climate change initiatives remains relatively small to global needs and may also be unpredictable – thus putting a huge question mark on the global level impact of climate change interventions.

The main thesis of the Fifth Overall Performance Study of the GEF, that success at the micro level is not leading to a change in trends at the macro level because of funding issues, is thus supported in the CIF and NICFI evaluations. Various elements are mentioned in the funding gap: the subsidies for non-sustainable use of natural resources which are substantially higher than funding of international action against climate change; the relative low amount of funding versus the identified global needs; and the unpredictability of funding in the coming years. The core of the micro-macro paradox is further substantiated in the older World Bank evaluations and the briefing note that the Evaluation Cooperation Group of the multilateral banks developed for COP17 in Durban in 2010.

2.4 Surviving the Negative Effects of Climate Change

While the onslaught of climate change continues unabated, the relevance and urgency of adaptation to changing circumstances has been increasing. While this is still questioned in developed countries where climate change deniers hold office, many if not all developing countries are working on national priority and action plans for adaptation to climate change. While support for adaptation did not figure prominently in the early years of climate action after the Earth Summit in 1992, it has come to the foreground and is now seen as of equal importance as mitigation in guidance of the UN Framework Convention for Climate Change. The international portfolio for support to countries on adaptation is not as mature as the portfolio on mitigation. Furthermore, international agreement on a comprehensive framework for adaptation – what adaptation is, what it would be composed of and how it should be measured – is still developing. While UNFCCC and the Intergovernmental Panel on Climate Change have done important work in providing a first understanding of adaptation and how countries can develop national priority plans for adaptation action, evaluations have not yet delivered a critical mass of evaluative evidence.

It could be argued that it is not necessary to look at adaptation from a global perspective. To adapt or not adapt is not something that happens on a global scale. While greenhouse gas emissions lead to climate change for the globe, adaptation is by definition more local – if one country is well adapted, it does not lead to better adaptation in its neighboring countries. Furthermore, while greenhouse gas emissions can be measured through technical processes, there is no similar mea-surement for adaptation. No single indicator for adaptation will suffice. We argue that adaptation has at least three distinct dimensions: (1) changes in social and economic development that ensure that the outputs and outcomes are sustainable from a climate change perspective; (2) preparedness for and dealing with natural disasters that may increase in intensity due to climate change; (3) resilience of populations and societies to tackle unexpected changes in the natural environment that they are living in. The first dimension increasingly overlaps with mitigation action. While mitigation may be primarily directed at reducing greenhouse gas emissions, the micro-macro paradox establishes clearly that for these actions to be ultimately successful, systems need to change and become environmentally sus-tainable. This is the route towards durable emission reductions, and it is also the route towards increased adaptive capacity. For this reason we see an increasing use of the same transformative mechanisms for adaptation as for mitigation.

Adaptation and mitigation are two different but linked dimensions in social, economic and environmental sustainability. Adaptation concerns the ways in which the social and economic domains are "ready" for change in the environmental domain, and includes resulting actions. Mitigation focuses on one particular way society and the economy use natural resources and aims to make this use environ-mentally sustainable. Adaptation perspectives in mitigation often are termed "cli-mate proofing" of actions; ensuring that the mitigation interventions will be resilient against climate change. Both adaptation and mitigation ultimately require action that transforms the interaction between the social, economic and environ-mental domains. One of us argued that sustainability is fundamentally an adaptation issue (van den Berg 2014, p. 34–35): "achieving a sustainable balance among civil society, the economy and the environment will require constant adaptation". In this light we include some of the evaluative evidence on adaptation in our discussion of transformative action.

2.5 Three Priority Areas for Transformative Action

The seven evaluations and their predecessors also provide much information and evaluative evidence on how transformative processes can be set in motion and what is essential for these processes. A coherent picture emerges of action at the country level, from civil society, the private sector and the government; action which requires legal and regulatory amendments and changes in markets and behaviour in society; of engaging with civil society which collaborates or is the main actor for behaviour change; of engaging with the private sector which introduces new solutions and technologies that could together with changed behaviour lead to market change and transformation. A crucial cross-cutting issue is whether activ-ities take gender, equity and inclusiveness into account, as they are essential to ensure the transformation will not just have an economic and environmental, but also a social impact. For this reason the next section of this chapter discusses briefly the interaction with civil society through the example of small grants provided to local communities, the introduction of new technology in collaboration with the private sector and the gender, equity and inclusiveness dimensions.

2.6 Civil Society Action Supported Through Small Grants

While civil society is active at all levels of governance, including the global level, it tends to be rooted in local organizations and action. Bottom-up action and representation are considered essential by many civil society organisations. Change in behaviour initiated by and in civil society often follows its own dynamics – some of it top-down, where governments impose rules and regulations, or behavior is modelled on the example of popular characters or opinion leaders, but often durable change is initiated at local levels and gradually (or quickly) spreading to the general population. Of the organisations evaluated the Global Environment Facility has supported local civil society initiatives through its Small Grant Programme (SGP). We turn now to the evaluations of this programme to look at whether this provides a promising avenue for civil society engagement in climate action.

The SGP was established in 1992 and implemented by the United Nations Development Programme (UNDP) on behalf of the Global Environment Facility (GEF). SGP provides small grants of up to US$50,000 to local communities as they take action on sustainable use and conservation of biodiversity, climate change, land degradation, international waters, sustainable forest- and chemical management. SGP has provided over 18,000 grants to communities in more than 125 countries. In its fifth operational phase (2011–2014), the Programme's aim included expanding its coverage to 136 countries. US$288.28 million was allocated to the SGP and total co-financing mobilized from diverse sources amounts to US$345.24 million (GEF/IEO and UNDP/IEO 2015, p. 1).

From 2013 to 2015, the GEF and UNDP Independent Evaluation Offices jointly evaluated SGP. One of the conclusions of the evaluation states that '*SGP continues to support communities with projects that are effective, efficient, and relevant in achieving global environmental benefits while addressing livelihoods as well as promoting gender equality and empowering women*' (GEF/IEO and UNDP/IEO 2015, p. xiii). The evaluation further notes that SGP's system ensures global policies are translated into action at the local level. The results at local level were impressive, with high percentages of projects that contributed to livelihoods, poverty reduction and gender.

In many countries, SGP achievements were replicated, upscaled and mainstreamed, sometimes to the extent of policy influence, into local and sometimes national development processes. Replication often takes place on a local scale only: other villages and communities copying what had been achieved in a specific SGP supported activity. Thus, successful introduction of conservation of mangrove forests in Senegal at the local community level was replicated in other villages. Mainstreaming happened less often, but an interesting example was an SGP grant in

Uganda to initiate behaviour change in local communities on the waste they produced and to take responsibility for this waste, rather than to expect the government to remove it. This behaviour change was then promoted throughout Uganda, supported amongst others by the World Bank (Examples from GEF/IEO 2015, p. 18). However, broader adoption at the national and regional levels tends to run into obstacles – success is most prevalent at local levels (GEF/IEO 2015, p. 19). The micro-macro paradox is therefore also visible in civil society involvement and action. Successes at the local level do not necessarily translate to national and regional levels, even though these successes evidently "extend beyond the project level" (GEF/IEO 2015, p. 19).

One can also see the SGP success with a bottom up approach as it contributes to numerous institutional and policy changes at the local, provincial, and national levels, and to building capacities within civil society and academic organizations to address global environmental concerns. Its success has resulted in a high demand for support (GEF/EO 2010, p. 18).

This is further demonstrated in a regional GEF project in the Pacific on Biodiversity Conservation[7] which aimed to introduce community based conservation approaches throughout the Pacific Islands. This approach, focusing on solving land-use problems between villages, while integrating livelihood issues in local conservation planning, is now in use throughout the Pacific and has been successfully adapted to local circumstances. However, evaluative evidence in Vanuatu shows that success in communities does not (yet) equal success at the national level, as the government has not been able to dedicate resources to institutionalize the new approach (and with the devastation caused by Cyclone Pam in 2015, it may take extra time before the approach can be integrated in its national policies).

Evaluative evidence thus shows that the bottom-up activism of civil society organisations and local communities, when supported with focused funding as provided by the Small Grants Programme of the GEF, can be successful and provide solutions that can be incorporated at national and even regional scale. However, the micro-macro paradox is also evident at this level and additional action is required to achieve broader adoption and systems change.

2.7 Introducing New Technologies Through the Private Sector

There is wide-spread agreement that climate action involves a substantial and transformative technological overhaul of production processes in the private sector. Innovation, together with the promotion, development and transfer of environmentally sound technologies, and uptake of these in the private sector is critical in enabling countries to combat climate change and to pursue their sustainable

---

[7]GEF ID 403.

development objectives. This can mean using renewable energy or transforming current equipment or technologies into something that is cleaner and more climate-resilient. This is reflected in the comprehensive evaluations: the introduction of technology scores high in the support provided by the organisations evaluated.

Since 1991, the GEF has been facilitating technology transfer to support developing countries through know-how, goods and services, equipment, as well as organizational and managerial procedures. The GEF has invested around US$250 million annually[8] in, among others, energy efficiency, renewable energy, emerging, low carbon and energy generating technologies and sustainable urban transport.

In July 2008, World Bank Executive Directors approved the establishment of the Clean Technology Fund (CTF), under the Climate Investment Funds (CIF). CTF is a US$5.6 billion fund that empowers the transformation in middle income and developing countries by providing resources to scale up the demonstration, deployment and transfer of low carbon technologies with significant potential for long-term greenhouse gas (GHG) emission savings.[9] Although implementation is still at its early stages, CTF investment plans, if successful, would boost renewable energy generation capacity or reduce national power consumption by 1–8 %. CTF funding for concentrated solar power, if successful, could boost total global capacity by more than 40 %.

In 2012, with financing from the GEF and in collaboration with the United Nations Environment Programme (UNEP), the Asian Development Bank established the Climate Technology Finance Center (CTFC). The Center is designed to promote transfer of and investment in climate technologies and to help mainstream climate technology considerations in development planning. Since inception, the Center has provided lessons on climate change initiatives to other multilateral development banks (MDBs) (ADB 2014).

While other organisations may not have dedicated instruments for technology transfer, they show a similar emphasis on innovation and introduction of technologies, especially in collaboration with the private sector. Evaluative evidence in the seven comprehensive evaluations focuses on the following issues.

Technologies That Work Best Tend to Be Already Tested Elsewhere A lot of technology transfer has been successful precisely because it was focused on well proven technologies. Replication was typically taken on by the private sector as a result of evidence showing that a technology was both cost-effective and profitable. Sound monitoring that demonstrates the benefits of a technology becomes even more important to its broader adoption (GEF/IEO 2014, p. 54). CTFC experience showed that many country governments do not give high priority to the introduction of relatively high-cost climate change risk reduction technologies. As a result, CTFC undertook a phased approach whereby it is required to first demonstrate

---

[8]Global Environment Facility (2016) Technology Transfer for Climate Change. https://www.thegef.org/gef/technology_transfer. Accessed 20 April 2016.

[9]Climate Investment Funds (2016) Clean Technology Fund. https://www-cif.climateinvestmentfunds.org/fund/clean-technology-fund. Accessed 20 April 2016.

benefits to countries (ADB/IED 2014, p. 21). The Report on Effectiveness of the Swiss International Cooperation in Climate Change (2014, IV) highlighted that groups of projects with strong scores for mitigation effectiveness were found to include projects that targeted the rehabilitation of hydropower systems and power systems with direct energy efficiency benefits and enabling impacts for renewable energy promotion, the strengthening of measuring, reporting and verification capacity and carbon market readiness, the use of knowledge sharing among cities and companies, and the rehabilitation and re-deployment of used Swiss trams to other countries.

A Fully Supportive Enabling Environment Is Necessary Wörlen's (2014) meta-analysis of mitigation interventions led to a systematic overview of all the barriers to change – providing a "theory of no change" – an explanation of why market change or transformation was not happening. The theory of no change demonstrated that introduction of technology will only be successful if all potential barriers for change have been tackled. The ICF evaluation showed that in more than half of CTF countries, policy, regulatory, and macroeconomic situations have the potential to slow down or limit transformation and replication. These countries have supportive policies in place that provide building blocks, but lack implementing regulations specifying key details of the regulatory environment, weakening the potential for immediate replication. Non-investment-grade credit ratings are also a limiting factor in some countries (ICF 2014, X). ADB's Climate Technology Finance Center (CTFC) also encountered difficulties during its design and launch. Barriers include financial constraints, insufficient knowledge base and expertise, and inadequacies of public policies, regulations, and enforcement (ADB/IED 2014, p. 21).

A Crucial Supporting Factor Is the Availability of Financing If loans for investment in new technology are unavailable, then this technology will not be widely adopted. The Fifth Overall Performance Study of the GEF (2014) showed that mainstreaming typically took place because of financial incentives provided by the national government to adopt the technologies (p. 54). The IDB/OVE evaluation notes that promoting the development of small-scale energy efficiency projects has proven to be more difficult, as small firms face high transaction costs and low financial returns from these investments (partly because of energy subsidies), and they require access to long-term financing (2014, p. 67).

The CIF evaluation could not see a clear path towards broader adoption of many technologies tested and demonstrated in CIF support, because these projects and programmes lacked a convincing theory of change that would explain how replication and market change and transformation would take place. This seems at least partly due to investment criteria, for example in CTF, that focus on quantifying GHG emission reductions rather than causal pathways to transformative change (ICF 2014, x). The focus on GHG emission reductions is visible in other evaluations as well – it points to the possibility that technology is easier judged on its contribution to climate change mitigation, without full recognition that any technology will only perform if the social, economic and environmental prerequisites are in place over time.

2.8 Gender, Equity and Inclusiveness

The Rio Declaration on Environment and Development,[10] adopted at the Earth Summit in 1992, introduced principles (10, 20, 21, 22) on participation and importance of specific groups (civil society, women, youth and indigenous groups) for sustainable development. The Sustainable Development Goals recognize that society and the economy need to ensure an equitable distribution of wealth, attention for gender perspectives and ensure inclusiveness of both civil society and the private sector, as well as government. Achieving this will ensure sustainable development in the social and economic domains. Without this balance, the balance of social and economic needs with the environmental domain will be meaningless.

Climate change impacts affect men and women, with the poorest being the most vulnerable. Seventy percent of the world's poor are women,[11] making them extremely affected. On the other hand, they also play a large and important role in tackling climate change. As impacts of climate change increase, work predominantly undertaken by women (i.e. food production, supplying household water, ensuring fuel for heating and cooking) is becoming increasingly more difficult. Coping strategies and their resilience give them a practical understanding of innovation and skills to adapt to changing realities, as well as contribute to finding solutions.

The GEF has recognized gender as highly important to achieve behavioural change that will lead to broader adoption of sustainable solutions to global environmental problems. In 2010, OPS4 highlighted that 'social and gender issues in GEF strategies and projects are not addressed systematically, and the GEF cannot rely completely on the social and gender policies of its Agencies.' (GEF/EO 2010, p. 30). As a response, the GEF developed its policy on gender mainstreaming and adopted it in May 2011. There has also been an increase in the proportion of projects that aim to mainstream gender. These improvements may be attributed to adoption of gender mainstreaming by several GEF agencies, of which the best international practices come from IFAD, UNDP and the World Bank. Despite the adoption and review of a gender policy and designation of a focal point OPS5 provides evaluative evidence that attention for gender in projects is often lacking. No less than 43 projects evaluated qualified themselves as "gender not relevant",

---

[10]United Nations Environment Programme (2003). Rio Declaration on Environment and Development. http://www.unep.org/documents.multilingual/default.asp?documentid=78& articleid=1163. Accessed 12 May 2016.

[11]United Nations Framework Convention on Climate Change (2014) Gender and Climate Change. http://unfccc.int/gender_and_climate_change/items/7516.php. Accessed 19 April 2016.

but provided evidence that gender turned out to be relevant after all. OPS5 concluded that "omitting attention for gender where it is needed may have led to unintended negative gender-related consequences". A baseline study undertaken by OPS5 revealed that many climate action projects were formulated by experts insufficiently aware of gender issues. On the good side it should be noted that the same study also revealed projects that tackle gender issues adequately (GEF/IEO 2014, p. 61).

When the Climate Investment Funds (CIF) started in 2008, they did not have an explicit gender focus – most countries did not include women's organisations in investment plan consultations. However, in 2009 and 2010, 15 % of the plans started declaring gender considerations. Some works remains to ensure that gender considerations are mainstreamed in CIF planning and carried through to investment projects in the field. In a positive step forward, the CIF hired a gender specialist to develop and implement an action plan to support collaboration among MDBs.

Attempts have been made throughout the NICFI portfolio to address gender issues in REDD+. However, it is stated by the evaluation that among partners, there is a lack of understanding of, and low general capacity to address gender. The strongest contribution has been through the UN-REDD programme, whereby numerous publications on REDD+ and gender have been produced.

For UN-REDD, the importance and need for gender mainstreaming is reflected in most of its policy and programmatic documents and guidelines. However, the implementation of gender mainstreaming activities at the country level is not taking place in a cohesive and systematic way throughout the programme. The evaluation (2014) stated that drivers of deforestation will be better addressed if gender considerations are integrated especially at the local level.

The track record on equity and inclusiveness is even less impressive. While equity and inclusiveness are essential dimensions of social, economic and environmental sustainability, they are perhaps too far removed from the often technical nature of the climate actions reviewed in the seven comprehensive evaluations. The Fifth Overall Performance Study of the GEF does not mention equity or inclusiveness, while the CIF evaluation only mentions equity in relation to investments and inclusiveness of stakeholders in consultations. There is indirect attention to the issues – for example in the attention for local livelihoods, involvement of indigenous peoples and civil society organisations. An example is to be found in the NICFI evaluation: since 2008, NICFI provided a total of NOK 1 billion or 9 % of its funding to civil society to generate needed knowledge, for advocacy (international and political), piloting and facilitating implementation (Frechette etc. 2014, xix). UN-REDD's evaluation stated that 'The Programme provides an enabling platform for Indigenous Peoples and civil society organisations to influence global discussions on REDD+. The ability of forest-dependent populations to influence REDD+ processes has so far proven to be more limited at the country level, and non-indigenous communities are not well represented in the programme, overall' (Frechette etc. 2014, vi).

While attention for gender, equity and inclusiveness is on the rise, the evaluative evidence is overwhelming that these dimensions have not yet been fully included in climate action. With the adoption of the Sustainable Development Goals, this has become more important and we hope to see evaluative evidence emerging the coming decade.

2.9 When Will We Achieve Systems Change?

The comparison of the seven comprehensive evaluations published in 2014, and some influential evaluations from the years before, leads to the conclusion that if we want to achieve transformational change, we need to ensure that the impact drivers working towards such a change are stronger than the impact drivers that cause climate change. An important part of the fight to mitigate climate change is therefore outside of climate change action: continue the fight against public subsidies for non-sustainable use of natural resources; take action to ensure that the costs of climate change are paid for by the "polluters", by industries and people who are causing climate change to happen. Until that time climate change action will consist of beautiful flowers in a walled garden: just a demonstration that we can have a beautiful planet, if only the winds blowing against us would not destroy these beautiful flowers if they emerge from the walled garden.

The conclusions we draw from the seven comprehensive evaluations are as follows.

1. The OPS5 conclusion that a high percentage of climate action is **effective** is supported by all other comprehensive evaluations that have been able to look at effectiveness;
2. For reduction of greenhouse gas emissions, the energy sector and energy policies hold greatest promise, and tackling energy efficiency issues is more effective than support for renewable energy, but the latter is effective as well, even if costlier.
3. Subsidies for non-sustainable use of natural resources (fossil fuels, agricultural practices, overuse of water resources, etc.) prevent the reductions in greenhouse gas emissions to have more than a marginal impact on climate change: i.e. the pace of climate change is slowed almost imperceptibly.
4. The micro-macro paradox is thus shown to exist: anything the international organizations and the bilateral donors do to prevent climate change continues to be effective in its own right, but powerless against the enormous spending power and damage done by subsidies for non-sustainable use of natural resources, with fossil fuel subsidies as the largest barrier to change.
5. To change the system, action from many partners, bottom-up and top-down, with full recognition of cross-cutting issues such as gender, equity, inclusiveness is needed, and evaluative evidence shows that pieces of the puzzle are known and can be effectively set up and used.
6. To change the system, an important input of technology is needed – the shift from fossil fuels to a low-carbon economy needs technological innovation and

change. The technologies to do this exist and need to be fostered and promoted: evaluative evidence shows that when they are introduced properly and supported by all actors, they can be effective fertilizers of change.

7. On adaptation to climate change many promising evaluative findings are emerging; what is lacking is a concentrated effort to gather the evaluative evidence and interpret it, learn at higher levels of aggregation and integrate adaptation into social, economic and environmental development as the essential ingredient that will ensure sustainability.

2.10   Recommendations for Future Evaluations

As argued in van den Berg (2013, p. 47), evaluations of climate change interventions, especially if they aim for a higher more comprehensive level of understanding what the interventions mean and what they achieve in the longer run, need to evaluate in the context of the continuing societal and economic winds that are causing climate change. Evaluators need to point out to policy makers and decision makers that what they promote with one hand, is more than sufficiently undone with a very active and much bigger other hand.

If the forces of destruction can be reduced or even halted, climate change action will become successful. How successful can currently not be established fully for all actions – there is some international agreement on measurement of reductions of greenhouse gas emissions, but this agreement needs to be further developed. Many countries, multilateral and bilateral organizations are currently using various measurement systems at the same time in different projects – this needs to be improved.

Evaluations, not just the seven comprehensive ones, but the many evaluations at the intervention level as well (many of which are highlighted in this book), provide an increasing body of knowledge that has been insufficiently explored for policy makers and decision makers on what works, where and when and for whom under what circumstances. This should focus on:

- How systems change can be effected through activities on key issues that will "tip" or "tilt" the system in the right direction;
- Identify the top-down actions that can and should be taken as they have proven to be effective; similarly look at bottom-up actions that can and should be supported;
- Present evidence on the difference in effectiveness between inclusive, gender sensitive, equity based approaches versus approaches that lack these perspectives;
- Contribute to a repository of evidence on which technologies under which circumstances, for whom, have proven to be effective in supporting more sustainable and low-carbon growth;
- Contribute to a repository of evidence on what works for whom, when, where, and under which circumstances on adaptation to climate change.

The evaluation departments of countries, bilaterals and international organizations, as well as philanthropic foundations and private and social enterprises should increase their collaboration, as the war on climate change requires a concentrated effort, rather than everybody focusing on their own constituency and their own accountability. Your children and your children's children will one day ask you, as an evaluator, what you have done to stop climate change. Your response should not be that you have provided evidence to your government or Board on how the money was spent, but on what is useful and a potential winner in our battle to keep the planet habitable for humankind.

References

ADB/IED. (2014). *Real-time evaluation of ADB's initiatives to support access to climate change.* Manila: Asian Development Bank Independent Evaluation Department. [Thematic Evaluation Study].

Arndt, C., Jones, S., & Tarp, F. (2010). *Aid, growth, and development: Have we come full circle?* (Working paper no. 2010/96). Helsinki: UNU-WIDER.

Dollar, D., & Prichett, L. (1998). *Assessing aid. What works, what doesn't, and why.* New York: Oxford University Press.

Frechette, A., Minoli de Bresser, & Hofstede, R. (2014). *External evaluation of the United Nations Collaborative Programme on reducing emissions from deforestation and forest degradation in developing countries (the UN-REDD Programme). Volume I – Final report.* Retrieved from http://tinyurl.com/zoavwkm on 17 Apr 2016.

Gaia Consulting Oy, Creatura Ltd., Zoi Environment Network. (2014). *Report on effectiveness 2014: Swiss international cooperation in climate change 2000–2012.* Bern: SDC and SECO.

GEF/EO. (2010). *OPS4: Progress toward impact. Fourth overall performance study of the GEF. Executive version.* Washington, DC: Global Environment Facility Evaluation Office.

GEF/IEO. (2014). *OPS5 fifth overall performance study of the GEF: Final report: At the crossroads for higher impact.* Washington, DC: Global Environment Facility Independent Evaluation Office.

GEF/IEO. (2015). *GEF country portfolio evaluation: Vanuatu and SPREP (1991–2012). Evaluation Report No. 98.* Washington, DC: Global Environment Facility Independent Evaluation Office.

GEF/IEO and UNDP/IEO. (2015). *Joint GEF-UNDP Evaluation of the Small Grants Programme.* Global Environment Facility Independent Evaluation Office, Washington DC and United Nations Development Programme, New York.

ICF Consulting and partners. (2005). *OPS3: Progressing toward environmental results. Executive version.* Washington, DC: Global Environment Facility Office of Monitoring and Evaluation.

ICF International. (2014). *Independent evaluation of the climate investment funds.* Washington, DC: World Bank.

IDB/OVE. (2014). *Climate change at the IDB: Building resilience and reducing emissions.* Washington, DC: Inter-American Development Bank Office of Evaluation and Oversight. [Thematic Evaluation].

LTS International, et al. (2014). *Real-time evaluation of Norway's international climate and forest initiative: Synthesising report 2007–2013.* Oslo: Norwegian Agency for Development Cooperation.

Mosley, P. (1987). *Foreign aid: Its defense and reform.* Lexington: University of Kentucky.

World Bank/IEG. (2009). *Climate change and the World Bank Group. Phase I: An evaluation of World Bank win-win energy policy reforms.* Washington, DC: Independent Evaluation Group of the World Bank.

World Bank/IEG. (2010). *Climate change and the World Bank Group. Phase II: The challenge of low-carbon development*. Washington, DC: Independent Evaluation Group of the World Bank.

World Bank/IEG. (2012). *Climate change and the World Bank Group. Phase III. Adapting to climate change: Assessing the World Bank Group experience*. [publication without date; first posted on the IEG website on October 25, 2012 – see http://goo.gl/SWcTUR. Accessed May 9 2016.

Wörlen, C. (2014). Meta-evaluation of climate mitigation evaluations. In J. I. Uitto (Ed.), *Evaluating environment in international development* (pp. 87–104). London: Routledge.

van den Berg, R. D. (2011). Evaluation in the context of global public goods. *Evaluation, 17*(4), 405–415.

van den Berg, R. D. (2013). Evaluation in the context of global public goods. In R. C. Rist, M.-H. Boily, & F. Martin (Eds.), *Development evaluation in turbulent times. Dealing with crises that endanger our future* (pp. 33–49). Washington, DC: The World Bank.

van den Berg, R. D. (2014). A global public goods perspective on environment and poverty. In J. I. Uitto (Ed.), *Evaluating environment in international development* (pp. 17–36). London: Routledge.

Part I
Policy

Chapter 3
Mainstreaming Impact Evidence in Climate Change and Sustainable Development

Rob D. van den Berg

Abstract This chapter examines the demand for impact evidence and concludes that this demand goes beyond the experimental evidence that is produced during the lifetime of an intervention through "impact evaluations" as currently the term is used by many in the evidence movement. The demand for evidence of longer term impact at higher levels requires inspiration from an older tradition of impact evaluation and rethinking how the full range of impact evidence can be uncovered in evaluations. This is especially relevant for sustainable development which calls for a balanced approach on societal, economic and environmental issues. Climate change is a good example of this and a theory of change approach serves to identify key questions over time, space and scale to ensure that impact evidence can be found and reported throughout the lifetime of projects, programmes and policies and beyond in ex post impact assessments. Such an approach leads to mainstreaming of impact questions and related evaluation approaches throughout project and policy cycles. This chapter will demonstrate that evidence can be gathered throughout the lifetime of a project and beyond, in different geographic locations from very local to global, at different levels from relatively simple one dimensional interventions to multi-actor complex systems, up to global scales. It will thus argue for mainstreaming impact considerations throughout interventions, programmes and policies and for evaluations to gather evidence where it is available, rather than to focus the search for impact and its measurement on one or two causal mechanisms that are chosen for verification through experimentation.

Keywords Impact • Evaluation • Sustainable development • Climate change • Evidence

R.D. van den Berg (✉)
King's College London and International Development Evaluation Association (IDEAS),
Leidschendam, The Netherlands
e-mail: rdwinterberg@gmail.com

J.I. Uitto et al. (eds.), *Evaluating Climate Change Action for Sustainable Development*, DOI 10.1007/978-3-319-43702-6_3

3.1   Re-instating an Older Impact Tradition?

The debate on what constitutes impact in evaluation continues, with many in the evidence based movement[1] focusing on "rigorous" experiments to measure and identify what works and what doesn't, versus participatory and democratic approaches enabling beneficiaries to state what would be relevant for them. It is important to note that both approaches, and many others, tend to focus on the here and now: what is relevant now, what works now and what doesn't. However, there is another tradition in impact evaluation which is often overlooked or ignored, which is the historical approach. Every once in a while a historical evaluation is done (Jerve et al. 1999), and every once in a while somebody asks attention for this approach (van den Berg 2005), but it cannot be said to have been a strong tradition, nor a tradition that made a big impression. Complaints have been that historical evaluation studies are very expensive, are perhaps more research than evaluation, take a lot of time and are not impressive as regards learning, because lessons from years ago may not be relevant to the present circumstances, let alone the future (see for example the controversies surrounding the Dutch historical evaluations of long term relationships with several countries in van Beurden and Gewald 2004, pp. 63–67). So it is with some enthusiasm that the development community turned to experimental impact evaluation, preferably integrated into the design of projects and executed during their lifetime, and hoped that this would turn up relevant evidence of what works that would provide lessons for the immediate future. However, what if the evidence of what works and what doesn't only reveals itself over time? What if the time horizon is in decades? What else are we to do but integrate historical approaches with other tools and methods?

Many problems in development are longer term in nature: to reduce absolute poverty, to reduce child-birth related death rates, to improve nutritional status, to integrate countries into the global economy, and so on – these are measured over decades and changes tend to happen relatively slowly. The Millennium Development Goals in general addressed global trends and impacts at higher scales. At these levels impact evidence can no longer be generated directly through experiments and other analytical tools such as meta-analysis, statistical analysis and modelling tend to take over. The Millennium Development Goals were monitored through statistical data. As 2014 report on the achievements of the Millennium Development Goals states: "reliable and robust data are critical for devising appropriate policies and interventions for the achievement of the MDGs and for holding governments and the international community accountable" (UN 2014, p. 6). However, especially when complex programmes and policies need to be improved, evaluations and research have to play their role, as they can provide answers to questions why a certain trend is occurring. For this reason the 2030 agenda for sustainable

---

[1] A movement that has its roots in evidence based medicine (see https://en.wikipedia.org/wiki/Evidence-based_medicine) and has spread to education, international development and other areas, where its characteristics may differ in some aspects.

development includes evaluation as a follow-up and review principle for the agenda (UN 2015, p. 37).

Evaluations at higher scales and at the global level are not often done and are difficult to design, implement and report on. Many problems could be mentioned, such as reliability and comparability of data, external validity of evidence of causality, but a particular problem that raises its head in relation to impact evidence, is the problem whether evidence at local levels and lower scales translates into evidence at global levels, at higher scales and over longer time periods. The first chapter of this book has dealt with this issue in detail. In 2013 I argued that a "micro-macro paradox", which points to successes at the micro level that seem not to be reflected in trends at the macro level, is particularly relevant to the linkage between environment and development and thus to sustainable development which aims to achieve a balance between society, the economy and the environment (van den Berg 2013, pp 41–43). Climate Change provides good examples for this. Many climate change related interventions are successful and achieve what they set out to do. However, the success of individual activities has not affected global climate change substantially. As the Intergovernmental Panel on Climate Change concluded in its 2014 report: "without additional mitigation efforts beyond those in place today, and even with adaptation, warming by the end of the twenty-first century will lead to high to very high risk of severe, widespread and irreversible impacts globally (*high confidence*)" (IPCC 2014, p. 77). Notice the use of the term impact for global phenomena.

3.2 Demand for Impact Evidence

Although the evidence movement has aimed to narrow down and reduce the meaning of the term "impact" as referring to what can be found through counterfactual testing, the term impact is an ordinary word in the English language, the meaning of which varies according to context. While science and in this case evaluation may prefer a precise definition and a narrow meaning of terminology, in general this will not change how terminology is used in conversation and debates. When the public demands to see proof of impact, they will use the term impact in an undefined way. To correct the public tends to be rather difficult if not impossible. The question thus emerges whether narrowing the definition of impact is helpful and whether another approach would not be more appropriate, which is to identify how the term is used, what kind of evidence would be required to meet the demand and to identify clearly what the advantages and disadvantages are of the tools and thus of the reliability, validity and credibility of the evidence.

A good example of the discrepancy between what works and does not work at the local level and whether "impact" is achieved according to the way the public thinks about it, is climate change. At the level of individual activities good, solid evidence is found on what works, especially on mitigation of climate change. Mitigation activities aim to reduce the level of greenhouse gas emissions and thus aim to reduce the inflow of carbon dioxide and other greenhouse gases into the atmosphere. If a sufficient number of these activities take place, it should be possible to stabilize or even reduce the concentration of greenhouse gas molecules in the air, which is currently about 400 particles per million. While individual activities may be quite successful in reducing emissions, the overall concentration of greenhouse gases in the atmosphere continues to increase. There are thus two kinds of impacts that the public is concerned about: do individual interventions work and lower the emissions, and is climate change stopped? The first question may be answered through counterfactual experimentation, modelling or through before/after measurements of greenhouse gas emissions. Nothing about this is as simple as it sounds. The calculation and measurement of greenhouse gas emissions is not yet based on full understanding, agreement on principles and validation through international norms and standards. For an overview of the issues and what the current state of the art is, see STAP (2013).

All the successes of achieving impact at project level have so far not been able to change the overall trend in climate change, which is that the global mean temperature continues to rise. When asking for evidence of impact, donors and the public want to know whether projects have an impact, whether the project delivers and the causal mechanism that it embodies works. But donors and the public also want to know whether this leads to changes at higher levels, beyond the direct influence of the project, and ultimately they would like to see climate change stopped or even reversed. The demand for impact evidence is legitimate at all levels and cannot be met by referring to impact evidence only at project level or in the context of one intervention or one causal mechanism. Understanding the range of questions on impact evidence will enable evaluators to focus on the key questions that need to be asked in evaluations and will enable them to identify the tools and methods that need to be used.

3.3   Theories of Change for Climate Change Mitigation

The standard approach to identify key questions in an evaluation is to look for the "theory of change" that identifies how the intervention is expected to achieve impact. In traditional impact evaluations this leads to an identification of the causal mechanism that is supposed to "work". In climate change, this is usually a combination of a technical mechanism and a behaviour mechanism: "if this new technology is adopted by people/institutions/countries it will lead to reduced greenhouse gas emissions and thus to a lower rate of global warming". Traditional impact evaluations tend to focus on what works to effectuate behaviour change. If the behaviour change occurs, the intervention "works" and should be promoted. If it does not work, it should be stopped.

Organisations like 3ie, devoted to promoting traditional impact evaluations, are very much aware that this simple version may lead to all kinds of perverse effects that need to be taken into account or looked at, and for this reason they advocate that impact evaluations should be "based on a thorough analysis of an intervention's theory of change"[2] as there may be other links in the causal chains that should be tested or taken into account. Adopting the new technology is a change of behaviour, but it could potentially lead to unintended consequences which may lead to an overall increase of greenhouse gas emissions, if energy use increases overall. Other changes in the context may make a specific behaviour change redundant, as for example where new markets emerge and take over functions that are done more efficiently through new technology. However, the focus remains on checking for evidence of the behaviour change, as this is the causal mechanism that can be checked in a traditional impact evaluation. Let us explore whether a deeper understanding of the theory of change would lead to different and new questions.

Let us take a typical mitigation intervention as an example: the introduction of a new technology that would lower greenhouse gas emissions. The Hilly Hydel project in India was a typical project funded by the Global Environment Facility and the Government of India, supported through UNDP, which took place from 1995 to 2003. This has been a particularly well evaluated project (see Ratna Reddy et al. 2006). It was the object of a case study for a major GEF study on local benefits generated through support for global benefits (GEFEO 2006), has an end-of-project evaluation including a counterfactual impact assessment (Ittyerah et al. 2005) and was further studied for the GEF impact evaluation of mitigation projects in emerging economies (GEFIEO 2013). For a total amount of $ 14.6 million this project led to the introduction of small hydroelectrical power plants in hilly regions in India, mostly in remote villages without access to the main grid. The reduction of greenhouse gas emissions was supposed to be achieved through using a renewable source of energy (hydro power) and reducing the need for wood as a source for fuel, thus leading to a secondary but important benefit: reduced deforestation. The outputs of the project were a national strategy and master plan for hydro electrical power generation, 20 stand-alone small hydel power generating water mills, upgrading of 100 existing water mills to incorporate power generation and institutional and human capacities to ensure sustainability. In general these outputs were achieved or surpassed – upgrading of no less than 143 water mills took place. All in all this led to direct greenhouse gas emission reductions of 1900 tons CO_2 equivalent per year. If the potential for installation of these small-scale hydroelectric water mills would be fulfilled throughout India, the total amount of reductions per year would calculate as 4 million tons CO_2 per year (GEFIEO 2013, table 24 p. 70).

The theory of change of the project focused on introducing a technology that was new for the villages in the hilly areas, that would lead to a source of energy that would be more reliable and would lead to a halt to deforestation because of energy needs, reduced greenhouse gas emissions as a result and given its benefits, would convince villages to invest in this kind of technology. This would lead to a change in the market for rural energy in hilly areas, where hydroelectricity would take the

---

[2]See From Influence to Impact. 3ie strategy 2014–2016, p. 2, found at http://www.3ieimpact.org/media/filer_public/2014/09/07/3ie_strategy_summary_final_rgb.pdf, on September 4, 2015.

place of wood burning and fossil fuel generators, also resulting in less pollution in these villages. The behavioural assumption was that villagers would be willing to spend more money on energy given the benefits in reliability of supply, reducing the need for wood and thus reducing deforestation, reducing pollution and saving time in searching for wood. The hydroelectric power plants would be made available through public-private arrangements, supported by the States and by the Federal Government, and legitimized and promoted through a national strategy. The theory of change provided a series of causal linkages that together would change the market for hydro-electric power in remote hilly areas and would lead to considerable reductions of greenhouse gas emissions. More challenging was the perspective that this would also lead to reduced deforestation and to biodiversity benefits (Ratna Reddy et al. 2006, 4071).

The demand for evidence of impact can be placed at various levels in this theory of change. First of all, the hydroelectric power plants are supposed to produce energy with greater efficiency in greenhouse gas emissions than other local energy sources: these emissions should be lower than the same levels of energy produced through burning wood and through fossil fuel generators. Technological expectations in this regard need to be met and one could argue that the first impact question would be whether the hydroelectric plants deliver what they promise. The second question is whether the village manages to integrate the hydroelectric mills into their society: will they maintain the mills, pay their energy bills and use this source of energy instead of reverting to wood and fossil fuels? This is the kind of behaviour question that is beloved in traditional impact evaluations. A third question concerns whether the shift towards hydro-electric power is leading to a change on the energy market in remote hilly areas. Have demonstration and the first verifiable outputs of the project led to an increased supply on this market; i.e. is the private sector offering hydroelectric technologies to villages? And if so, is there a demand for this? Are villages actively taking this up for consideration when looking at their energy options? And is the financial sector willing to provide loans for investments to the communities or villages? A fourth impact question is then whether the market has changed – if it has changed – locally, regionally or nationally. These questions need to be looked at from three different perspectives: time, space and scale.

3.4   Key Questions Related to Time, Space and Scale

Especially with a global issue like climate change the demand for impact evidence ranges from "what works here and now" to "has it contributed, or will it contribute, to stop climate change". The first is very local, time and scale bound, just looking at whether a specific mechanism works as it is supposed to. The second looks at the planet, at scenarios that go into the future and that are at the highest (global) scale. Both are relevant questions and need to be answered.

This translates into issues of time, space and scale. It is quite clear that a project of $ 14.6 million cannot change the national energy market for remote hilly areas overnight. This takes time; in fact the impact assessment done at the end of the project asked for "adequate time" to pass and for a stable situation to be achieved before impact is assessed (Ittyerah et al. 2005, p. xv). And if individual projects need adequate time to have an impact, it follows that market change can only be observed and measured over even longer stretches of time. Longer time lapses are well known in environmental circles and on environmental impact, as Hildén (2009) and Rowe (2014, 54–55) have pointed out, but they tend to be less associated with market change. The slow pace of market change is more often observed with impatience, raising the question why no change is happening, which led Wörlen (2014) in her study of climate mitigation evaluations to reformulate the "theory of change" approach to a "theory of no change approach" that focuses on a better understanding of market barriers and how they can be overcome.

In general environmental boundaries do not follow jurisdictional boundaries. One ecosystem may spread over several countries, and one country may have several ecosystems. Rowe (2012) asked attention for the fact that location may differ conceptually and practically between a social and economic system that is targeted for change and an ecosystem that is influenced through the same intervention or action. But this is not only an issue of different locations of systems, but also of scope of an intervention: it may be focused on a direct impact in the villages in which it is implemented, while other areas are still outside the scope of the project or have not yet been approached by suppliers, or invited to participate by State or Federal government.

It is an issue of scale when impact needs to be observed at several levels: that of energy supply and demand, of greenhouse gas emissions related to energy, of greenhouse gas emissions including deforestation and alternative sources of energy, of livelihood and financial resources issues in the villages, of hilly rural areas in general, and perhaps somewhat more removed, whether greenhouse gas emissions in India are positively influenced by what happens in remote hilly areas. The last does not seem likely, and it may lead to a feeling of disenchantment – if it does not help India, it does not help the world, and it does not stop climate change.[3] But that was the reason the project was co-funded by the Global Environment Facility in the first place!

Scale is not easily defined. It seems clear that while interventions or actions move from one actor to multiple, from one location to many, from a "local" to a "national" or even "global" level that moving up scales is involved, but scales can also be understood in terms of different dimensions or sectors. Kennedy et al. (2009) recognises jurisdictional and management dimensions as different scales, and Bruyninckx (2009) asks attention for overlap and discrepancies between social, economic, environmental and spatial scales. Yet even though there is no universal agreement on how scales should be defined or what their boundaries are,

---

[3] And a good overall conclusion on the project was formulated by Ratna Reddy et al. (2006, 4078): the overall impact of the project appears to be slightly positive or neutral in a majority of key indicators. Certainly not a major contribution to reduced greenhouse gas emissions as hoped for.

there is widespread agreement that to mainstream, replicate, reproduce, upgrade or upscale interventions to higher levels is an essential perspective in understanding causal pathways from the micro-level to higher level goals.

Garcia and Zazueta (2015) argue that at higher scales interventions should be interpreted and looked at from a systems perspective. Individual components and elements do not a system make, but when they start interacting, they tend to take on characteristics of a system, which can have its own dynamics and shifts and changes. Arguably markets operate as systems and market change is systemic change: subtle changes in supply, demand and enabling environment can lead to "tipping points", after which slow, reversible change becomes irreversible, or the point in time at which a new technology (such as hydel power) becomes mainstream.

In conclusion key questions related to **time** lead to the realisation that impact can be measured at each moment in time – ex ante as impact assessment, through modelling and calculations, real time through monitoring, experimental design, trend analysis etc. and ex post through various evaluations and studies. Key questions related to **space** make us realise that impact differs per area and that areas have different impacts. Key questions related to **scale** point to the need to mainstream, replicate, upscale and broaden the scope of interventions before impact can be achieved at higher levels.

3.5   Using Time and Space to Identify Approaches

In principle the three dimensions of time, space and scale can be used to build a three dimensional matrix in which the theory of change of an intervention, programme or policy can be represented. This will enable the evaluator to identify where a particular demand for impact evidence needs to be placed, and what would be appropriate analytical tools to evaluate impact. Figure 3.1 presents a matrix of time and space aspects. The time dimension goes from ex ante (designing and formulating a new intervention) to important moments in real time (from inception to mid-term to end-of-project) to ex post and identifies ex post evaluation approaches. Red "balloons" signify evaluation approaches; blue ones monitoring and data analysis, whereas a green balloon identifies a research approach. Of course evaluations use and analyse monitoring data, and often use research tools and methods. Figure 3.1 just presents a possible configuration of what is dominant in the matrix from an evaluation perspective. The space dimension goes from local through national and regional to global, but has an extra row for ecosystems, which overlap with other rows.

The ex ante column is occupied by ex ante evaluation and impact assessment, which is a lively community of practice that uses various methods and tools to come to conclusions on the potential impact that different scenarios may have throughout time. These impact assessments tend to use modelling as their preferred tool and may present several scenarios that would lead to different impacts. The ex post

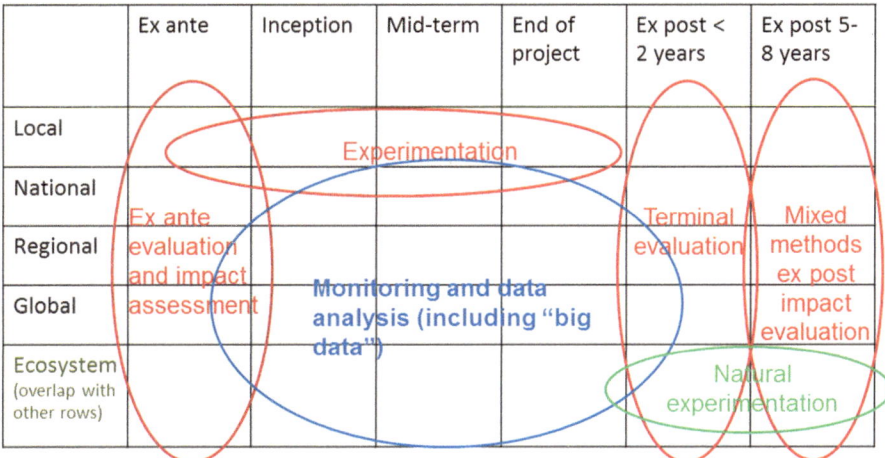

Fig. 3.1 The time and space dimensions of demand for impact evidence (Source: Author)

evaluation community tends to keep its distance from the ex ante evaluators, as there is widespread concern that any involvement of ex post evaluators in ex ante evaluation will lead to a conflict of interest when the activity needs to be evaluated later on. If design and implementation characteristics were decided upon because of an ex ante evaluation's outcomes, an ex post evaluator would in fact be required to evaluate his or her own judgments in the ex ante evaluation. In actual practice the two communities of practice hardly mingle. Ex ante evaluators have their own conferences and their own literature and good practice standards. What Fig. 3.1 shows is that they are the first to delve into the question of impact and aim to provide evidence, even if hypothetical at that stage, for what an intervention would set out to do.

During implementation monitoring and evaluation often become management tools. If the project needs to be steered through difficult circumstances and react adequately to changes, it needs to set up an adequate monitoring system, either collecting its own data or using data from available statistical services. Relatively new is the inclusion of real time evaluation, which on impact tends to take the form of randomized controlled trials that need to be included in the design of the project and need to be adhered to during implementation, in order to come to valid conclusions about the causal mechanism tested out. Other evaluations during implementation (such as mid-term evaluations) tend to look at processes and efficiency and are not represented in this matrix. Randomized controlled trials tend to be "local" in nature; rarely will we see RCTs at the national level and even more rarely at the regional level, as they would become very costly to reach a sufficient level of data (large "n") to allow for conclusions at that level.

In the ex post columns we tend to see two varieties of evaluation that provide impact evidence. First of all, end-of-project evaluations may present results of experiments or provide data on impact; usually these evaluations also contain important information on expected "progress toward impact" (van den Berg 2005) and whether the conditions have been set to enable longer term impact. The last column presents ex post evaluations 5–8 years after the project has ended. These are almost invariable historical evaluations, using a historical approach to trace whether the results of the project have contributed to observed changes in trends, markets, societies, economies and the environment. These evaluations tend to advertise themselves as "theory of change" oriented and using mixed methods and triangulation of evidence to come to conclusions on impact. They have less of a problem to move beyond countries to regions and the global level, not because the evidence is stronger after 5–8 years, but because they are more flexible in approach and are more pragmatic and adaptable in using data sources and linking data where possible. This sounds opportunistic, but there are many scientifically sound methods and tools that can be combined and triangulated, as amply demonstrated by Stern et al. (2012) and Garcia and Zazueta (2015).

3.6   Using Time and Scale to Identify Approaches

Another cut-through of the three-dimensional matrix of time, space and scale would be to combine time and scale. Figure 3.2 presents this matrix. The time dimension is of course the same as in the time-space matrix, but has been simplified a bit, for example presenting one row for ex-post rather than two. The scale dimension provides various perspectives of scale. From interventions focused on one causal mechanism, such as a project focusing on changing customer behaviour on the energy market through price setting, to multiple interventions within one project, of for example public-private partnerships, social change movements, capacity development efforts, to a perspective on an enabling environment that through rules and regulations, taxation, knowledge dissemination and other incentives tries to redirect a market or change behaviour, to market change and transformation, the interventions become more complex and challenging to evaluate. At the far right I have included climate change, and again this environmental scale overlaps with others, posing a special problem that two evaluends need to be recognized in an evaluation that includes environmental objectives (see Rowe 2012).

Again we see randomized controlled trials and quasi-experimental approaches focusing mostly on one intervention, as to control for combinations of interventions will become very costly. Ex ante research will deliver counterfactual assessments of how different scenarios will perform at all scales. A relatively new method such as Qualitative Comparative Analysis (QCA) is currently often used for case studies of more complex interventions and the enabling environment. Markets are of course the subject of economic research and for evaluations especially market research to assess whether a new product or approach has a chance on the market dominates in the market columns and ideally before the new intervention starts. At the programme and policy levels, ex post impact evaluations may look at triangulation

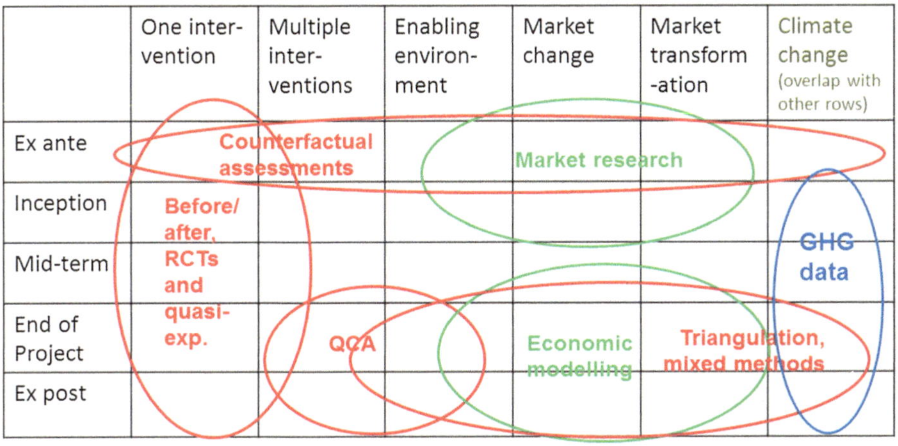

	One inter-vention	Multiple inter-ventions	Enabling environ-ment	Market change	Market transform-ation	Climate change (overlap with other rows)
Ex ante	Counterfactual assessments			Market research		
Inception	Before/ after,					GHG data
Mid-term	RCTs and					
End of Project	quasi-exp.	QCA		Economic modelling	Triangulation, mixed methods	
Ex post						

Fig. 3.2 The time and scale dimensions of demand for impact evidence (Source: Author)

of different sources of evidence (including monitoring data on Greenhouse Gas emissions) and use a mixed methods approach (as advocated by Bamberger 2012).

3.7 Using Space and Scale to Identify Approaches

When the two dimensional cut of space and scale are taken out the three dimensional time, space and scale matrix, at first sight a less well covered picture emerges, with some clear gaps where currently no favourite tool or method for evaluation seems to be in use. Figure 3.3 presents the rows used in Fig. 3.1 with the scales used in Fig. 3.2. I have focused methods and tools on what they are mostly used for and where their recognized strength is. Randomized controlled trials dominate in providing impact evidence on local interventions that focus on one causal mechanism. When we are looking at multiple causal mechanisms and interventions moving beyond national boundaries to regional collaborations, quasi-experimental methods and QCA become more or less dominant. Social network analysis is a particularly powerful analytical tool that could help in complex interventions with many partners, including the enabling environment that supports actors in participating in societal or economic action. The Delphi methodology has been used to evaluate market change, as market experts may be able to identify why changes have occurred and what would have happened without changes in the enabling environment or if certain technologies would not have become available. Research methods such as modelling take over on the right side of the matrix. The gap in the lower left hand corner of the matrix could potentially be an expression of costs: it would be prohibitively expensive to do global or ecosystem wide randomized controlled trials, while theory of change oriented

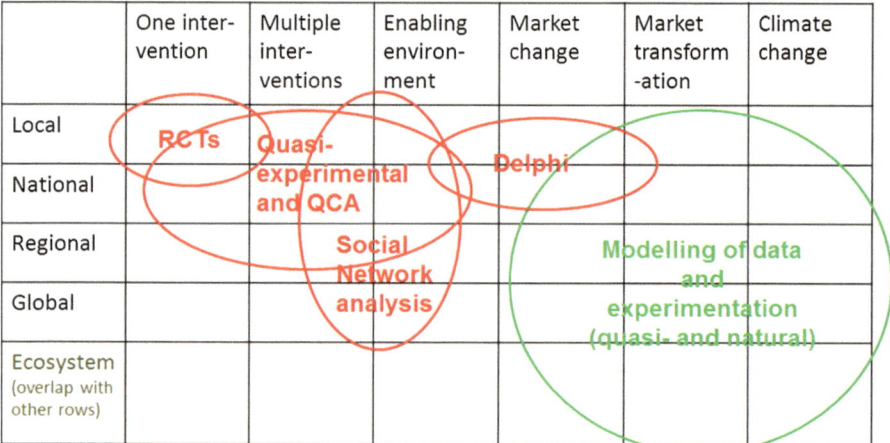

Fig. 3.3 The space and scale dimensions of demand for impact evidence (Source: Author)

mixed methods evaluations may see it as a waste of money to focus on one causal mechanism only.

Potentially meta-evaluations and meta-analysis could go a long way towards covering some of the gaps, as has been advocated by the evidence movement through so-called systematic reviews. However, there are methodological problems with these reviews. They tend to focus on a specific question and go through a huge number of studies and evaluations to see whether they provide evidence on that specific question. Many studies turn out not to have evidence for that question and thus are not used. Another issue is that these systematic reviews tend to not accept evidence that is gathered outside the narrow range of methods that are considered by the evidence movement to be sufficiently rigorous.[4] More recently realist perspectives have started to become more fashionable in meta-evaluations, which broadens the range of evidence that is accepted. An example can be found in Chap. 13, 'What do evaluations tell us about climate change adaptation' of this book.

3.8   Conclusions

There is a famous scene in the British comedy Fawlty Towers which provides a good metaphor of how impact evidence may be treated by a narrow interpretation of evidence based politics. In this particular episode of Fawlty Towers the hotel manager Basil Fawlty puts some money on a horse in the hope of substantial

---

[4]See for example http://www.whatworksgrowth.org/resources/the-scientific-maryland-scale/. See also the discussion in https://en.wikipedia.org/wiki/Hierarchy_of_evidence.

earnings, and desperately wants to keep this secret from his wife Sybil. But the Spanish waiter in the hotel, Manuel, discovers what goes on. Basil asks Manuel to deny, if Sybil would question him, that he has any knowledge of this. When Basil is discovered by Sybil in suspicious circumstances with a lot of money, he needs to proof that he came by this money through legal means, and he asks Manuel to vouch for him. Manuel looks at Basil, grins, and in a proud performance exclaims: "I know nothing". After a few seconds he repeats, with added emphasis: "I know nothing", thus sealing Basil's fate. The evidence based movement came to the foreground and argued for randomized controlled trials and counterfactual impact evaluations by claiming that old fashioned evaluations could be thrown in the wastepaper basket, and that there was a serious gap in evidence that needed to be filled. On international cooperation the evidence on what works and what doesn't was, to adapt Manuel's phrase: "we know nothing". However, an analysis of the dimensions of time, space and scale demonstrate that randomized controlled trials are particularly good at covering a few of them, and that in many cases evaluators will need to explore other methods and tools to provide evidence on impact. As a result of the narrow scope of evidence that is accepted by the evidence movement, they will have difficulty in explaining to policymakers, boards and parliaments that what they want to see evidence on cannot be provided through randomized controlled trials.

The three dimensional matrix of time, space and scale provides a systemic ordering of demand for impact evidence, and inspiration for how this can be uncovered through various evaluation techniques. It underscores the wide range of scientific tools and approaches as discussed in the Stern report (2012). Further analysis is needed. No doubt more scientific tools exist and can be placed in the matrix. It could be developed as a heuristic tool to identify key evaluation questions and approaches. It also demonstrates that impact evidence is available throughout the cycle of projects, programmes and policies and that demand for impact evidence can be throughout the lifetime of a project and will get to higher levels and scales after the project has ended.

In the case of climate change mitigation, the matrix provides a better understanding why impact is visible at project level and in markets directly influenced (and hopefully changed) by the project, but that impact at the global level is illusive, not visible, and has not led to the desired change in trends. Especially where goals are formulated at the highest level the matrix may be useful in providing a systematic understanding why impact cannot (yet) be demonstrated at that level.

My suggestion is to further develop the matrix as an analytical tool to:

1. Better identify the demand for impact evidence: is it on whether a specific causal mechanism works, or is it whether the problem that needs to be addressed is becoming solved, or whether global, regional or national trends are moving in the right direction, and if so, how that is linked to the intervention.
2. When the demand is identified, how would this translate to key evaluation questions that focus on the right moment in time, at the right location and at the appropriate scale?

3. Given these questions, the appropriate evaluation approaches and tools and methods can be found to address them.
4. Lastly, by framing the evidence in time, space and scale the evaluation can better explain why evidence is generated in the way that is chosen, and why other methods (such as randomized controlled trials in the case of complex interventions, or mixed methods case studies in the case of a straight-forward intervention that is localized and focuses on testing one causal mechanism).

The Centre for Development Impact in Brighton will continue to work on this tool and aims to further develop it along these lines.

References

Bamberger, M. (2012). *Introduction to mixed methods in impact evaluation.* InterAction [etc.] [Impact Evaluation Notes: No.3 August 2012].

Bruyninckx, H. (2009). Environmental evaluation practices and the issue of scale. In M. Birnbaum & P. Mickwitz (Eds.), *Environmental program and policy evaluation. New Directions for Evaluation, 122,* 31–39.

Garcia, J. R., & Zazueta, A. (2015). Going beyond mixed methods to mixed approaches: A systems perspective for asking the right questions. *IDS Bulleting, 46*(1), 30–43.

GEFEO (2006). *The role of local benefits in global environmental programs.* Washington, DC: Evaluation Office, Global Environment Facility.

GEFIEO (2013). *Climate change mitigation impact evaluation. GEF support to market change in China, India, Mexico and Russia.* Washington, DC: Independent Evaluation Office, Global Environment Facility [Unedited version, downloaded from gefieo.org in August 2015].

Hildén, M. (2009). Time horizons in evaluating environmental policies. In M. Birnbaum & P. Mickwitz (Eds.), *Environmental program and policy evaluation: Addressing methodological challenges. New Directions for Evaluation, 122,* 9–18.

IPCC. (2014). Climate change 2014: Synthesis report. Contributions of working groups I, II and III to the fifth assessment report of the Intergovernmental Panel on Climate Change. [Core Writing Team, R.K. Pachauri and L.A. Meyer (eds.)]. Geneva: IPCC.

Ittyerah, A. C., Choudhary, R., Narang, S., Choudhary, S. D. (2005). *Terminal evaluation and impact assessment of the UNDP/GEF project – IND/91/G-31 – optimizing development of small Hydel Resources in the Hilly Regions of India.* S.l., s.n.

Jerve, A. M., et al. (1999). *A leap of faith: A story of Swedish aid and paper production in Vietnam – the Bai Bang project, 1969–1996.* Stockholm: SIDA.

Kennedy, E. T., Balasubramanian, H., & Crosse, W. E. M. (2009). Issues of scale and monitoring status and trends in biodiversity. In M. Birnbaum & P. Mickwitz (Eds.), *Environmental program and policy evaluation: Addressing methodological challenges. New Directions for Evaluation, 122,* 41–51.

Ratna Reddy, V., Uitto, J. I., Frans, D. R., & Matin, N. (2006). Achieving global environmental benefits through local development of clean energy? The case of small hilly hidel in India. *Energy Policy, 34,* 4069–4080.

Rowe, A. (2012). Evaluation of natural resource interventions. *American Journal of Evaluation, 33,* 384.

Rowe, A. (2014). Evaluation at the nexus: Principles for evaluating sustainable development interventions. In J. I. Uitto (Ed.), *Evaluating environment in international development.* London: Routledge.

STAP. (2013). *Calculating greenhouse gas benefits of the global environment facility energy efficiency projects, version 1.0*. s.l.. Science and Technical Advisory Panel (STAP), March 2013.

Stern, E., Stame, N., Mayne, J., Forss, K., Davies, R., & Befani, B. (2012). *Broadening the range of designs and methods for impact evaluations* (Working paper 38). London: DFID.

United Nations. (2014). *The millennium development goals report 2014*. New York: United Nations.

United Nations. (2015). *Transforming our world: The 2030 agenda for sustainable development*. New York: United Nations [A/Res/70/1], downloaded from https://sustainabledevelopment.un.org/post2015/transformingourworld/publication on 3 Sept 2015.

van den Berg, R. D. (2005). Results evaluation and impact assessment in development co-operation. *Evaluation, 11*(1), 27–36.

van den Berg, R. D. (2013). Evaluation in the context of global public goods. In R. C. Rist, M.-H. Boily, & F. Martin (Eds.), *Development evaluation in turbulent times: Dealing with crises that endanger our future*. Washington, DC: The World Bank.

van Beurden, J., & Gewald, J.-B. (2004). *From output to outcome? 25 years of IOB evaluations*. Amsterdam: Aksant.

Wörlen, C. (2014). Meta-evaluation of climate mitigation evaluations. In J. I. Uitto (Ed.), *Evaluating environment in international development* (pp. 87–104). London: Routledge.

Chapter 4
Pathway to Impact: Supporting and Evaluating Enabling Environments for Research for Development

Tonya Schuetz, Wiebke Förch, Philip Thornton, and Ioannis Vasileiou

Abstract The chapter presents a research for development program's shift from a Logframe Approach to an outcome and results-based management oriented Monitoring, Evaluation and Learning (MEL) system. The CGIAR Research Program on Climate Change, Agriculture and Food Security (CCAFS) is designing an impact pathway-based MEL system that combines classic indicators of process in research with innovative indicators of change. We have developed a methodology for evaluating with our stakeholders factors that enable or inhibit progress towards behavioral outcomes in our sites and regions. Our impact pathways represent our best understanding of how engagement can bridge the gap between research outputs and outcomes in development. Our strategies for enabling change include a strong emphasis on partnerships, social learning, gender mainstreaming, capacity building, innovative communication and MEL that focuses on progress towards outcomes.

It presents the approach to theory of change, impact pathways and results-based management monitoring, evaluation and learning system. Our results highlight the importance of engaging users of our research in the development of Impact Pathways and continuously throughout the life of the program. Partnerships with diverse actors such as the private sector and policy makers is key to achieving change, like the attention to factors such as social learning, capacity building,

---

T. Schuetz
Independent Consultant, Munich, Germany
e-mail: schuetztonya@gmail.com

W. Förch
Deutsche Gesellschaft für Internationale Zusammenarbeit (GIZ) GmbH, Private Bag X12 (Village), Gaborone, Botswana
e-mail: W.Foerch@cgiar.org

P. Thornton (✉)
CGIAR Research Program on Climate Change, Agriculture and Food Security (CCAFS), International Livestock Research Institute (ILRI), Nairobi, Kenya
e-mail: P.THORNTON@cgiar.org

I. Vasileiou
World Bank, Washington, DC, USA
e-mail: ivasileiou@worldbank.org

© The Author(s) 2017
J.I. Uitto et al. (eds.), *Evaluating Climate Change Action for Sustainable Development*, DOI 10.1007/978-3-319-43702-6_4

networking and institutional change when generating evidence on climate smart technologies and practices. We conclude with insights on how the theory of change process in CGIAR can be used to achieve impacts that balance the drive to generate new knowledge in agricultural research with the priorities and urgency of the users and beneficiaries of these research results.

Evaluating the contribution of agricultural research to development has always been a challenge. Research alone does not lead to impact, but research does generate knowledge which actors, including development partners, can put into use to generate development outcomes. In CCAFS we are finding that a theory of change approach to research program design, implementation and evaluation is helping us bridge the gap between knowledge generation and development outcomes.

Keywords Results-based management • Impact pathway • Monitoring • Learning and evaluation • Theory of change

4.1 Introduction

Global poverty has been reduced over the past 25 years. The developing regions overall saw a 42 % reduction in the prevalence of undernourished people between 1990–1992 and 2012–2014 (FAO 2015). Despite major investments of the international community in reducing poverty and food insecurity, an estimated 805 million people were chronically undernourished in 2012–2014 (FAO 2015), almost all of whom live in developing countries. There are large regional differences in terms of the progress that has been made against poverty and hunger: in South Asia it has been limited, and in sub-Saharan Africa it has actually gone backwards since 1990–1992 (FAO 2015). There is much to be done to reach the targets for 2030 as articulated in the Sustainable Development Goals (UN 2015a). Research for development (R4D) has played a significant role in reducing food insecurity over the last decades and will continue to play a critical role moving forward.

R4D is a set of applied research approaches that aim to directly contribute towards achieving international development targets through innovation. In this, there is a wide range of understanding of the concept. In this chapter we focus on agricultural research for development as operationalized by CGIAR. The underlying assumption is that research within R4D is done with broader development outcomes in mind, e.g. demand-led prioritization of research, participatory and action research and stakeholder involvement (Harrington and Fisher 2014).

Agricultural R4D has a long history. CGIAR was founded in 1971 as a response to address global hunger in India, Pakistan and other South Asian countries. The adoption of improved agricultural practices and technologies developed by CGIAR, including high-yielding rice and wheat varieties, fertilizers, pesticides and irrigation, has proven to be a powerful instrument of the Green Revolution in fighting hunger in that part of the world. CGIAR currently comprises 15 international agricultural research centers that collectively aim to increase agricultural productivity, reduce poverty and enhance environmental sustainability. Renkow and Byerlee (2010) and Raitzer and Kelly (2008) reviewed evidence of impact across the centers and concluded that there have been strong positive impacts of CGIAR research relative to investment. Another way to describe CGIAR's success is to show a world without it (Evenson and Gollin 2003): focusing on the impact of crop improvement research from 1965 to 1998 provided counterfactual scenarios of the global food system: developing countries would produce 7–8 % less food; their cultivated area would be 11–13 million hectares greater at the expense of primary forests and other fragile environments; and 13–15 million more children would be malnourished.

However, agricultural R4D has not realized its full potential: the world food system continues to face challenges of persistent food insecurity and rural poverty in many parts of the developing world. The adoption of improved agricultural technologies and practices by farmers has often been less than expected, when considering their demonstrated benefits, primarily due to a supply-led approach to their development and dissemination, with limited attention paid to context specificity, to farmer's priorities beyond increased agricultural productivity, and to the socio-economic, political and institutional contexts within which smallholder farmers operate. Many studies have shown that 'scientifically proven' technologies alone are not the only key to get to impact. If a technology gets adopted or adapted, it is often not so much because of its quality and suitability but because of good social management and implementation processes (Hartmann and Linn 2008; Pachico and Fujisaka 2004). New challenges like population growth and climate change are adding complexity to the mission of CGIAR and other R4D organizations.

Within this context, this chapter aims to describe the journey towards a new R4D approach based on theory of change (TOC) and impact pathway thinking for program implementation, monitoring, learning and evaluation (MEL). It illustrates lessons of broad applicability regarding results-based management (RBM) and adaptive management approach to tackling complex development challenges through R4D. The key messages are summarized in Box 4.1. The chapter starts by describing a case study within CGIAR, where TOC combined with IPs and learning-based approaches were employed to build an outcome-focused RBM approach to R4D. It then analyses the main findings, focusing on program design and systems for planning and reporting, as well as a MEL framework within an impact pathways approach. The chapter concludes with lessons for required institutional change as well as for MEL practitioners, researchers and policy makers.

Box 4.1: Key Messages

Overall, RBM can offer many elements and approaches to help with strategic program design, but it needs to be adapted to the specific context of a program, institution, or organization. It requires some enabling conditions and an environment to support an outcome-focused R4D program.

Key lessons and enablers:

- Buy-in from the top, healthy balance between given structures but allowing for creativity in designing processes.
- Investing in facilitation and process – and bringing the three elements of MEL together is key and requires resources (time and money).
- Flexible condition, rigid system to allow adaptive management and learning (liberating structures).
- The 'three thirds' principle: one third partnerships, ownership and buy-in externally from partners, one third capacity enhancement at all levels internally and externally, and one third cutting-edge science.
- System support – building an online platform and working towards a one-stop-shop (database).

4.2   Background

CGIAR is a global agricultural research partnership for a food secure future. Its science is carried out by 15 research centers with 10,000 scientists working in 96 countries and a host of partners in national and regional research institutes, civil society organizations, academia, development organizations, and the private sector (CGIAR 2015a). Its work contributes to the global effort to tackle poverty, hunger and major nutritional imbalances, and environmental degradation. The 15 CGIAR Centers have different foci and operate semi-autonomously in pursuing their specific research agendas, ranging from promoting the productivity of specific crops, livestock, and fish commodities to production systems in specific agro-ecologies and research on policies natural resource management (Raitzer and Kelly 2008).

CGIAR was formed in 1971 to foster technical solutions to agricultural productivity constraints affecting developing countries (Renkow and Byerlee 2010). Research tended to focus on creating outputs, was often technology focused and supply driven; success was measured by peer-reviewed publications, citations and science products. Criticism has been mounting over the last decades, as the limitations of the output delivery model became evident: outputs do not automatically translate into impact. It was often assumed that communication and development specialists would repackage research findings after the researcher produced them and that farmers would realize the value of new technologies and happily adopt

Fig. 4.1 Early change theorists (Found in Duncan Green's 'From Poverty to Power' blog)

"I think you should be more explicit here in step two."

them to increase agricultural productivity (Fig. 4.1). CGIAR itself has long recognized these weaknesses and embarked on a far-reaching reform process in 2010.

The challenges of demonstrating wide-reaching impact through R4D are compounded by a rapidly growing human population, climate change and other complexities of our time. The human population has almost doubled from 3.8 billion in 1971 to 7.3 billion in 2014 (UN 2015b). With an expected extra two to three billion people to feed over the next 40 years, this will require targeted research efforts to achieve not just growing 70 % more food but making 70 % more food available on the plate to keep up with rapidly rising demand (WWAP 2012). Climate change is already affecting agriculture in many developing countries, and the effects will become increasingly challenging in the future. Higher temperatures, shifting disease and pest pressures, and more frequent and severe droughts and flooding will affect agricultural production and place increasing pressure on water and other natural resources (IPCC 2013). Climate change impacts are increasing the vulnerabilities of populations that are already struggling with food insecurity and poverty, even in the relatively conservative scenario of a global 2-degree temperature rise (Thornton et al. 2014a).

The increasing complexity of the challenges, particularly with regard to their impacts on poor and vulnerable populations, requires a rethinking of our approach to R4D. CGIAR has taken on this challenge by broadening its portfolio of major new initiatives for strategic research. A first round of some half-a-dozen 'Challenge Programs' were mandated to develop new R4D models over a period of up to 10 years, starting in 2002 (CGIAR 2015b). Box 4.2 describes one example of these programs, focusing on water and food.

Box 4.2: Challenge Program on Water and Food

The CGIAR Challenge Program on Water and Food (CPWF) piloted new ways of increasing the resilience of social and ecological systems through better water management for food production. From 2002 to 2013, the program supported more than 120 research projects in ten of the world's largest river basins (Hall et al. 2014; Harrington and Fisher 2014). The program early on developed IPs and theories of change for its R4D river basin programs. From a Monitoring and Evaluation (M&E) perspective this included results-based and adaptive management as well as learning-oriented approaches. The insights and knowledge gained from CPWF's 12 years of work are being integrated into another CGIAR Research Program on Water, Land and Ecosystems.

In a second round, from 2010 onwards, 16 CGIAR Research Programs (CRPs) were set up in a 5-year first phase (CGIAR 2015c). The major reorientation of the R4D portfolio was in the move from an output focus to an outcome focus. Success was now to be measured in terms of the CRP's contribution to behavioral changes, manifested in changes in knowledge, attitudes and skills and practices of a wide set of non-research next users, including development practitioners, farmers and policy makers.

Through approaches such as results-based management, theories of change and impact pathways, the term outcome came into focus. Organizations such as the International Development Research Center (IDRC) were early developers of M&E tools to capture and measure outcomes through their 'Outcome Mapping' methodology (Earl et al. 2001). Within CGIAR, 'Participatory Impact Pathways Analysis' (PIPA) (Douthwaite et al. 2003, 2007) was developed under the CPWF to unpack processes and mechanisms in the realm of outcomes.

Towards the end of the first phase, 4 of the 16 CGIAR research programs were tasked to develop a comprehensive, suitable and lean results-based management approach for research for development, initially for a period of 12 months. The following section describes how the CGIAR Research Program on Climate Change, Agriculture and Food Security (CCAFS) developed and implemented its RBM trial and highlights the main lessons learnt.

4.3 Approach

This section describes the approach to results-based management taken by one research for development program of CGIAR, CCAFS. The description is combined with theoretical and practical references to development agencies that started experimenting with results-based management some 10 years earlier. Figure 4.2 illustrates CCAFS' approach to implementing results-based management with a theory of change (TOC) approach along defined impact pathways, focusing on outcome delivery. The TOC defines several activities, such as developing the impact pathways for thematic research and regional work, trialing RBM with a subset of projects, training key partners in the impact pathways building, and analytical systems support. These led to tangible outputs, e.g. a finalized ex-ante set of impact pathways with coherently defined outcome targets, workshop reports and learning notes, facilitation guidelines (CCAFS 2015a), a RBM MEL strategy (CCAFS 2015b), and an online platform. This involved the engagement with and involvement of identified key next-users such as CGIAR Consortium Office, program partners, and fellow researchers, with the idea that these outputs would both be useable and an incentive to overcome existing barriers in the system. It was also envisaged that the outputs would facilitate changes in their practice: for example, working towards implementing more efficient and effective R4D, and proactively changing organizational norms. Moving from outcomes to impact in Fig. 4.2 requires several steps that are not elaborated because this is beyond the scope of this chapter.

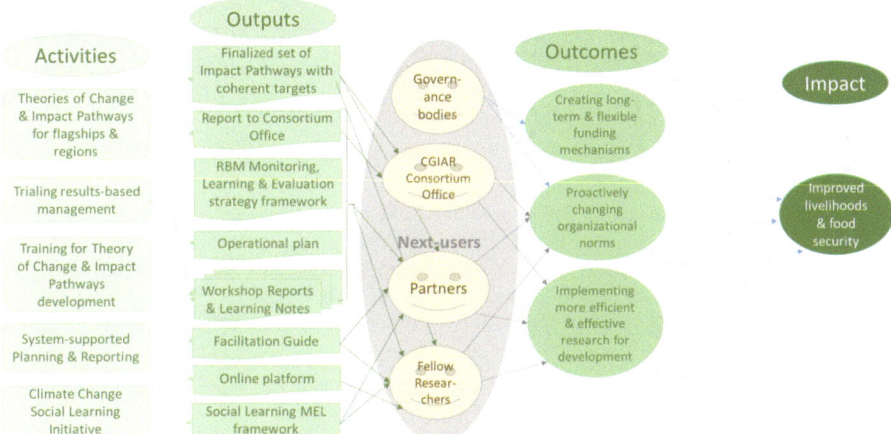

Fig. 4.2 CCAFS' theory of change for its results-based management approach and components

Box 4.3: About the CGIAR Research Program on Climate Change, Agriculture and Food Security (CCAFS)[1]

The CGIAR Research Program on Climate Change, Agriculture and Food Security (CCAFS) is a strategic partnership of CGIAR and Future Earth led by the International Center for Tropical Agriculture (CIAT). CCAFS brings together the world's best researchers in agricultural science, development research, climate science and Earth System science to identify and address the most important interactions, synergies and trade-offs between climate change, agriculture and food security. For more information see ccafs.cgiar. org

As an R4D program working on addressing the complexities of climate change, agriculture and food security, the main goal of CCAFS is to improve the livelihoods of the most vulnerable and poor people in target countries in Asia, Africa and Latin America. While CCAFS is at the cutting edge of generating demand-driven science products, it also plays a bridging role: to transform credible scientific evidence and results into development outcomes. A key underlying principle the CCAFS management team subscribes to is the "Three-Thirds Principle": one third of effort engaging with partners to decide what needs to be done and how; one third on doing the actual research, often in partnership; and one third on sharing results in appropriate formats and strengthening capacity of next users to utilize the research to achieve outcomes and impact. Deep engagement with stakeholders with the support from a wide network of partners to get science-based solutions to practical problems is fundamental to the CCAFS approach (CCAFS 2014).

CCAFS has been one of the programs at the forefront of testing and paving the way for moving a multi-million dollar R4D program from a logframe approach to an outcome-focused approach. Additionally, it has put in place a MEL mechanism for programmatic RBM, including elements of adaptive management.

---

[1]CCAFS started as a Challenge Program (2009–2011), and then became a CRP with Phase 1 (2012–2014) and an Extension Phase (2015–2016). The proposal for Phase 2 (2017–2022) is currently under development. We acknowledge support to CCAFS from the CGIAR Fund Council, European Union, and International Fund for Agricultural Development. We also acknowledge the inputs of many people in the work and activities described here.

4.4   Getting to the Right Mix

CCAFS was clearly committed to an outcome-focused R4D program from its inception. It became increasingly clear that a logframe approach (LFA) was not the most ideal way of doing R4D differently. In particular, when moving from a Challenge Program to a CGIAR Research Program with increasing complexities of partnerships, engagement and CGIAR integration, the limits of a logframe became apparent (as described in section 4.5.1). The program's vision of contributing towards development outcomes increasingly required a different approach: one that acknowledged the importance of stakeholder engagement and capacity development. As a result, monitoring the annual contribution of CCAFS and its partners towards development outcomes becomes increasingly complex.

While a wide range of MEL approaches and methodologies with an outcome focus exist (e.g. PIPA, Outcome Mapping, Outcome Harvesting), none provides a blue-print solution that can just be rolled out. The approaches were designed to address the particular needs of a specific program or organization. Thus, to adapt these approaches to a new program, it is key to select the right mix of elements creating a conceptual framework in support of the program's specific TOC and MEL requirements. Springer-Heinze et al. (2003) advocate a holistic approach to impact evaluation and program monitoring with quantitative and qualitative elements, based on an impact pathway that can accommodate different stakeholder views, allows for reflection, and emphasizes capacity of research organizations. Cummings (1997) compares RBM, LFA and Project Lifecycle Management and would welcome more discussions and learning among the different approaches. According to Bazeley (2004), 'The "mixing" may be nothing more than a side-by-side or sequential use of different methods, or it may be that different methods are being fully integrated in a single analysis'. Applying a mix of methodologies in a programmatic MEL framework raises certain terminological, definitional, paradigmatic and methodological issues, including over-interpretation of numbers, single dimensionality, and disregarding 'outliers' from the analysis (Bazeley 2004). However, mixed methods also provide opportunities to address the respective shortcomings of any single method as applied in practice.

CCAFS in its early years worked with various logframe elements in planning and reporting. Limitations of the more traditional LFA resulted in experimentation with elements of TOC that were integrated within the logframe, in order to more effectively capture the complexity of activities, partners and anticipated outcomes of the program. The limitations of this single method approach resulted in CCAFS deciding to operationalize a modular MEL approach, described in the next section. The findings and analysis section explains CCAFS's approach over time. With the limitations in mind CCAFS is aiming for a more holistic approach in line with Springer-Heinze et al. (2003).

4.5   Findings and Analysis

4.5.1   Moving Away from a Logframe

In line with funding agency requirements, CCAFS also initiated its programmatic management approach on the basis of a logframe (see Table 4.1 for an example). Annual milestones were defined that were largely focused on producing scientific outputs and evidence of their achievement, which would then lead to developmental impact.

R4D programming over the last few decades has commonly been based on a logframe approach (LFA). The LFA was initially developed for United States

Table 4.1 Excerpt from the CCAFS annual logframe (2011) as an example from Theme 4 (Integration for Decision Making), while outcomes and impacts are reported against in the medium-term plan (2010–2012) (CCAFS 2011)

Milestones output targets	Performance indicator	Means of verification	Assumptions	Partners
Objective 4.1 Explore and jointly apply approaches and methods that enhance knowledge to action linkages with a wide range of partners at local, regional and global levels				
Outcome 4.1: Appropriate adaptation and mitigation strategies mainstreamed into national policies in at least 20 countries, in the development plans of at least five economic areas (e.g. ECOWAS, EAC, South Asia) covering each of the target regions, and in the key global processes related to food security and climate change				
Output 4.1.1 For each region, coherent and plausible futures scenarios to 2030 and looking out to 2050 that examine potential development outcomes under a changing climate and assumptions of differing pathways of economic development; developed for the first time in a participative manner with a diverse team of regional stakeholders				
Milestone 4.1.1.1 Capacity built among three regional teams of diverse stakeholders trained in scenarios approaches and engaging with policymakers in their countries/regions and in global climate change processes and with the Earth System Science Partnership community; Methodological briefs, papers	Regional scenarios partners actively participating in regional food security debates and global climate change processes. Number of partners using/citing scenarios; number of regional partners trained in scenarios participating in regional food security debates and global climate change processes	CCAFS and partner websites and reports; newspaper and other media reports	Partners remain engaged and help communicate scenario research results widely and to inform key decision makers	Regional Agricultural Research Organizations; Regional policy organizations; International NGOs; Regional NGOs; Private Sector; Farmers Organizations; Regional Meteorological Organizations;

Note: While at planning outcomes and impacts were described, the reporting was against the given categories in the table, budgets were spread across regions and program crosscutting items

Agency for International Development (USAID) in 1970 and adopted by a range of international organizations, including agricultural R4D (Schubert et al. 1991). The approach has been widely required by funding agencies and has thus been used for project planning, management and evaluation and adheres to a relatively rigid framework. It tends to prescribe a hierarchy of objectives converging on a single goal, a set of measurable and time-bound indicators of achievement, checkable sources of information, and assumptions of other impinging factors (Gasper 2000). In the R4D context, the underlying assumption is that development agencies, communication units, ministry staff and other people who could use the findings are able to source the scientific evidence, understand it, know how to implement and apply it, and convey this to people who they think need them. In this case both research and development have their mandates, responsibilities and clearly defined boundaries.

While this has been a useful approach for several decades, it is debatable whether it is entirely suitable for ensuring the use of research results and their translation into outcomes. Crawford and Bryce (2003) note that although much of the literature promotes the use of the LFA for the purposes of M&E, it has proven inadequate and evidence for its usefulness is lacking. The LFA does not pay enough attention to involving key stakeholders in a joint process, emphasizing the stakeholder networks needed to achieve impact, providing managers with the information needed both to learn and to report to their funding agencies, and establishing a research framework to examine the critical processes of change that projects seek to initiate and sustain (Douthwaite et al. 2008).

CCAFS has gone through several iterations of the logframe that was employed for planning and reporting (CCAFS 2015c). In 2010, a limited version was used (CCAFS 2010) while more elements were added in the following years. Planning and reporting elements were pre-determined to some extent by requirements from CGIAR, though for internal purposes additional elements were added in response to the limitations that were identified from year to year.

4.6 Testing the Waters with Theory of Change and Results-Based Management in CCAFS

In addition to the use of logframe elements within the CCAFS planning and reporting system, at program design stage CCAFS also explicitly included a research theme entitled 'Knowledge to Action' in its portfolio (Jost et al. 2014a). The team was experimenting with strategies of getting from research outputs to development outcomes. This theme was tasked with research, not with creating an operational mechanism for CCAFS per se. It was only in year 3, when CCAFS started working in two new target regions, that opportunities presented themselves to trial a TOC approach within this new component of the R4D portfolio (Jost and Sebastian 2014; Jost et al. 2014b). Very early on it became clear that a new way of thinking needs to take effect for the whole program in order to plan for and capture outcomes more effectively and include engagement and capacity enhancement as key strategic elements (Thornton et al. 2014b). As a consequence, the 'Knowledge into Action' theme was then mainstreamed into the whole CCAFS program with its four research and five regional programs.

The opportunity to trial an alternative approach of RBM was taken up enthusiastically (Thornton et al. 2014b; Jost et al. 2014c). Theory of change, impact pathways and results-based management offer practical mechanisms to potentially enhance program design and its monitoring, learning and evaluation and help CCAFS to create an operational program management framework that is better suited to deal with the complexities at hand.

Closely linked to this, CCAFS and partners also started experimenting with learning-based approaches within R4D recognizing the need to include mechanisms that challenge business as usual and support institutional learning and innovation to ensure that research contributes to development outcomes, see Box 4.4.

Box 4.4: Why Learning
Learning-based approaches are useful in supporting transformational change across institutions and stakeholders. One such approach is social learning. We understand social learning to be a facilitated process of planning, implementing, reflecting, and adapting. It can effectively foster an institutional learning culture and pave the way for climate resilient food systems and sustainable development outcomes. For more information see Kristjanson et al. (2014), Gonsalves (2013), and Harvey et al. (2013) and ccafs.cgiar. org/social-learning-and-climate-change.

4.7   Trialing Results-Based Management in CCAFS

CCAFS decided to trial a RBM approach for one of its research themes, *Policies and institutions for climate-resilient food systems*, fast-tracking the extension phase for this particular theme. A new portfolio of six multiannual regional projects was set up and these were each tasked from the beginning with designing their project using a TOC approach (Schuetz et al. 2014a). TOC are key elements of CCAFS' approach to RBM.

There is no single definition of a TOC and no set methodology, as the approach assumes flexibility according to its respective user needs (Vogel 2012). A TOC provides a detailed narrative description of an impact pathway (a logical causal chain from input to impact, see Fig. 4.3) and how changes are anticipated to happen, based on underlying assumptions by the people who participated in describing these trajectories. As such they provide an ex-ante impact assessment of a program's anticipated success. TOC is at its best when it combines logical thinking and critical reflection; it is both process and product (Vogel 2012).

Fig. 4.3 Theory of change logical causal chain

RBM builds on the same logical causal chain and is more explicit about output-use. Within R4D output-use refers to strategies that directly engage the next-users in the research process, e.g. through stakeholder platforms and user-oriented communication products. At the turn of the century, many development and funding agencies, including USAID, Department for International Development, IDRC, UNDP and the World Bank, reformed their performance management systems and M&E approaches towards a RBM approach (Binnendijk 2000; Bester 2012; Mayne 2007a, b). At the time, these organizations faced a number of common challenges: how to establish an effective performance measurement system, how to deal with analytical issues of attributing impacts and aggregating results, how to ensure a distinct yet complementary role for evaluation, and how to establish organizational incentives and processes that will stimulate the use of performance information in management decision-making (Binnendijk 2000). These early experiences with RBM have informed further development of the approach.

Early on, IDRC has attempted to unpack the in-between area of outcomes and were at the forefront of developing means to measure outcomes through the Outcome Mapping methodology (Earl et al. 2001). To show that R4D contributes to the desired behavioral changes, i.e. outcomes, that enable long-term positive impacts is a particular challenge, as it requires more qualitative monitoring than dealing with quantitative means of measuring alone (Young and Mendizabal 2009; Springer-Heinze et al. 2003). Evaluators generally agree that it is good practice to first formalize a project's TOC, and then monitor and evaluate the project against this 'logic model' (e.g. Chen 2005). The TOC is a mental model made explicit by involving as many people as possible in its design. Key principles of the Participatory Impact Pathways Analysis also include reflecting on these models, regularly validating the assumptions that were made, and adjusting program management accordingly (Douthwaite et al. 2013).

Within the CCAFS RBM trial projects, this TOC approach to project planning helped position the R4D agendas further along the IP (Schuetz et al. 2014a). Projects expanded their skill sets by bringing on board non-research partners that would help implement output-to-outcome strategies and thus create more clearly defined causal logical chains (Fig. 4.3; Schuetz et al. 2014b, c). This is not to take over the work of development agencies, but it is to ensure that research findings are maintained in their content and get contextualized to be best fit for purpose (see Table 4.2 for a comparison of key difference between research, development and R4D). The RBM trial projects have thus challenged the common thinking that good science and publications are enough and by themselves will lead to impact – rather, they are necessary but not sufficient.

Table 4.2 Comparison between research, research for development and international development

Criteria/ elements for RBM, TOC, IPs	Research	R4D	International development
Organizational formats	Research centers with a key scientific focus	Interdisciplinary research programs around a development challenge and partnership approach	NGOs, development aid agencies, UN agencies
Mandate and performance focus	Outputs	Outcomes	Impacts
Responsibility for achieving impact	Provision of solid science and technologies	Strong partnerships	Implementation
Type of communication, knowledge management	More traditional/ corporate communications	Communications for development, engagement	People communications
Type of partners	International, regional, national research partners	International and national research partners, and development agencies	Local/ district implementing agencies, central/national governments
Program evaluation	Focused on quantitative measuring of publications, quality of journals, citations,	Forward looking external evaluation, learning-based approach, contribution (not attribution), balanced quantitative with qualitative measures	Focused on traditional impact assessment, quantitative measuring including baselines
Timeframe for achieving outcomes/impact	Often not considered	Achieving outcomes at scale within 5 years and impact within 15 years	Long term impacts 10–20 years at large scale
Languages of products	International standards	Both international and locally appropriate languages	Both international and locally appropriate languages

4.8   Building Capacity and Learning Within the Program for Theory of Change Approach

The RBM trial project teams were thrown in at the deep end. Used to a more traditional LFA, they were tasked with shifting to a TOC and learning-based approach for planning their projects within the trial. It was quickly apparent that capacity to plan projects using this new approach had to be built within CCAFS (and wider CGIAR).

Using TOC approaches within R4D requires the strengthening of capacities of scientists to do research differently and work with non-research partners for impact, but also of institutions to facilitate such a shift. Several authors highlight the importance of building capacity for institutional learning (Hall et al. 2003; Horton and Mackay 2003; Eade 1997; Springer-Heinze et al. 2003). Eade (1997) emphasizes a capacity-building approach, training of staff in a variety of relevant skills, and the dynamic and long-term nature of the process when looking at types of social organization of NGOs engaged in development theory and practice. Johnson et al. (2003) show that participation of non-research stakeholders early on in the research process is important, as it can inform institutional learning in research organizations to change priorities and practices. It can also enhance the relevance of agricultural technologies and the capacity of these stakeholders to design their own action research processes (Johnson et al. 2003). Horton and Mackay (2003) outline the links between M&E, learning and institutional change and highlight the importance of institutional learning as a means to develop the capacities of the organization and of individual researchers, as well as empowering non-research partners as key stakeholders in the process.

CCAFS worked with expert facilitators and trainers from PIPA to implement a 1-week training course on using TOC for project and program planning (Alvarez et al. 2014). Participants were chosen strategically so that capacity would be available in the CGIAR Centers at the point in time when CGIAR proposals would need to be developed following the TOC principles. In addition to project representatives, CCAFS science officers representing all themes and regions participated, in order to build in-house capacity of TOC champions. The training, in combination with TOC facilitation guides (version 1: Jost et al. 2014d; version 2: Schuetz et al. 2014d) and learning notes (CCAFS 2015a), helped highlight the opportunities (and constraints) of rolling out RBM to a whole R4D program. An online community of practice (and wikispace) was established and allowed for continued documentation and exchange of experiences.

4.9 CCAFS' Results-Based Management Trial: Insights from Researchers and Partners

CCAFS' approach to RBM is centered on adaptive management, regular communications between program and projects, and facilitated learning within projects. Besides periodic virtual meetings, trial participants were surveyed for a more in-depth and standardized reflection, and for capturing lessons and achievements from their experience (Schuetz et al. 2014b, c). These lessons also formed the basis for the progress report to CO (Thornton et al. 2014c). Ten months into the RBM trial, the progress report summarized project participant experiences, as well as the programmatic perspective.

From the programmatic level, reflections and lessons by the CCAFS Program Management Committee have been documented in the CO progress report, as well as in the series of learning notes (CCAFS 2015a). It was a great learning experience to have an RBM trial with the six projects and to be allowed to test and tryout what is required to make the shift from a LFA to an approach that is much more people-focused, learning-focused and outcome-focused. The approach to developing the IPs was simplified over time, mostly in relation to a reduction in the type and number of indicators and level of complexity so that the wider group of people who were expected to work with them would continue to buy in to the approach (Schuetz et al. 2014d).

The survey results show that there are many people within CGIAR Centers and CCAFS partners who are willing to take on the challenge to develop new ways of collaborating and working beyond delivering outputs towards outcomes (Schuetz et al. 2014b). From the survey, the RBM trial team found that the projects had made considerable progress, but also that making fundamental shifts in the way of working take time and (initially at least) additional resources. It requires iterative and continuous processes. Staffing, or the profile of project team members, and project team composition are emerging as key factors for success. Project staff has acknowledged that they may require additional skills beyond disciplinary expertise, such as skills in coordination, facilitation, engagement, communications, and participatory and learning-oriented M&E. The RBM trial team is using the findings from the survey to explore how additional support can be provided in such areas as engaging with stakeholders and using RBM.

4.10   Rolling Out Results-Based Management for CCAFS as a Whole

Opportunities for changing the programmatic approach to project planning, implementation and MEL emerged when CCAFS was approaching the end of its first phase in 2014. The mandate to implement an RBM trial came at a perfect time – it was initiated in advance so that it could inform the planning of the CCAFS extension phase (2015–2016), as well as Phase 2 proposal development (2017–2022). With a time lag of several months between the RBM trial and CCAFS as a whole, the program planning process and TOCs were developed and defined for all four research and five regional programs as a first step to putting together the new program portfolio (Schuetz et al. 2014e). Figure 4.4 provides an illustration of one research theme's impact pathway component with its regional elements, indicators and outcome targets.

Experience in CPWF also shows that an intense process is required to finalize the program portfolio and allow for the appropriate triangulation and harmonization between thematic perspective, regional context and individual project proposals to ensure programmatic coherence, cohesion and its relevance and potential for impact (Hall et al. 2014; Biswas et al. 2008). This requires intense bilateral virtual preparation between research and regional teams, facilitated face-to-face time (e.g. in the form of workshops or writeshops), and follow-up work. Intensive workshops bring together project leaders, key national and regional partners and core program

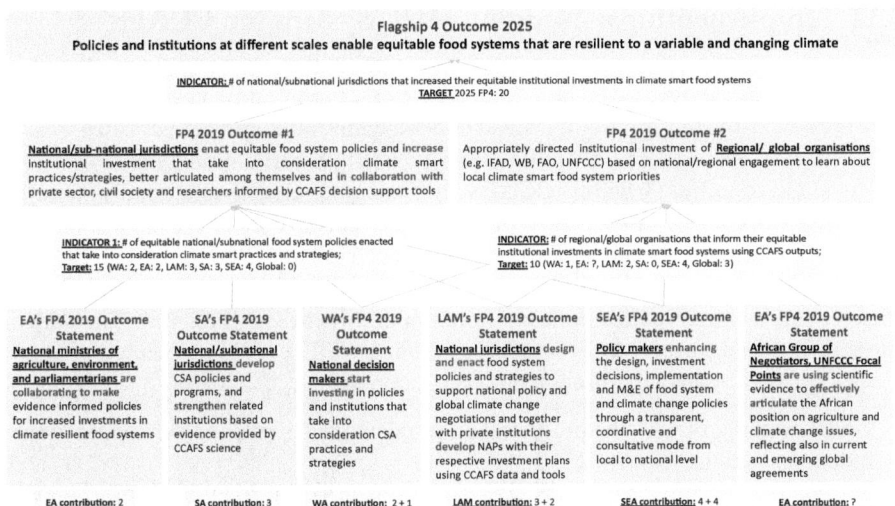

Fig. 4.4 Illustration of a CCAFS thematic IP component (Drawn from the Flagship Program on Policies and Institutions for Climate Resilient Food Systems)

staff within a respective regional or thematic focus. The workshops/writeshops can bring together selected projects in a region as a team that will continue to work together over a period of time. It is key that the agenda is designed to hone the individual project IPs towards a coherent and cohesive regional and global R4D program that complements other ongoing initiatives and contributes to the given development goals.

While it took a considerable amount of effort, the iterative development of the CCAFS TOCs and impact pathways was done in a resource-efficient way. It started off mostly virtually and intensely facilitated, building on CCAFS Phase 1 engagement and regional priorities, and was completed in five regional face-to face meetings with key next-users and stakeholders within the CGIAR research community (Schuetz et al. 2014e, f). Building on the learning-based approach to developing a suitable TOC approach for CCAFS, a series of learning notes was written to document the RBM trial experiences and the rolling out of the approach to the whole program (CCAFS 2015a). The TOC development and facilitation process, and guidance documentation were revised to make them leaner, more contextualized and easier to implement (Schuetz et al. 2014d). The TOC building process is one key component in the CCAFS MEL system that was developed to support the new approach in a comprehensive manner (CCAFS 2015b).

4.11   Implementing a Modular MEL System for CCAFS

Building on the above, a CCAFS Monitoring, Learning and Evaluation Strategy was approved by the program's management committee and its advisory board (Schuetz et al. 2014g). The overall goal of the CCAFS MEL strategy is to develop an "evaluative culture" within CCAFS that encourages self-reflection and self-examination, seeks evidence, takes time to learn, encourages experimentation and change so that MEL becomes an integrated mechanism. The strategy includes a conceptual framework, guided by overall program principles for partnership, engagement and communications and a modular system (see Fig. 4.5). The added value of the framework has been adapted from UNDP's (2007) expected competencies for their managers through an RBM approach:

- Understanding of why the program and projects are believed to contribute to the outcomes sought – the TOC.
- Setting meaningful performance expectations/targets for key results (outputs and outcomes).
- Measuring and analyzing results and assessing the contribution being made by the program to the observed outcomes/impact.
- Deliberately learning from this evidence and analysis to adjust delivery and, periodically, modify or confirm program design, i.e. have an adaptive management in place.
- Reporting on the performance achieved against expectations – outcomes accomplished and the contribution being made by the program.

Fig. 4.5 CCAFS modular MEL system

A modular system can best meet the demands of the program as a whole and its projects, as well as the wider CGIAR system (see below; Thornton et al. 2014d). Some elements are prescribed by CGIAR governance bodies, including the carrying out of baselines, independent impact assessments, and periodic external evaluations. Programmatic flexibility exists within the day-to-day operational MEL, as a system is required that allows enough flexibility and adaptability to be applied to the different types of projects and programs.

CCAFS has identified the following modules to guide its MEL system (Schuetz et al. 2014g):

Harmonization of TOCs: the framework for this modular approach is set through the TOC development across CCAFS thematic and regional operations, describing how CCAFS flagships, regions and projects anticipate changes in next-user behavior and practices, and their role in it. Investment in the development, harmonization and use of IPs and more elaborated TOC: (1) ensures that CCAFS plan of work is targeted at achieving outcomes and requires that tasks addressing the 'use of outputs' are built into each activity plan; (2) strategically encourages communication and collaboration among colleagues within research, regions and projects and guides exchanges across disciplines and regions; and (3) revisits the trajectory of CCAFS contributions to change and uses them as an ex-ante impact assessment.

Indicators & Baselines: In preparation for a harmonization process, as described above, indicators and outcome target numbers to which the program and projects will be held accountable were defined by the regional and research program leaders. The regionally and thematically aggregated targets were then checked against what individual projects proposed to contribute towards an agreed set of target values. Additionally, a programmatic baseline at site level was conducted at the beginning of the program to be able to compare achievements against these later on, with respect to behavior and practice change of farmers. Furthermore, projects are responsible for conducting specific baselines to monitor progress over time within their respective thematic and regional foci.

Reflexive spaces & activities: These need to be built in systematically to ensure that the key elements of adaptive management are operationalized. Adaptive management provides for flexibility and corrective actions to strengthen predictive capacity, which is essential when working in a constantly changing, complex environment. In working with TOCs, we make assumptions as to how we anticipate change will happen, but we know that change does not always happen as predicted, and so reflexive spaces are critical for allowing us to make well-documented and well-justified adjustments in response to the insights gained through our work.

Planning and reporting support: First, an online planning and reporting platform (P&R) collects project information at project inception, so that projects population the system once, and build on this for follow-up planning, reporting and learning. Project teams are guided in their TOCs/IPs-building from the beginning and use this as basis for monitoring annual progress. Thematic and regional programmatic goals/frameworks are prefilled by the program team, while projects map their individual contributions into these. Second, an MEL support pack provides practical mechanisms and tools to ensure a balanced quantitative and qualitative monitoring.

Assessment and bonus: Feedback loops, spaces for justification of changes and learning are weaved into the P&R to allow for systematic and strategic adaptive management throughout. Evaluation and synthesis are done from the regional and thematic perspectives after project reporting, to facilitate reporting to funding agencies, but also to minimize double counting of outcome target numbers and facilitate learning and knowledge brokerage across the program portfolio and beyond. Evaluation criteria include traditional output focused criteria, as well as progress towards outcomes, partnership and learning. Incentive mechanisms are being introduced, recognizing that these do not always have to relate to budgetary bonuses.

Institutional transformation and learning: Through feedback loops and reflexive spaces the program's evaluative learning-oriented culture is also built into the system to ensure that the program is not only capturing 'are we doing the right thing?', 'are we doing it right?', but also 'how do we know we are getting it right?' (Kristjanson et al. 2014; van Epp and Garside 2014).

Chapter 14 (*Adaptation Processes in Agriculture and Food Security*: *Insights from Evaluating Behavioral Changes in West Africa*) of this book describes an example of how this has been operationalized in a regional program of CCAFS.

4.12 Implications for Policy, Practice and Research

In this section we list some practical implications for a research-for-development organization that is considering moving to an approach based on RBM and TOC (Schuetz et al. 2015).

Working along TOCs and impact pathways has **major implications for M&E**. It implies a move to contribution rather than attribution, to acknowledge the role and inputs of partners and other actors both in achieving outcomes and in providing evidence for those outcomes. Building in triple-loop learning can make a major contribution to reflection and to supporting adaptive management, so that project teams can better deal with uncertainty. At the same time, not everything can be measured; this highlights the need for narratives that can complement and support more quantitative information.

As part of creating a program enabling environment, **embracing the three thirds principle** facilitates investment into solid science, critical partnerships, ownership and buy-in by partners, and capacity enhancement at all levels both internally and externally. CCAFS has been pushing the boundaries of R4D and has been serious about taking on the expanded CGIAR mandate to deliver outcomes, see Fig. 4.6.

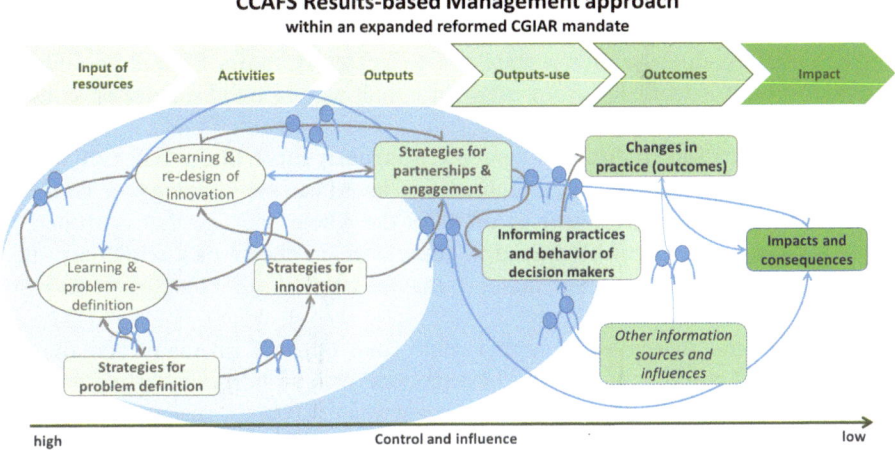

Fig. 4.6 R4D within an expanded CGIAR mandate

The three thirds principle implies **different budgeting and funding structures**, so that appropriate levels of resources are allocated to capacity building, communications and engagement with the wide range of different partners likely to be needed. These elements need to be budgeted for explicitly within a project life-cycle, rather than as an after-thought. At the same time, there is still much work to be done on how to monitor outcomes effectively, evaluate the real share of contribution towards the observed change, and assess value for money. Similarly, delivery of outcomes, especially at scale, may take time for research-for-development programs. Longer funding cycles could be expected to facilitate this considerably.

The CCAFS experience has highlighted several **operational principles for programmatic RBM**. First, there is a need to focus on people and users, on utilizing M&E as a tool to help achieve outcomes, and on accountability – it is the people within organizations that make behavioral and practice changes happen. Second, there should be an emphasis on learning through M&E activities. Robust knowledge needs to be generated that can feed into developmental policy and investment decision making, and this in turn requires a cumulative and catholic approach to choice of impact assessment methods at different levels (Maredia 2009). Third, adaptive management needs to be encouraged, as a key element of RBM. As a tool that is based on learning processes, it can improve long-run management outcomes. The challenge in using it is to find the balance between gaining knowledge to improve management in the future and achieving the best short-term outcome based on current knowledge. Fourth, the development and implementation of an online platform is a great investment for capacity development. Planning, reporting and evaluation procedures need to be as simple as possible while still providing (most of) the information needed for effective and timely management.

Sharing findings along the way is a good way to foster the inclusive involvement of as wide a range of stakeholders as possible in project planning and implementation. Encouraging researchers to get early drafts of findings out to potential users for feedback from early on is one way to build a learning culture and to encourage open-mindedness.

Rigid application of just one specific approach most likely will not work. Whether it is the adoption of a technology, an M&E methodology, a learning approach or a scientific result, it is often not the whole package that is attractive to users but specific pieces. We need to allow users to cherry pick while ensuring that the relevant linkages remain intact so that the context is not lost for others who may want other cherries.

Solutions that are good enough rather than optimal. In many domains of knowledge and practice there is no best practice or option, particularly when the problem is complex and resources are constrained. CCAFS made considerable changes once it had started to implement an approach based on TOC and impact pathways, and in time moved towards a leaner and simpler model. Time will tell if some of the details inevitably lost in this process will need to be added back in, but the notion of "good enough" systems needs to be a key guiding principle.

Addressing tensions across scale. CCAFS is still in the process of embedding TOCs for the different organizational units of the program, in order to provide a flexible framework that allows for aggregation of output, outcomes and targets across the different units. For example, targets need to be framed locally with users and beneficiaries, and voiced in such a way as to allow the flexibility to deal with uncertainty and emerging priorities and opportunities. New investments of time and effort may be needed to identify and work with non-traditional partners to promote behavioral change in shared IPs.

Providing value for money. Many funding agencies now require that grantees demonstrate value for money. The Deutsche Gesellschaft für Zusammenarbeit states that its 'work is systematically geared towards results, the yardstick by which we measure the success of our work. We want to help achieve tangible positive changes on the ground' (GIZ 2015). Some have critiqued the whole notion of payment by results as applied to development and research-for-development on the basis that it provides perverse incentives that actually diminishes cost-effectiveness (see Chambers 2014). As noted above, there is much work still to do on appropriate measurement mechanisms, but this does not diminish the need to demonstrate accountability.

Balancing science and outcomes. Research is often curiosity-driven, and traditional indicators of success center on peer-reviewed publications in high-profile academic journals. In today's highly competitive research environment another crucial success factor relates to fundraising: the ability to write and win competitive research proposals. Neither of these motivations for research is guaranteed to deliver development outcomes. For CGIAR and its research programs, it is still early days, but preliminary results suggest that "successful RBM" relates to effective and efficient research leading to outcomes, with a minimum of perverse incentives. The building of an IP with a narrative TOC forces researchers to give some thought to what lies between solid science, great technologies, and their positive developmental impact. A mix of an outcome-focused TOC with people and partners at the core, and a RBM approach that allows us to monitor, reflect, evaluate, and learn, are key elements for a programmatic MEL strategy – coupled with great science.

4.13 Conclusion

Requests by funding agencies for a move towards outcome-oriented research programs are having considerable impact on the way in which research is conceived, planned, implemented and evaluated. A key requirement for such work is flexibility – the flexibility to adjust so that the outcome orientation works as a support mechanism and enabler rather than a one-size-fits-all straitjacket without any space for innovation, serendipity and creativity. The shift to a R4D approach based on TOC is fostering massive change, much of it for the better, in our view. However, it also comes with considerable challenges. Defining the necessary changes, and developing new processes and mechanisms, need time and resources, which are often grossly underestimated and inadequately planned for. Some of these challenges arise because of the nature of research: the results are not known from the start, unlike in engineering where the outcomes are generally much less uncertain. Another challenge is that CGIAR is a R4D organization, not a development organization, and it is still in the process of sorting out how to balance the need to do great science with the need for impact. We need to avoid the results-based focus being to the disadvantage of the science, and development being seen to be in competition with the science. Rather, they need to be seen as complementary, enabling, and liberating.

References

Alvarez, S., Jost, C., Schuetz, T., Förch, W., Schubert, C., & Kristjanson, P. (2014). *Lessons in theory of change from the introductory training on theories of change, impact pathways and monitoring & evaluation* (CCSL Learning Brief No. 10). Copenhagen: CGIAR Research Program on Climate Change, Agriculture and Food Security (CCAFS). http://hdl.handle.net/10568/52992.

Bazeley, P. (2004). Issues in mixing qualitative and quantitative approaches to research. In R. Buber, J. Gadner, & L. Richards (Eds.), *Applying qualitative methods to marketing management research* (pp. 141–156). Basingstoke: Palgrave Macmillan.

Bester, A. (2012). *Results-based management in the United Nations Development System: Progress and challenges.* A report prepared for the United Nations Department of Economic and Social Affairs, for the Quadrennial Comprehensive Policy Review Final Report.

Binnendijk, A. (2000). *Results-based management in the development co-operation agencies: A review of experience. DAC Working Party on Aid Evaluation Report.* OECD. www.oecd.org/development/evaluation/dcdndep/31950852.pdf.

Biswas, A., Palenberg, M., & Bennet, J. (2008). *Report of the first external review of the water and food challenge program review.* CGIAR Science Council. www.sciencecouncil.cgiar.org/sys tem/files_force/ISPC_CPER_WaterFood.pdf?download=1

CCAFS. (2010). *Activity workplans and budgets for 2010.* www.ccafs.cgiar.org/publications/ ccafs-activity-workplans-and-budgets-2010

CCAFS. (2011). *Annual workplan 2011 for theme leaders and regional program leaders.* cgspace. cgiar.org/bitstream/handle/10568/10214/workplan2011themeleaders1.pdf

CCAFS. (2014). *Annual report 2014.* ccafs.cgiar.org/research/annual-report/2014

CCAFS. (2015a). *How we work: Impact pathways.* ccafs.cgiar.org/impact-pathways

CCAFS. (2015b). *Planning, monitoring, learning and evaluation.* ccafs.cgiar.org/planning-moni toring-learning-and-evaluation

CCAFS. (2015c). *Key strategic documents.* ccafs.cgiar.org/key-strategic-documents

CGIAR. (2015a). www.cgiar.org

CGIAR. (2015b). *CGIAR challenge programs.* www.cgiar.org/our-strategy/challenge-programs/

CGIAR. (2015c). *CGIAR research programs.* www.cgiar.org/our-strategy/cgiar-research-programs/

Chambers, R. (2014). *Perverse payment by results: Frogs in a pot and straitjackets for obstacle courses.* Blog. participationpower.wordpress.com/2014/09/03/perverse-payment-by-results-frogs-in-a-pot-and-straitjackets-for-obstacle-courses/

Chen, H. T. (2005). *Practical program evaluation: Assessing and improving planning, implementation, and effectiveness.* Thousand Oaks: Sage Publications.

Crawford, P., & Bryce, P. (2003). Project monitoring and evaluation: A method for enhancing the efficiency and effectiveness of aid project implementation. *International Journal of Project Management, 21*(5), 363–373.

Cummings, F. H. (1997). Logic models, logical frameworks and results-based management: Contrasts and comparisons. *Canadian Journal of Development Studies, 18*(S1), 587–596.

Evenson, R. E., & Gollin, D. (Eds.). (2003). *Crop variety improvement and its effects on productivity.* Wallingford: CABI Publishing.

Douthwaite, B., Kuby, T., van de Fliert, E., & Schulz, S. (2003). Impact pathway evaluation: An approach for achieving and attributing impact in complex systems. *Agricultural Systems, 78* (2), 243–265.

Douthwaite, B., Alvarez, S., Cook, S., Davies, R., George, P., Mackay, R., & Rubiano, J. (2007). Participatory impact pathways analysis: A practical application of program theory in R4D. *The Canadian Journal of Program Evaluation, 22*(2), 127–159.

Douthwaite, B., Alvarez, S., Thiele, G., & Mackay, R. (2008). *Participatory impact pathways analysis: A practical method for project planning and evaluation.* Colombo: CPWF. http://hdl. handle.net/10568/33649.

Douthwaite, B., Kamp, K., Longley, C., Kruijssen, F., Puskur, R., Chiuta, T., Apgar, M., & Dugan, P. (2013). *Using theory of change to achieve impact in AAS* (AAS working paper). aas.cgiar. org/sites/default/files/publications/files/AAS-theory-of-change-impact.pdf

Eade, D. (1997). *Capacity-building: An approach to people-centred development.* Oxford: Oxfam. policy-practice.oxfam.org.uk/publications/capacity-building-an-approach-to-people-centred-development-122906

Earl, S., Carden, F., & Smutylo, T. (2001). *Outcome mapping, building learning and reflection into development programs.* Ottawa: IDRC Publications.

FAO. (2015). *Climate change and food systems: Global assessments and implications for food security and trade.* Rome: FAO. www.fao.org/documents/card/en/c/2d309fca-89be-481f-859e-72b27a3ea5dc/.

Gasper, D. (2000). Evaluating the 'logical framework approach' towards learning-oriented development evaluation. *Public Administration & Development, 20*, 17–28.

GIZ. (2015). *Results-based monitoring.* https://www.giz.de/rd-en/aboutgiz/518.html

Gonsalves, J. (2013). *A new relevance and better prospects for wider uptake of social learning within CGIAR* (CCAFS Working Paper No. 37). Copenhagen: CGIAR Research Program on Climate Change, Agriculture and Food Security (CCAFS).

Hall, A., Sulaiman, V. R., Clark, N., & Yoganand, B. (2003). From measuring impact to learning institutional lessons: An innovation systems perspective on improving the management of international agricultural research. *Agricultural Systems, 78*(2), 213–241.

Hall, A., Bullock, A., & Adolph, B. (2014). *Forward-looking review of the CGIAR challenge program on water and food.* Colombo: CPWF. http://hdl.handle.net/10568/41729.

Harrington, L. W., & Fisher, M. J. (Eds.). (2014). *Water scarcity, livelihoods and food security, research and innovations for development.* New York: Routledge.

Hartmann, A., & Linn, J. F. (2008). *Scaling up: A framework and lessons for development effectiveness from literature and practice* (Wolfensohn Center for Development Working Paper 5). Brookings Institute.

Harvey, B., Ensor, J., Garside, B., Woodend, J., Naess, L. O., & Carlile, L. (2013). *Social learning in practice: A review of lessons, impacts and tools for climate change* (CCAFS Working Paper No. 38). Copenhagen: CGIAR Research Program on Climate Change, Agriculture and Food Security (CCAFS).

Horton, D., & Mackay, R. (2003). Using evaluation to enhance institutional learning and change: Recent experiences with agricultural research and development. *Agricultural Systems, 78*, 127–142.

IPCC. (2013). In T. F. Stocker, D. Qin, G.-K. Plattner, M. Tignor, S. K. Allen, J. Boschung, A. Nauels, Y. Xia, V. Bex, & P. M. Midgley (Eds.), *Climate change 2013: The physical science basis. Contribution of working group I to the fifth assessment report of the intergovernmental panel on climate change.* Cambridge, UK: Cambridge University Press.

Johnson, N., Lilja, N., & Ashby, J. A. (2003). Measuring the impact of user participation in agricultural and natural resource management research. *Agricultural Systems, 78*, 127–142.

Jost, C., & Sebastian, L. (2014). *Workshop on mapping out a CCAFS R4D agenda and strategy for Southeast Asia.* Copenhagen: CCAFS. http://hdl.handle.net/10568/35586.

Jost, C., Kristjanson, P., Vervoort, J., Alvarez, S., Ferdous, N., & Förch, W. (2014a). *Lessons in theory of change: Monitoring, learning and evaluating knowledge to action* (CCSL Learning Brief No. 9). Copenhagen: CCAFS. http://hdl.handle.net/10568/42446.

Jost, C., Sebastian, L., Kristjanson, P., & Förch, W. (2014b). *Lessons in theory of change: CCAFS Southeast Asia research for development workshop* (CCSL Learning Brief No. 8). Copenhagen: CCAFS. http://hdl.handle.net/10568/42447.

Jost, C., Kristjanson, P., Alvarez, S., Schuetz, T., Foerch, W., Cramer, L., & Thornton, P. (2014c). *Lessons in theory of change: Experiences from CCAFS* (CCSL Learning Brief No 6). Copenhagen: CGIAR Research Program on Climate Change, Agriculture and Food Security (CCAFS). http://hdl.handle.net/10568/35184.

Jost, C., Alvarez, S., & Schuetz, T. (2014d). *CCAFS theory of change facilitation guide.* Copenhagen: CGIAR Research Program on Climate Change, Agriculture and Food Security (CCAFS). http://hdl.handle.net/10568/41674.

Kristjanson, P., Harvey, B., Van Epp, M., & Thornton, P. K. (2014). Social learning and sustainable development. *Nature Climate Change, 4*, 5–7.

Maredia, M. K. (2009). *Improving the proof – evolution of and emerging trends in impact assessment methods and approaches in agricultural development* (IFPRI discussion paper). Washington, DC: IFPRI.

Mayne, J. (2007a). *Best practices in results-based management: A review of experience a report for the United Nations Secretariat volume 1: Main report.* wpqr4.adb.org/lotusquickr/copmfdr/pagelibrary482571ae00516aa5.nsf/0/92A8DA95E6556A1A482576AF0029BB29/$file/john_mayne_un_1.pdf.

Mayne, J. (2007b). Challenges and lessons in implementing results-based management. *Evaluation, 13*(1), 187–109.

Pachico, D., & Fujisaka, S. (Eds.). (2004). *Scaling up and out: Achieving widespread impact through agricultural research* (Vol. 3). Cali: CIAT.

Raitzer, D., & Kelly, T. (2008). Benefit-cost analysis of investment in the international agricultural research centers of the CGIAR. *Agricultural Systems, 96*, 108–123.

Renkow, M., & Byerlee, D. (2010). The impacts of CGIAR research: A review of recent evidence. *Food Policy, 35*, 391–402.

Schubert, B., Nagel, U. J., Denning, G. L., & Pingali, P. L. (1991). *A logical framework for planning agricultural research programs*. Manila: IRRI.

Schuetz, T., Crame, L., Foerch, W., Jost, C., Alvarez, S., Thornton, P., & Kristjanson, P. (2014a). *Summary for the CCAFS flagship 4 projects kick-off meeting 28–29 January 2014: Result-based management trial*. http://hdl.handle.net/10568/35407

Schuetz, T., Schubert, C., Thornton, P., Förch, W., & Cramer, L. (2014b). *Lessons and insights from the CCAFS results-based management trial. Summary from CCAFS Flagship 4 RBM trial projects survey 2014*. CGIAR Research Program on Climate Change, Agriculture and Food Security. Report cgspace.cgiar.org/bitstream/handle/10568/53103/FP4_RBM%20Survey%20results_Final.pdf

Schuetz, T., Förch, W., Schubert, C., Thornton, P., & Cramer, L. (2014c). *Lessons and insights from CCAFS results-based management trial* (CCSL Learning Brief No 12). Copenhagen: CGIAR Research Program on Climate Change, Agriculture and Food Security (CCAFS). https://cgspace.cgiar.org/handle/10568/56629

Schuetz, T., Förch, W., & Thornton, P. (2014d). *Revised CCAFS theory of change facilitation guide*. Copenhagen: CGIAR Research Program on Climate Change, Agriculture and Food Security (CCAFS). http://hdl.handle.net/10568/56873

Schuetz, T., Förch, W., Thornton, P., Wollenberg, L., Hansen, J., Jarvis, A., Coffey, K., Bonila-Findji, O., Loboguerrero Rodriguez, A. M., Martínez Barón, D., Aggarwal, P., Sebastian, L., Zougmoré, R., Kinyangi, J., Vermeulen, S., Radeny, M., Moussa, A., Sajise, A., Khatri-Chhetri, A., Richards, M., Jost, C., & Jay, A. (2014e). *Lessons in theory of change from a series of regional planning workshops* (CCSL Learning Brief No 11). Copenhagen: CGIAR Research Program on Climate Change, Agriculture and Food Security (CCAFS). Link: http://hdl.handle.net/10568/52990

Schuetz, T., Förch, W., Thornton, P., Wollenberg, L., Hansen, J., Jarvis, A., Coffey, K., Bonilla-Findji, O., Loboguerrero Rodriguez, A.-M., Aggarwal, P., Sebastian, L., Zougmoré, R., Kinyangi, J., Jost, C., & Jay, A. (2014f). *Lessons in theory of change from a series of regional planning workshops*. Copenhagen: CCAFS. Full report: http://hdl.handle.net/10568/52315

Schuetz, T., Förch, W., & Thornton, P. (2014g). *CCAFS monitoring and evaluation strategy*. Copenhagen: CGIAR Research Program on Climate Change, Agriculture and Food Security (CCAFS). http://hdl.handle.net/10568/41913

Schuetz, T., Förch, W., & Thornton, P. (2015). *CCAFS reporting and evaluation in a results-based management framework* (CCSL Learning Brief No 15). Copenhagen: CGIAR Research Program on Climate Change, Agriculture and Food Security (CCAFS). http://hdl.handle.net/10568/67362.

Springer-Heinze, A., Hartwich, F., Henderson, J. S., Horton, D., & Minde, I. (2003). Impact pathway analysis: An approach to strengthening the impact orientation of agricultural research. *Agricultural Systems, 78*(2), 267–285.

Thornton, P. K., Ericksen, P. J., Herrero, M., & Challinor, A. J. (2014a). Climate variability and vulnerability to climate change: a review. *Global Change Biology, 20*(11), 3313–3328.

Thornton, P., Förch, W., Cramer, L., Vasileiou, J. C., & Kristjanson, P. (2014b). *Lessons learned from the Flagship 4 results-based management trial*. Copenhagen: CCAFS. http://hdl.handle.net/10568/35188

Thornton, P., Schuetz, T., Förch, W., & Campbell, B. (2014c). *The CCAFS flagship 4 trial on results-based management: Progress report*. cgspace.cgiar.org/bitstream/handle/10568/52261/CCAFS%20RBM%20trial%20report%202014_FINAL.pdf

Thornton, P., Förch, W., Schuetz, T., Barahona, C., Abreu, D., & Tobón, H. (2014d). *Guiding principles or "propositions" on monitoring and evaluation in CCAFS.* Copenhagen: CGIAR Research Program on Climate Change, Agriculture and Food Security (CCAFS). http://hdl.handle.net/10568/51830

UN. (2015a). *Transforming our world: The 2030 agenda for sustainable development.*

UN. (2015b). *World population prospects. The 2015 revision.* esa.un.org/unpd/wpp

UNDP. (2007). *Evaluation of results-based management at UNDP, achieving results.* Evaluation Office. http://web.undp.org/evaluation/documents/thematic/rbm/rbm_evaluation.pdf

Van Epp, M., & Garside, B. (2014). *Monitoring and evaluating social learning: A framework for cross-initiative application* (CCAFS Working Paper no. 98). Copenhagen: CGIAR Research Program on Climate Change, Agriculture and Food Security (CCAFS).

Vogel, I. (2012). *'Theory of Change' in international development.* Review Report for the UK Department of International Development.

WWAP. (2012). *The United Nations world water development report 4: Managing water under uncertainty and risk.* Paris: UNESCO.

Young, J., & Mendizabal, E. (2009). *Helping researchers become policy entrepreneurs* (ODI briefing paper No. 53). www.odi.org/sites/odi.org.uk/files/odi-assets/publications-opinion-files/1730.pdf.

Chapter 5
Lessons from Taking Stock of 12 Years of Swiss International Cooperation on Climate Change

Monika Egger Kissling and Roman Windisch

Abstract A stronger focus on results achieved in international cooperation on climate change has become common in the Swiss Agency for Development and Cooperation SDC (www.eda.admin.ch/sdc) and the State Secretariat for Economic Affairs SECO (www.seco.admin.ch). In 2014 these agencies have commissioned an assessment on the effectiveness of more than 400 of their climate change interventions over the timeframe of 12 years (2000–2012). This paper presents the methodological approach of the assessment and its results. In a second step and most importantly, it summaries the challenges and lessons learnt of commissioning and conducting such a stock-taking exercise in the field of climate change. These lessons are addressed to evaluators, practitioners and policy makers. In general, the paper concludes that preparing such a report on the effectiveness of the international cooperation in climate change is indeed a very challenging exercise. More specifically, the paper argues that firstly many more efforts are needed from evaluators to identify best methodological practices in dealing with such a mass of information, the wide and highly diverse portfolio and a lack of good quantitative and qualitative data. Secondly, practitioners need to invest more in project design and in monitoring in order to provide accurate data as a basis for sound assessment. Finally, policy makers should be well aware of the significant investments needed for such assessments as an instrument of accountability. This paper thus contributes to the debate among interested stakeholders on the need for better results measurement and results reporting in international cooperation on climate change.

M. Egger Kissling (✉)
Evaluation and Corporate Controlling Division, Swiss Agency for Development and Cooperation, Berne, Switzerland
e-mail: egger.consulting@bluewin.ch

R. Windisch
Quality and Resources Division WEQA, Swiss State Secretariat for Economic Affairs SECO, Berne, Switzerland
e-mail: roman.windisch@eda.admin.ch

© The Author(s) 2017
J.I. Uitto et al. (eds.), *Evaluating Climate Change Action for Sustainable Development*, DOI 10.1007/978-3-319-43702-6_5

Keywords Climate change • Swiss International Cooperation • Effectiveness • Measurement and reporting • Lessons

5.1 Introduction

A stronger focus on results has become common among international development agencies over the last decade. This is also the case for Switzerland and its two development agencies, the Swiss Agency for Development and Cooperation SDC[1] (Federal Department of Foreign Affairs) and the State Secretariat for Economic Affairs SECO[2] (Federal Department of Economic Affairs, Education and Research). For SDC and SECO it is important and of great interest to understand what worked and which interventions were effective, which interventions have not produced tangible results and what the reasons for success or failure are. Consequently, Switzerland regularly produces thematic Reports on the Effectiveness of the Swiss International Cooperation. Following effectiveness reports on Water (2008) and on Agriculture (2010),[3] the third Report on Effectiveness (2014) was dedicated to Climate Change. Taking stock of results achieved in international cooperation on climate change is a challenging exercise. One has to deal with a mass of information, a broad and highly diverse portfolio and a variety of actors. The consultants had to build on poorly developed methodologies and few internationally recognized standards for measuring climate change adaptation. They were also confronted with the lack of explicit climate baseline data and the difficulties in attributing (and aggregating) the effects of mitigation measures to Swiss interventions. Informing the parliament and the greater public on the results in a synthesized but still relevant manner on the basis of a comprehensive and highly technical report was another demanding task.

The main reason for those significant challenges was the fact that the assessment of the International Cooperation portfolio of 423 climate change relevant projects covering the timeframe 2000–2012, was a pioneer undertaking. Switzerland was one of the first bilateral donors commissioning such an assignment. Consequently, this assessment is of specific originality and can be considered as a pioneer venture of a bilateral donor in putting the climate lens on a longstanding development cooperation portfolio.

The authors' perspective is that of a donor administration. In this chapter the results of the assessment are briefly presented. However, the chapter is mainly focused on the process and presents the lessons of commissioning and conducting the stock taking on 12 years of Swiss International Cooperation on Climate Change. It also presents lessons on how to improve the evaluability of climate change

---

[1]www.deza.admin.ch

[2]www.seco.admin.ch

[3]Available under https://www.eda.admin.ch/deza/de/home/resultate_und_wirkung/wirkungs-_und_jahresberichte.html and http://www.seco-cooperation.admin.ch/themen/01033/01130/05122/index.html?lang=en

relevant programs and maximize climate change effectiveness from a practitioner's perspective. Finally the authors also present related conclusions and lessons learnt for policy makers.

5.2   Purpose

The purpose of initiating a "Report on the Effectiveness of the Swiss International Cooperation in Climate Change" was primarily **accountability**. The report aimed to provide mainly the members of the Swiss Parliament and the interested Swiss public with an accountable and transparent assessment of the climate change relevant interventions financed through public funds in the period 2000–2012. The report further accounts for the use of additional funding for climate change relevant interventions which aimed at raising Swiss ODA contributions to 0.5 % of gross national income (GNI).[4] These additional Swiss ODA contributions had been classified as Fast Start Financing (FSF) under the United Nations Framework Convention on Climate Change (UNFCCC).

The scope of the evaluation is focusing exclusively on the effectiveness of the portfolio. Thus the assessment is not an evaluation sensu stricto. The other OECD-DAC evaluation criteria (relevance, efficiency, impact, sustainability) have not been assessed and the report has not produced any recommendations.

Furthermore, it is important to highlight that the assessment has a clear focus on *the climate change effectiveness* of the portfolio, rather than assessing its overall results and achievements in relation to poverty alleviation which is regularly scrutinized in other studies and evaluations. Its findings on climate change effectiveness can therefore not be used to imply anything to the over-arching poverty reduction objectives and results of the Swiss International Cooperation. However, the impact of climate change on development is evident. People in developing countries are also more vulnerable to the negative consequences of climate change due to widespread poverty and lower resilience and coping capacities. Therefore, it seems apparent that climate change adaptation and mitigation measures have positive impacts on poor populations.

The evaluation assessed the effectiveness of SDC's/SECO's projects along the following general questions:

- How have climate change relevant interventions achieved their objectives and proven to be successful and effective in terms of mitigation and adaptation?
- To what extent have climate change relevant projects proven to be successful and effective in contributing to a low carbon development in the partner countries?

---

[4]Refer to the message for the increase of funds for the official development aid available under http://www.seco-cooperation.admin.ch/org/00515/00516/index.html?lang=en. This documents is available only in German and French×

- To what extent have climate change relevant projects proven to be successful and effective in contributing to a climate-resilient development in the partner countries?
- What obstacles, difficulties and challenges have undermined the desired success and effectiveness of climate change relevant interventions and which measures were undertaken to address them?

The evaluation was commissioned to Gaia Consulting Oy Ltd, Helsinki/Finland, in consortium with Zoï Environment Network, Geneva/Switzerland and Creatura Ltd, Bath/UK through an open tendering process. The tender document included the task to develop a suitable methodology, using different techniques and tools, which allow for the assessment of the project results and the production of aggregated result statements at portfolio level. Gaia consortium was required to document methodology, assessment, results and conclusions in a fully technical report. The contract included also the production, based on the technical report, of a public report for dissemination, using modern communication techniques, including the production of a video. The consortium was therefore charged to present solid evidence-based results in an attractive manner for different targeted audiences. By reporting on and accounting for the achieved results in Climate Change, the report also contributed to the institutional learning at SDC and SECO.

5.3 Methodology

In preparation for the Terms of Reference of the mandate, SDC and SECO had already undertaken some analytical work in order to specify the scope and volume of the assessment. Firstly, every project within the whole portfolio of Swiss International Cooperation was rated ex post on its climate change relevance (the extent to which its main objectives contribute to climate change mitigation and climate change adaptation), resulting in a portfolio of 508 individual, climate change relevant projects. Within this portfolio 283 projects with a total value of CHF 975 million were implemented by SDC, and 140 projects with a total value of CHF 346 million by SECO. A number of these projects were initiated before 2000, and some projects were still ongoing by the time the evaluation was finished. The total budget of climate change related commitments for this period amounted to CHF 1.32 billion, around 5 % of the overall ODA funding provided by Switzerland during these years. Secondly, the intervention logic on portfolio level was reconstructed, resulting in the definition of seven different result chains defining concrete outputs, outcomes and impacts (see Fig. 5.1).

The intervention logic sets the frame for formulating a theory of change for each of the three areas of interventions (Enabling Framework, Mitigation and Adaptation). They are closely linked to the intended results at outcome/impact level formulated in Fig. 5.1.

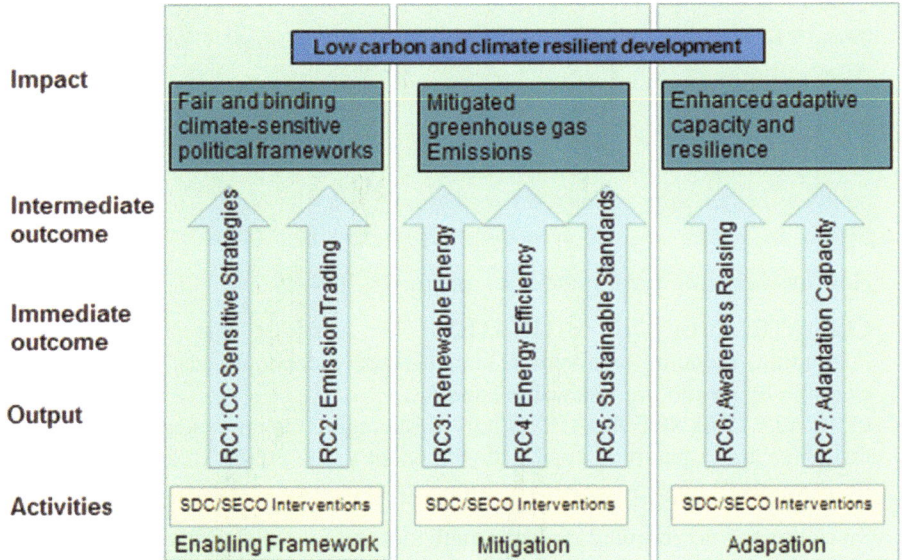

Fig. 5.1 **Intervention logic of Swiss International Cooperation in climate change**. *RC*=result chain (Source: SDC/SECO, Tender Document, Report on Effectiveness of the Swiss International Cooperation in climate Change Mitigation and Adaptation Interventions 2000–2012, 2013-04-09, p. 8)

Theory of Change for the Area of Intervention 'Enabling Framework'
Switzerland's engagement for Enabling Frameworks contribute to the development of fair and binding climate-sensitive political frameworks on international level and in partner countries. It ensures that negotiations on strategies on growth and development are built on principles of "green and low carbon growth" and on "building climate resilience of systems and people".

Theory of Change for the Area of Intervention 'Climate Change Mitigation'
Switzerland's engagement for climate change mitigation reduces GHG Emissions in partner countries by facilitating the access and use of low carbon technologies in the production processes and energy systems. It also supports the sustainable use of natural resources through the use of norms and standards as well as best practices in agriculture, forestry and water management.

(continued)

Theory of Change for the Area of Intervention 'Climate Change Adaptation'
Switzerland's engagement for climate change adaptation enhances the adaptive capacity and resilience in partner countries through a combination of interventions allowing to secure and improve living conditions and livelihoods of people affected by climate change.

How the terms are used in the assessment:

- CLIMATE CHANGE MITIGATION = Avoiding the unmanageable. Preventing, reducing or avoiding human-made greenhouse gas emissions, for example by promoting renewable energies.
- CLIMATE CHANGE ADAPTATION = Managing the unavoidable. Increasing resilience and capacity to cope with and adapt to the effects of climate change, for example by improving early warning systems for extreme weather events.

Since there is no accepted standard methodology for the summative assessment on portfolio level as requested in the mandate, the consultants applied an innovative and adaptive approach to develop a suitable methodology. The finally applied methodology covered the following three steps:

- Portfolio appraisal: In a first step, the consultants conducted an independent appraisal of the portfolio, reviewing and developing an understanding of the nature of all 508 projects, exploring the quality of available data, validating the proposed climate change relevance of the projects and identifying suitable clusters in reference to the proposed result chains. This resulted in a portfolio of 423 assessable projects, categorized into six thematic clusters (energy, cleaner production, natural resources, hazards, livelihoods, knowledge) and the funding and grants to organizations as a separate cluster. Furthermore six countries including 30 projects (five in each country) were identified for field visits and in-depth studies. The selection had to consider the following criteria:

  - Thematic balance: The selected projects had to include and balance interventions in the three Areas of Intervention (Enabling Framework, Adaptation and Mitigation).
  - Geographical balance: The selected projects had to include and balance interventions in priority countries from different continents including the former Soviet Republics/countries from Eastern Europe.
  - Institutional balance: The selected projects had to include and balance projects of SDC and SECO and reflect bilateral and multi-bilateral funding schemes.
  - Performance balance: The selected projects had to represent strengths and weaknesses/successes and challenges of the Swiss International Cooperation in Climate Change.
  - Time balance: The selected projects had to represent the whole observation period, considering the increasing relevance of climate change in Swiss International Cooperation over time.

- A further critical requirement was the availability of an adequate documentation of the selected projects.
- Finally, some projects should have produced visible effects that would allow an attractive visualization of achievements.

- Key questions: The Climate Change Report on Effectiveness investigates the achieved results of the selected 423 climate-relevant projects carried out by the SDC and the SECO in the areas of climate change adaptation and climate change mitigation. The key questions for the analysis were: What contribution did the projects make towards improving people's ability to cope with the negative effects of climate change (adaptation)? What contribution did the projects make towards reducing greenhouse gas emissions (mitigation)? To what extent did Switzerland's engagement for enabling frameworks contribute to the development of fair and binding climate-sensitive political frameworks at the international level and in partner countries?
- Detailed investigations: The second step comprised more detailed investigations of these 30 projects during field visits to the selected six countries (Nepal, South Africa, Peru, Mongolia, Serbia and Albania). The desk study of 31 additional projects (including an in-depth study of projects in Vietnam) ensured the balanced coverage across the various themes and modalities within the portfolio. The project documentation included planning and reporting documents such as Credit Proposals, Annual Reports, Progress Reports as well as Evaluation Reports. The detailed investigations involved direct interviews with knowledge holders at project level.
- Portfolio analysis: The third step was to analyse the complete portfolio of 423 projects, and to determine adaptation and/or mitigation effectiveness scores, with the aim of estimating the overall effectiveness of each thematic approach and of the whole portfolio. This assessment drew on the portfolio appraisal, detailed project reviews, questionnaires, interviews, and focus group discussions. Overall effectiveness scores for the 423 projects for which sufficient information was available were distributed across all themes. These scores were either 'tentative' or 'confirmed' and both represented the reviewer's judgement on the project's effectiveness, from 'extremely strong' (score 7) to 'none'. Tentative scores were based on the information presented in key documents, informed by similar projects that have been reviewed through in-depth assessments, as well as sectoral specific reputation of the implementing partner. Confirmed scores were based on the findings of the 61 detailed desk and field studies, and replaced the tentative scores in each of these cases. The distribution of effectiveness scores in the sample of confirmed scores (n = 61) was compared with that in the larger sample of tentative scores (n = 362), and the distributions were found to be significantly correlated. Though not as perfect as an in-depth study of all 508 projects would have been, the use of tentative scores in the overall assessment was necessary. The portfolio was far too diverse to yield meaningfully representative results or aggregate results statements for the whole portfolio.

5.4   Results

The assessment concluded that, on average, the 423 projects of Swiss international cooperation analyzed show "moderate to strong" effectiveness in reducing greenhouse gas emissions and in increasing people's ability to cope with the impacts of climate change. Approximately 40 % of the portfolio was found to be strongly or very strongly effective, both in climate change mitigation (114 projects) and adaptation (121 projects). Around 50 % of the total portfolio budget was allocated to interventions assessed as moderately effective (198 projects) in terms of climate mitigation or adaptation. Only 10 % of the projects showed little or no climate benefit.

Despite the geographical and cultural diversity of over 70 partner countries within the portfolio, no difference in effectiveness between the different geographical regions were identified. It has been found that climate effectiveness improved over time, with higher effectiveness scores of projects implemented after 2007. Thus the share of adaptation projects rated as highly and very highly effective increased from 23 to 66 % between the projects implemented in the periods 2000–2006 and 2007–2012. For mitigation, the increase was from 36 to 54 %. Furthermore, recent projects in the portfolio integrated climate change more explicitly into project design and the quality of design of these specific interventions improved. Finally, the creation of the SDC Global Programme on Climate Change and the development of a new thematic priority "Fostering climate-friendly growth" in SECO are signs of increased strategic importance and institutional awareness on climate change.

The stock taking exercise did not identify factors of success that are specific for high climate change mitigation or adaption effectiveness. It rather concluded on general success factors such as a comprehensive project design, high stakeholder commitment and ownership, good project management to be a precondition for highly satisfactory results achievement. At the same time, the report identified several domains that proved having predominantly positive results. For mitigation they include the rehabilitation of hydropower plants, improving energy efficiency, promoting renewable energy and cleaner production, and improved ecosystem management. Multi-stakeholder forest management projects, biotrade-based conservation and organic farming projects create in addition to mitigation results, in many cases, also important adaptation benefits. In the adaptation field, Swiss-funded interventions in the areas of risk management, disaster risk reduction (including early warning) and insurance are providing real benefits to large numbers of people in various parts of the world. Swiss contributions to several multilateral institutions show high effectiveness overall (both for mitigation and adaptation). For example, the results achieved through the Forest Carbon Partnership Facility (FCPF), the Partnership for Market Readiness and the UNFCCC Adaptation Fund are clearly noted, with Switzerland contributing to the results through its funding alongside expertise and strategic guidance.

5.5 Challenges and Lessons Learnt

5.5.1 In General

Preparing such a report on the effectiveness of the international cooperation in climate change is indeed a challenging exercise: one has to deal with a mass of information, with a wide and highly diverse portfolio and with a variety of actors; moreover developing a method for assessing adaption is a crucial challenge. Another demanding task is to inform a larger public on the results presented in a comprehensive and highly technical report in a synthesised but still relevant manner. The elaboration of the report has shown clear limits that must be balanced with too high expectations. Given the lack of comprehensive and reliable data as well as efficient and agreed methodologies to collect quantitative data, mainly in the field of adaption, there is a risk that the results are either too generic at a portfolio level or that "show cases" are reduced to a few examples.

5.5.2 For Evaluators

- *Resources*: The numerous challenges for evaluators in a complex exercise start with the allocation of sufficient resources for such a pioneering assessment. The expectation to conduct a pure accountability exercise in a most efficient way often leads to the allocation of insufficient resources. The absence of well-developed methodologies, the huge size of data and information to be assessed in a large portfolio, combined with the expected lack of direct evidence of climate effectiveness has to be taken into account.
- *Expectations*: The expectations have to be aligned with the size of the invest-ment. The ex-post reconstruction of baselines and the assessment of quantitative results is an intensive and time consuming process. If rigorous climate change related quantitative and qualitative data are not available in final reports or evaluations of the assessed projects, it is illusive to think that an assessment covering a portfolio of several hundred projects is able to fill that gap and to produce aggregated quantitative data, for example on mitigated GHG emissions. When producing data on proxies or qualitative assessments, the expectations must be realistic, not to say modest.
- *Independence* is one of the most important principles in evaluations focusing on accountability. Ensuring this independence of the consultants in such an inno-vative approach is however challenging. It could either undermine a constructive exchange between the consultant and the commissioner if implemented too strictly, leaving the consultants too isolated. Or it could lead to in-depth involve-ment and micro-management by the commissioner especially if there are dispa-rate perceptions on how to approach and address the upcoming challenges in developing the methodology from the very beginning.

- *Scope and focus*: The challenges for the consultants in commissioning an assessment with such a narrow scope are twofold. Firstly, consultants might tend to expand their assessment to other OECD DAC evaluation criteria such as relevance, efficiency or sustainability. In particular in a case where the climate change relevant portfolio under review is predefined by the commissioner, the consultants might refuse to accept this climate change earmarking by the mandatory without additional re-verification and assessment.

   Secondly, the focus on accountability for effectiveness as well as the renouncement to develop recommendations also demands a clear management of expectations toward the project managers. The intensive involvement of responsible project managers often leads to the expectation that the scope of the assessment can be widen individually and that a report on effectiveness also produces recommendations. The SDC/SECO reports on effectiveness treat learning clearly as a secondary objective and the formulation of recommendations is not part of the evaluation.

- *Method*: From a clear methodological point of view, the main challenge lies in the late introduction of climate change earmarking SDC and SECO's interventions, the fact that climate change benefits are co-benefits in most projects and that results relevant for accountability toward the public are only achieved with a significant time-lag. Earlier interventions implemented before the introduction of the OECD Rio Markers in 2006 for adaptation and 2010 for mitigation (see References) often do not have an explicit focus on climate change mitigation and adaptation. As a consequence, they often lack clear climate change related objectives, indicators and baselines. Nonetheless, they have potentially produced significant results in terms of climate change mitigation or adaptation and are worth to be included in a report on effectiveness. As mentioned above, the complexity and the resources needed to assess their effectiveness is however much higher in comparison with newer project that have systematically integrated climate change into their results framework (with respective indicators and targets) and consecutive monitoring and evaluation activities.

It is important to notice that the portfolio assessed for this analysis embraced projects and initiatives that were not explicitly making reference to climate change. Initially the projects and programmes implemented mainly during the earlier period were neither fully geared towards nor openly declared as climate change relevant interventions. Only over time, some of them were gradually oriented towards climate change and declared as such. The introduction of the OECD Rio Markers between 2006 and 2010 supported a clear earmarking of climate change relevant projects. Finally the Bill to Parliament on ODA 0.5 % in 2010 specifically earmarked some of its funds to tackle climate change. As a consequence, the precise tracking of climate change relevant interventions was far more difficult for the first half of the period 2000–2012 and many projects had to be classified ex post.

The challenge of time-lag between the implementation of a project and the presence of measurable results at outcome and impact level is particularly relevant for climate change. A report on effectiveness is a very challenging undertaking for a topic that is high on the political agenda for a short period only. The results that are of interest for the wider public materialize only with a certain time lag and are not available with the first 3–5 years of a project. In fact, more time between the intervention and the evaluation would be needed in order to assess whether the adaptation measures have been effective and contributed to increased resilience, or whether mitigation measures finally resulted in the expected reduction of greenhouse gas emissions. This also leads to the question if a highly diverse portfolio covering a timeframe of 12 years can be assessed with the same methodology.

Finally the methodological challenges to assess effectiveness also depend on the topic. For adaptation interventions they are considerably higher than for those projects in the field of mitigation. Contrary to the field of mitigation, no clear metric and reliable baseline data exist for measuring adaptation and a lot of the measures are rather more of a qualitative than quantitative nature. Therefore, it is more difficult to find adequate indicators, which can measure effectiveness of the interventions. Thus, more time needs to be invested in the development of baselines and measurement, reporting, and verification systems. Moreover, aggregating and scoring will remain difficult.

Finally the influence and effectiveness of projects working on the policy level to create a better climate change framework is difficult to measure. The assessment concentrated here on interviews with Swiss experts engaged in policy dialogue in international institutions and initiatives and on the institutions' results reporting.

5.5.3   For Practitioners/Program Managers

- *Results reporting*: As in other areas of intervention, the common difficulties in assessing results statements at outcome and impact level have been experienced in the climate change assessment. It revealed several lessons in term of Result Based Management (RBM) and monitoring for project managers of climate change relevant project. Despite the fact that the reorganization process of SDC (2008–2012) has focused on result orientation and that results based management within SECO has been improved, it appears still premature to expect comprehensive and well-documented result reporting on all interventions. This is especially the case for the earlier projects under review. Given the fact that the design of projects in terms of climate change has improved significantly over time, it can be expected that a similar assessment in a few years would be more successful in gathering quantitative and qualitative results, thus allowing for an aggregation at higher level. However, this depends on the development of result frameworks with smart and standardized indicators across the portfolio. Consequent baseline studies and the onset and rigorous monitoring during implementation are further preconditions.

    This does not mean that gathering results on climate change effectiveness will become an easy task. The above-mentioned measures need significant resources. Consequently the expectation on a quantitative monitoring of GHG emission reductions needs to be clarified explicitly at the beginning of each project. Moreover, if there is a real demand for clear results on portfolio level, the investments in RBM and M&E need to be approved in order to set the ground for reports on effectiveness that assess results based on evidence.

- *Mainstreaming*: Although the OECD Rio markers have obliged project developers to systematically consider climate change relevance and benefit within their project, there is further need for more systematic mainstreaming of climate change adaptation and mitigation into development projects. Explicitly mentioning the climate change components and objectives does potentially increase the awareness and ownership at the level of implementing partners, stakeholders and beneficiaries which will positively contribute to the effectiveness of the programs.

- *Common understanding*: A common understanding between donors and implementing partners on the relevance of climate change within a project is crucial in order to ensure transparent reporting on achieved results. Many donors have been criticized in the past for not applying the Rio markers, in particular the climate-related ones, in a coherent manner. In the framework of this results assessment, Switzerland has conducted an exhaustive revision, has gained a valuable overview and improved its skills in reliable coding of its climate change portfolio.

- *Synergies between adaptation and mitigation*: The tender document initially proposed a clear separation between a climate change mitigation and adaptation portfolio. The assessment revealed, however, that climate change adaptation and mitigation are often interlinked. A clear separation does miss potential synergies. One should try to reach for multipurpose results in the design of the projects. The report therefore encourages a systematic integration of climate change adaptation into development as a more promising approach in order to achieve sustainable and resilient development, instead of trying to clearly identify "additionality" of adaptation actions. Adaptation and mitigation synergies could be increased, in particular in the natural resource management sector, but also e.g. in hydropower, by addressing the issues more systematically during planning and establishing the adequate measurement, reporting and verification (MRV) systems.

5.5.4 For Policy Makers

- *Joint forces for better cost-benefit*: Policy makers should be better informed on the investment needed for producing reports on effectiveness and be aware about the difficulties and challenges in terms of quality, accessibility and availability of data and the development of adequate methods. A discussion on the need for rigorous results measurement on the "end" side and in consequence the need to invest in rigorous results planning systems on the "entry" side mainly raises questions on priorities and resources (human, financial, time). National parliaments could think about joining forces with other donors for initiating joint results assessments on selected topics of the international cooperation in order to have a better cost-benefit-balance and benefit from mutual learning.

- *International debate*: With regard to international commitments to a global climate deal (Paris 2015) it is critical to sharpen the international understanding on the results measurement related to climate change sensitive investments and to decide on the level of ambition. Strengthening the climate change capacity (policy, planning, and programming) in partner countries are also one precondition to achieve mutual accountability in this sector.
- *Swiss CC know-how for development*: For climate targeted projects, SDC and SECO could focus/concentrate even more on areas where Switzerland has proven technical expertise, such as renewable energy (in particular hydropower), disaster risk reduction and disaster risk/weather insurance, energy efficiency in buildings and small and medium enterprises, air quality, and ecosystem management.

5.6 Conclusions

Based on the experience from this pioneering assessment the key conclusions from the donor's perspective are the following:

- Be precise in the terms and methods: Clearly say what is meant by effectiveness and what the results are about. Avoid vague terms such as "climate benefit" or "climate effectiveness".
- It is difficult to isolate the effect of a single donor's intervention in mitigation. The attribution of climate change projects in the field of mitigation to a single bilateral donor is methodologically questionable, especially without clear baseline data.
- In the field of climate change adaptation quantitative data are often lacking and it is important to appreciate qualitative data in an adequate manner.
- It is difficult to report on policy influencing at international and regional level and to link the effects from being at the table of negotiations with concrete changes in people's life.

The overall conclusion is that this pioneering undertaking of producing the Report on Effectiveness in Climate Change did not allow identifying best methodological practices how best taking stock of climate change projects and programs. This chapter is much more an appeal to be precise, realistic, authentic and transparent in the communication of the methodological challenges and of the results. Let's take the reports on effectiveness as a chance to enter into an open dialog with interested stakeholders, mainly with the national parliament. Let us explore the opportunity and utility to undertake effectiveness assessments jointly with other development agencies and join forces and resources for further improving the approach, the methods and the common learning from effectiveness reports.

References

Documents available on the "Report on the Effectiveness of the Swiss International Cooperation in
    Climate Change":
    Swiss Development Cooperation, Swiss State Secretariat for Economic Affairs. (2014). *Report
    on Effectiveness 2014: Swiss International Cooperation in Climate Change 2000–2012*. http://
    www.seco-cooperation.admin.ch/themen/01033/01130/05122/index.html?lang=en. Accessed
    29 Apr 2016.
Information on the legal basis and federal decisions for Swiss International Cooperation available
    under:
    The Swiss Confederation. (2010). *Message concernant l'augmentation des moyens pour le
    financement de l'aide publique au développement*. http://www.seco-cooperation.admin.ch/org/
    00515/00516/index.html?lang=en. Accessed 29 Apr 2016.
Documentation on SDC/SECO accountability reports/reports on effectiveness available:
    Swiss Development Cooperation, Swiss State Secretariat for Economic Affairs. (2008). *Report
    on the effectiveness of Swiss development cooperation in the water sector*. http://www.seco-
    cooperation.admin.ch/themen/01033/01130/05122/index.html?lang=en. Accessed 29 Apr
    2016.

Chapter 6
An Analytical Framework for Evaluating a Diverse Climate Change Portfolio

Michael Carbon

Abstract The Climate Change Sub-programme (CCSP) of the United Nations Environment Programme (UNEP) has four components: Adaptation, Mitigation, REDD+ and Science and Outreach. It cuts across all UNEP divisions located in Nairobi and Paris, and relies a lot on partnerships to drive its work and scale up its impact. The CCSP evaluation conducted by the UNEP Evaluation Office over the period 2013–2014, aimed at assessing the relevance and overall performance of the CCSP between 2008 and 2013. The complexity, geographical spread and rather weak results framework of the CCSP, coupled to rather limited evaluation resources and a shortage of evaluative evidence, required the Evaluation Office to develop an innovative analytical framework and data collection approach for this evaluation. It combined three areas of focus (strategic relevance, sub-programme performance and factors affecting performance), five interlinked units of analysis (UNEP corporate, sub-programme, country, component and project level), a Theory of Change approach and an appropriate combination of data collection tools. This chapter discusses the overall evaluation approach and process, followed by a summary of lessons learned which could be useful for future similar exercises.

Keywords Programme evaluation • Complex evaluation • Theory of change • Climate change • UNEP

6.1 Introduction

The United Nations Environment Programme (UNEP) has been working on climate-related issues for more than 20 years,[1] but UNEP has a formal Climate Change Sub-programme (CCSP) only since the Medium-Term Strategy (MTS) for

---

[1]UNEP 2010, Climate Change Strategy for the UNEP Programme of Work 2010–2011. Web link: http://www.unep.org/pdf/UNEP_CC_STRATEGY_web.pdf

M. Carbon (✉)
Evaluation Office, United Nations Environment Programme, Nairobi, Kenya
e-mail: michael.carbon@outlook.com

© The Author(s) 2017   95
J.I. Uitto et al. (eds.), *Evaluating Climate Change Action for Sustainable Development*, DOI 10.1007/978-3-319-43702-6_6

2010–2013. According to the MTS 2010–2013[2] UNEP's CCSP objective is *"to strengthen the ability of countries to integrate climate change responses into national development processes"*. UNEP is expected to support countries and institutions to meet the challenges of climate change by promoting ecosystem-based approaches to adaptation, up-scaling the use of and facilitating access to financing for clean and renewable energy and technologies, and capitalizing on the opportunities of reducing emissions from deforestation and forest degradation. UNEP is also working to improve awareness and understanding of climate change science for policy decision-making. As such, the UNEP CCSP is organized around four components: Adaptation, Mitigation, REDD+, and Science and Outreach. Each component has its own Expected Accomplishments (direct results expected from UNEP's interventions) achieved through Programme of Work Outputs (different products and services delivered by UNEP).

In UNEP, Sub-programmes cut across the divisional structure of the organization and the CCSP is the most cross-cutting of all sub-programmes in UNEP. For instance, the Division for Technology, Industry and Economics, based in Paris, is accountable for delivering the Mitigation component and the Division of Environmental Policy Implementation, based in Nairobi, manages the majority of projects under the Adaptation and REDD+ components. The Division for Early Warning and Assessments, based in Nairobi, is accountable for the delivery of certain assessments and assessment capacity building under the Science and Outreach component. The structural complexity and geographical spread of the CCSP posed specific challenges for the evaluation, as described below.

The CCSP heavily relies on partnerships to drive the work. These partnerships are important both for global efforts, such as the preparation of annual global reports that help establish norms and track progress in achieving them, as for efforts at the regional and country level. Partners often bring complementary technical skills and provide access to decision making fora. Since UNEP is a non-resident agency, it must also rely on operating through partners at the country level. Cooperation with government and other local partners is necessary because the country projects/pilots serve the double purpose of developing and testing concepts and tools, but also to build country ownership and capacity to use them to promote in-country replication. Also this posed challenges for the evaluation, in particular in terms of attribution of Sub-programme results to UNEP.

---

[2]UNEP 2009, United Nations Environment Programme Medium-term Strategy 2010–2013: Environment for Development. Web link: http://www.unep.org/PDF/FinalMTSGCSS-X-8.pdf

6.2   Scope of the Evaluation

In accordance with the UNEP Evaluation Policy, all Sub-programmes are evaluated on a rotating basis every 4 years.[3] They are part of a larger evaluation architecture that include project, sub-programme and UNEP-wide, Medium Term Strategy evaluations. Sub-programme evaluations are conducted by the UNEP Evaluation Office in consultation with the relevant UNEP Divisions. While the Evaluation Office reports to the UNEP Executive Director, its evaluations are conducted in an independent manner and evaluation findings are reported without interference. However, the Evaluation Office does not enjoy financial independence and its limited financial and human resources are sometimes a major challenge.[4]

The CCSP evaluation aimed at assessing the relevance and overall performance of UNEP's work related to climate change from 2008 to 2013 according to standard evaluation criteria (relevance, effectiveness, efficiency, sustainability and impact). The evaluation assessed whether, in the period under review, UNEP was able to strengthen the ability of countries to integrate climate change responses into national development processes, by providing environmental leadership in the international response to climate change and complementing other processes and the work of other institutions. The evaluation was an in-depth, independent exercise conducted by a multidisciplinary team of consultants and Evaluation Office staff, with oversight from the UNEP Evaluation Office. The author was in charge of overall design, management and quality assurance of the evaluation process and participated in interviews and country visits.

The evaluation tried to answer the following key questions:

- Are the Sub-programme objectives and strategy relevant to the global challenges posed by climate change, global, regional and country needs, the international response and UNEP's evolving mandate and capacity in this area?
- Has UNEP achieved its objectives in the area of climate change? Have projects been efficiently implemented and produced tangible outputs as expected? Are the required external factors in place so that the CCSP outputs can lead the expected outcomes and, ultimately, to sustainable, large-scale impact?
- What are the key factors affecting sub-programme performance, such as portfolio design and structure; human and financial resources administration; collaboration and partnerships; and monitoring and evaluation?

The evaluation covered the four components of the CCSP. However, because the Science and Outreach component was largely implemented within projects belonging to the first three components, the Evaluation Team decided to treat Science and

---

[3]UNEP 2009, Evaluation Policy. Web link:
  http://www.unep.org/eou/StandardsPolicyandPractices/UNEPEvaluationPolicy/tabid/3050/Default.aspx
[4]UNEG 2012, Professional peer review of the evaluation function, United Nations Environment Programme. Web link: www.uneval.org/document/download/1527

Outreach as a cross-cutting issue rather than a stand-alone component. The portfolio under review included 57 projects and programmes classified by UNEP as belonging to the CCSP and that were either on-going or had been initiated after 1 January 2008. A little over half (32) of these projects were completed at the time of the evaluation, 20 were on-going and the remaining 5 were inactive or had an unknown status. Within this portfolio, there were a number of interventions known as "umbrella projects", which included several, independent sub-projects contributing to the same Expected Accomplishment or (set of) Programme of Work Outputs. If all sub-projects were counted, the total evaluation portfolio comprised about 88 interventions. Their spread over the different thematic components was as follows: 60 % were mitigation, 23 % were adaptation, 5 % were REDD, and 9 % science and outreach. The remaining combined both mitigation and adaptation objectives.

6.3 Challenges to the Evaluation

A rapid assessment of the evaluability of the sub-programme during the inception phase had brought to light several challenges the evaluation was bound to face. First, it was expected to assess a large, highly diverse and dispersed project portfolio, spread over four components, managed by various branches across the organization based in different duty stations. Second, a review of strategic documents had revealed serious issues with the results framework of the sub-programme, namely its internal logic, the results levels at which Expected Accomplishments and Programme of Work Outputs were pitched and the changes in results statements, indicators and targets every 2 years. Table 6.1 presents the results framework for the mitigation component as an illustration. Third, the assessment of strategic relevance would prove quite challenging considering the rapidly changing political and institutional context such as new decisions immerging from UNFCCC COPs and others.

At the same time, the evaluation would need to cope with very limited evaluative evidence. For instance, monitoring of progress at the sub-programme level was limited to output milestones and weak outcome indicators. Project reporting was donor-specific, incomplete and focused on activities and outputs and, over the period covered by the evaluation, less than one quarter of the projects in the portfolio under review had been independently evaluated due to resource limitations and a lack of pressure from senior management and Member States. In addition, this ambitious evaluation had to be carried out with a very limited budget, which allowed the recruitment of only three consultants for a relatively short period of time.

These challenges were, however, not specific to the Climate Change Sub-programme evaluation. Similar issues were encountered by previous sub-programme evaluations, requiring the Evaluation Office to develop an

Table 6.1 Results framework of the mitigation component of the Climate Change Sub-programme

Programme of work 2010–2011		Programme of work 2012–2013	
Expected accomplishment	Programme of work output	Expected accomplishment	Programme of work output
EA(b) Countries make sound policy, technology, and investment choices that lead to a reduction in greenhouse gas emissions and potential co-benefits, with a focus on clean and renewable energy sources, energy efficiency and energy conservation	1. Technical and economic assessments of renewable energy potentials are undertaken and used by countries in making energy policy and investment decisions favouring renewable energy sources	EA(b) Low carbon and clean energy sources and technology alternatives are increasingly adopted, inefficient technologies are phased out and economic growth, pollution and greenhouse gas emissions are decoupled by countries based on technical and economic assessments, cooperation, policy advice, legislative support and catalytic financing mechanisms	1. Economic and technical (macroeconomic, technology and resource) assessments of climate change mitigation options that include macroeconomic and broad environmental considerations are undertaken and used by countries and by major groups in developing broad national mitigation plans
	2. National climate technology plans are developed and used to promote markets for cleaner energy technologies and hasten the phase-out of obsolete technologies		2. Technology-specific plans are developed through public-private collaboration and used to promote markets for and transfer of cleaner energy technologies and speed up the phase-out of obsolete technologies in a manner that can be monitored, reported and verified
	3. Knowledge networks to inform and support key stakeholders in the reform of policies and the implementation of programmes for renewable energy, energy efficiency, and reduced greenhouse gas emissions are established		3. Knowledge networks and United Nations partnerships to inform and support key stakeholders in the reform of policies, economic incentives and the implementation of programmes for renewable energy, energy efficiency and reduced greenhouse-gas emissions are established, supported and used to replicate successful approaches
	4. Macro-economic and sectoral analyses of policy options for, fostering low greenhouse gas emissions, including technology transfer, are undertaken and used		
	5. Sustainability criteria and evaluation tools for biofuels development are refined globally and applied nationally		
	6. Public/private partnerships are promoted and best practices are applied leading to energy		

(continued)

Table 6.1 (continued)

Programme of work 2010–2011		Programme of work 2012–2013	
Expected accomplishment	Programme of work output	Expected accomplishment	Programme of work output
	efficiency improvements and greenhouse gas emission reductions		
EA(c) Improved technologies are deployed and obsolescent technologies phased out, through financing from private and public sources including the Clean Development Mechanism and the Joint Implementation Mechanism of the Kyoto Protocol	1. Barriers are removed and access is improved to financing for renewable and energy efficient technologies at the national level through targeted analysis of costs, risks and opportunities of clean energy and low carbon technologies in partnership with the finance sector	EA(c) Countries' access to climate change finance is facilitated at all levels and successful innovative financing mechanisms are assessed and promoted at the regional and global level	1. Financing barriers are removed and access to financing is improved for renewable and energy-efficient technologies through public-private partnerships that identify costs, risks, and opportunities for clean energy and low-carbon technologies
	2. Clean Development Mechanism projects are stimulated through market facilitation and the application of relevant tools, methodologies and global analyses, including on environmental sustainability		2. Use of the Clean Development Mechanism and other innovative approaches to mitigation finance is stimulated through analyses and the development and application of relevant tools and methodologies, including on environmental sustainability and measuring, reporting and verification compatibility
	3. National institutional capacity for assessing and allocating public funding and leveraging private investment for clean energy is strengthened		3. Institutional capacity for assessing and allocating public funding and leveraging private investment for clean energy is strengthened and new climate finance instruments are developed and applied by financiers, lenders and investors
	4. New climate finance instruments are launched and investments in clean energy are made by first-mover financiers and lenders and investors		
	5. Financial institutions adopt best climate, environmental and sustainability practices		

Sources: UNEP Biennial Programme of Work and Budget for 2010–2011; UNEP Biennial Programme of Work and Budget for 2010–2011

innovative analytical framework and data collection approach for sub-programme evaluations.[5] These were further refined for the CCSP evaluation and are discussed in the following sections, followed by a summary of lessons learned which could be useful for future similar exercises.

6.4   Analytical Framework of the Evaluation

The evaluation assessed the Climate Change Sub-programme in **three areas of focus**, corresponding to three distinct but strongly related clusters of evaluation questions (see Table 6.2). First, the evaluation assessed the strategic relevance and

Table 6.2 Areas of focus and examples of evaluation questions

Areas of focus	Examples of evaluation questions
Strategic relevance	Are the sub-programme objectives and strategy relevant to the global challenges posed by climate change; global, regional and country needs; the international response; and UNEP's evolving mandate and capacity in this area?
	How are the respective strategies of the CCSP components designed to ensure relevance in their respective thematic areas and how do their efforts address crosscutting areas (DRR, land-use, etc.)?
Sub-programme performance	Has UNEP achieved its expected accomplishments in the area of climate change?
	Have projects been efficiently implemented and produced tangible outputs as expected?
	Are the main drivers present and are the key assumptions valid so that the outputs delivered by the sub-programme can lead to sustainable, higher-level changes at outcome and impact level?
Factors affecting performance	What were the key factors affecting sub-programme performance?
	How well were the overall sub-programme and its project portfolio designed and structured?
	Are organizational arrangements adequate, and what is the quality of management within the operational units?
	Have human and financial resources been optimally deployed to achieve sub-programme objectives?
	What role did partnerships play in achieving sub-programme objectives and are these optimally developed?
	How well were sub-programme activities and achievements monitored and evaluated?

Source: UNEP Evaluation Office 2014/2015, Evaluation of the UNEP Sub-programme on Climate Change

---

[5]UNEP Evaluation Office 2011, 2010–2011 Evaluation Synthesis Report, pp. 54–60. Web link: http://www.unep.org/eou/Portals/52/Reports/2010-2011_Synthesis%20Rpt(E).pdf

appropriateness of sub-programme objectives and strategy. It analysed the clarity and coherence of the CCSP's vision, objectives and intervention strategy, within the changing global, regional and national context, and the evolving overall mandate and comparative advantages of UNEP. Second, the evaluation assessed the overall performance of the CCSP in terms of effectiveness (i.e. achievement of outcomes), sustainability, up-scaling and catalytic effects. It also reviewed the potential or likelihood that outcomes were leading towards impact. Which outcomes were assessed, was determined by a reconstruction of the sub-programme's Theory of Change (see below). Third, the evaluation examined the factors affecting performance in more detail: intervention design issues, organizational aspects, partnerships etc. that affected the overall performance of the sub-programme.

These areas of focus were not addressed in sequence but simultaneously as they are strongly linked to each other and dynamic as shown in Fig. 6.1. For instance, elements of strategic relevance of UNEP's involvement in Climate Change determine the scope and scale of the sub-programme and shape the kinds of products, services and delivery mechanisms are used to reach core objectives. Decisions surrounding strategic relevance of the CCSP thereby also influence the administrative, management and implementation structure, and other factors that affect performance. Sub-programme performance, in turn, affects funding availability and programme orientation. Progress made on expected accomplishments and impact also changes the priority needs of countries and other stakeholders, justifying strategic adjustments to sub-programme objectives and strategies.

Fig. 6.1 Three interlinked areas of focus of the evaluation (Source: Author)

Fig. 6.2 Five concentric units of analysis of the CCSP (Source: UNEP Evaluation Office 2014/2015, Evaluation of the UNEP Sub-programme on Climate Change)

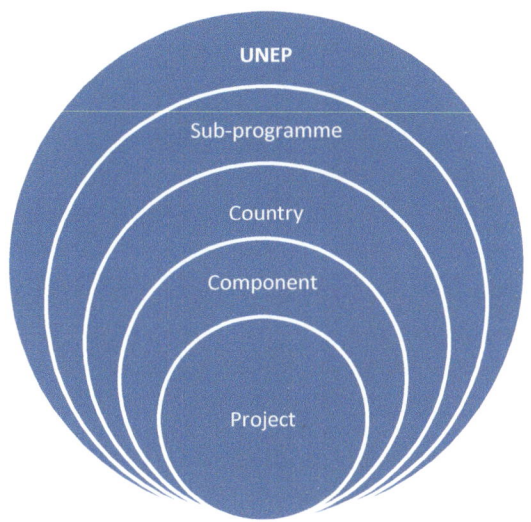

As illustrated in Fig. 6.2, the evaluation was conducted at **five units of analysis**. The two upper units are UNEP corporate and the Sub-programme itself. Considering the vast number and high variety of interventions, and highly diverse institutional arrangements and other factors affecting performance under the CCSP, neither UNEP or the sub-programme as a whole were the most practical and straightforward level at which to conduct analysis. They were also not the best level at which to attribute performance and uncover lessons learned.

Therefore, three lower units of analysis were used, which, combined, would provide sufficient information and analysis for the assessment of the sub-programme as a whole. The main unit of analysis was the sub-programme component (adaptation, mitigation etc.). At that level, performance could be most easily attributed to the line managers and partners delivering against the Expected Accomplishments of each component. The components were also the best units of analysis for learning, as they were usually better defined and delimited, and less complex than the sub-programme as a whole, but still provided the opportunity to see linkages between interventions either within or between main areas of intervention.

Another useful aggregated level of analysis was the country, where it was possible to obtain insights on the linkages (complementarities and synergies) between projects within a component, between the different components of the CCSP, and also between the CCSP and other sub-programmes within one, confined geographical and political space. The evaluation team visited six countries selected on the basis of geographical spread (spanning the regions of Latin America, Africa, Europe, West Asia, and Asia and the Pacific), presence of the sample projects (see next paragraph) and diversity of UNEP support on climate change in the country. A country case study was prepared for each visited country.

The lowest unit of analysis was the individual project. This was the most appropriate level to unveil factors affecting performance, but as the resources for the evaluation were limited only a sample of projects could be looked at in sufficient depth. The evaluation team prepared rapid reviews of 19 projects – about one third of the entire portfolio. Projects were selected on the basis of four criteria: thematic area (adaptation, mitigation or REDD), project size (based on estimated cost), project scope (global, regional or national) and maturity.

The evaluation made use of a **Theory of Change** (ToC) approach to address several evaluation questions. A ToC depicts the logical sequence of desired changes (also called "causal pathways" or "results chains") to which an intervention, programme, strategy etc. is expected to contribute. It shows the causal linkages between changes at different results levels (outputs, outcomes, intermediate states and impact), and the actors and factors influencing those changes. Initially inspired by guidance provided by the Global Environment Fund[6] the UNEP Evaluation Office has been systematically using a ToC approach in project and sub-programme evaluations since 2009.

The ToC for each component of the CCSP, and then for the CCSP as a whole, was reconstructed based on a review of strategic documents and UNEP staff interviews, and using best practice in determining correct results levels. Figure 6.3 presents the overall reconstructed ToC for the CCSP. The reconstructed ToC helped identify the expected outcomes of UNEP's work on Climate Change and the intermediary changes between outcomes and desired impact. Thus, it allowed to cluster outputs and define summary direct outcome statements cutting across components, which would prove very useful to frame data collection and synthesize findings on sub-programme effectiveness.

The reconstruction of the ToC also helped to determine the key external factors affecting the achievement of outcomes, intermediary states and impact, namely the drivers that UNEP could influence through awareness raising, partnerships etc., and the assumptions that were outside UNEP's control. As these were key determinants of the likelihood of impact, upscaling and sustainability of the sub-programme, it was important to identify them early on so that adequate information on their status could be collected in the course of the data collection phase.

The reconstructed ToC was also used to assess the internal logic and coherence of the formal results framework of the sub-programme. Therefore, the formal results framework comprised of the Sub-programme objective, Expected Accomplishments and Programme of Work Outputs was compared with the reconstructed ToC, and differences between the two were pointed out. For instance, in the formal results framework the results levels at which Expected Accomplishments and Programme of Work Outputs had been set were inconsistent between and within

---

[6]GEF Evaluation Office 2009, Fourth overall performance study of the GEF: The ROtI Handbook: Towards Enhancing the Impacts of Environmental Projects, Methodological Paper #2. Web link: https://www.thegef.org/gef/sites/thegef.org/files/documents/M2_ROtI%20Handbook.pdf

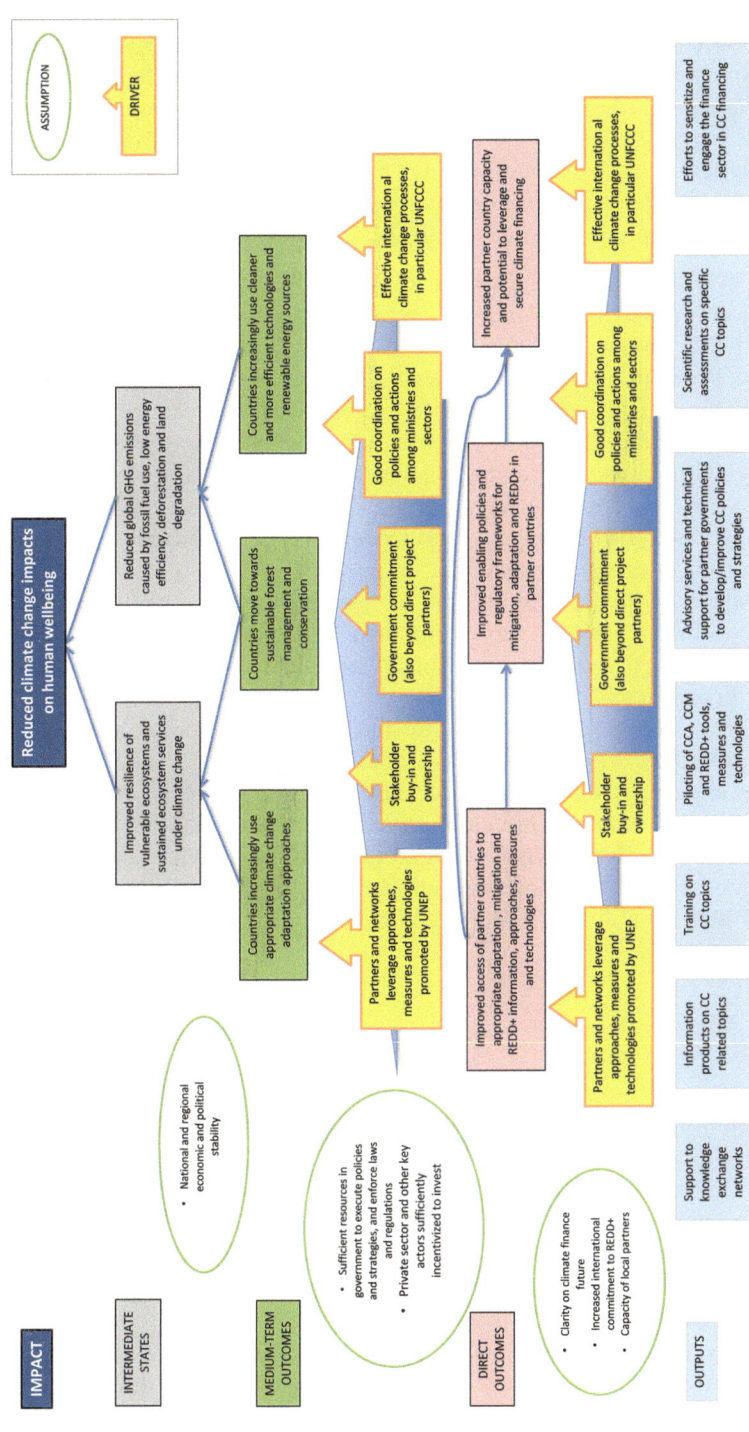

Fig. 6.3 Reconstructed theory of change of UNEP's climate change sub-programme (Adapted from: UNEP Evaluation Office 2014/2015, Evaluation of the UNEP Sub-programme on Climate Change)

components, some cause-to-effect relationships were either non-existent or had been overlooked, and several key drivers and assumptions had been neglected.

As explained above, attribution of large-scale, global changes to UNEP's work was difficult due to the largely normative nature of UNEP's work. Casual pathways from UNEP outputs to impact on the environment and human living conditions tended to be very long, with many external factors coming into play all along the causal pathways. The reconstructed ToC was used to **assess the likelihood of impact** by considering four distinct elements:

- UNEP's effectiveness in achieving its expected direct outcomes. This included verification of progress on output delivery and, most importantly, of the extent to which UNEP outputs led to increased stakeholder capacity, for instance: enhanced access to information and technological know-how, enabling policies and regulatory frameworks, or increased access to climate change finance.
- The validity of the ToC. The purpose was to prove the causal connection between UNEP direct outcomes and results higher-up the causal pathways. This was done by applying logic, through interviews with key stakeholders, and through analysis of evaluative evidence of progress towards impact at the country or lower geographical levels.
- The presence of drivers and validity of assumptions. The evaluation had to collect adequate evidence, mostly through desk review and key informant interviews, to verify the presence of an adequate enabling environment in supported countries.
- Early signs of large-scale progress on medium-term outcomes, intermediate states and impact. In itself this was not evidence of UNEP's contribution to higher-level changes, but was still necessary to inform stakeholders about global trends. Also, if UNEP's contribution to direct outcomes had been established, the ToC was very likely to be valid, and the required drivers were present and assumptions were valid, then the likelihood of UNEP's contribution to impact was very high even though it remained unquantifiable.

6.5   Data Sources

The evaluation team conducted a **comprehensive desk review** spread over the inception and main evaluation phase. During the inception phase, it helped to reconstruct the ToC of the components and the sub-programme as a whole, and to refine key areas of analysis and the evaluation approach highlighting evaluation challenges and information gaps. During the main evaluation phase, it was essential to collect information on achievements, impact, sustainability and upscaling, and the main factors affecting performance, while also leaving room for unanticipated results. The evaluation team conducted an in-depth analysis of CCSP key documents: background documents on climate change science and technology, the UNFCCC process and Climate Change finance, UNEP strategy and planning documents, evaluation reports (by the UNEP Evaluation Office and UNEP partners), project design documents and progress reports etc.

The evaluation team also conducted a large number of **interviews** with UNEP staff and managers at headquarters, concerned divisions and branches, in regional offices and country offices. Country visits were organized to conduct interviews with government officials, NGOs, development partners, and recipients of UNEP technical and/or financial support, which enabled the evaluation team to deepen its analysis and understanding of key internal and external factors affecting performance. The six country visits allowed the evaluation team to gauge how beneficiaries and other key stakeholder perceived programme effectiveness, sustainability and likelihood of impact. The country visits also helped the evaluation team to assess synergies and complementarities between UNEP climate change interventions, and also to address cross-cutting issues such as gender.

The evaluation further conducted a **staff and partner perception survey**. The purpose of the survey was to collect perceptions on sub-programme relevance and effectiveness and key factors affecting performance such as communication and coordination between divisions, inclusiveness within UNEP in determining work plans and budgets, human and financial resources devoted to the CCSP and its components, engagement with partners, monitoring and reporting systems etc. The survey was conducted online using the SurveyMonkey platform. Responses were received from 56 UNEP staff and managers – the response rate was acceptable at about 40 %. Only three partners responded to the survey – a response rate of less than 15 %.

6.6   Evaluation Process

As a first deliverable, the evaluation team produced an inception report based on an initial desk review and introductory interviews within UNEP. It included a more detailed presentation of the evaluation background (global context, programme framework, institutional arrangements and project portfolio); a draft Theory of Change of the sub-programme components; and the evaluation framework (a detailed description of the methodology and analytical tools that the evaluation would use to answer the evaluation questions). The inception report was first reviewed by the Evaluation Office and then shared for comments with the Sub-programme Coordinator and the heads of functional units involved in the sub-programme.

The data collection phase for the evaluation was expected to take place over a relatively short timeframe from January to April 2013. However, some country visits had to be rescheduled due to unavailability of key persons or conflicting schedules within the evaluation team, prolonging the data collection until June 2013. The evaluation team prepared country case studies and component working papers, which went through several rounds of comments from the Evaluation Office (for quality assurance) and UNEP stakeholders (for fact checking). The main report was drafted by November 2013, but also required a series of reviews by the Evaluation Office and subsequent revisions, so that it was shared within UNEP for comments as late as February 2014. Considering that the period covered by the evaluation ended on 31 December 2012, there was a time lag of more than 1 year between much of the information collected for the evaluation and the distribution of its first draft report. During the first half of 2014, comments were received from UNEP staff and data from the UNEP Programme Performance Report 2012–2013 was incorporated where appropriate to make the report as up-to-date as possible. Because the consultants' team had been disbanded by mid-2014, finalisation of the report was done internally in the Evaluation Office. The report was finally published in January 2015.

6.7   Lessons Learned on the Evaluation Approach

This evaluation has shown the importance of developing an appropriate analytical framework, well suited for the scope and complexity of the object of evaluation. The analytical framework and evaluation approach used for the UNEP Climate Change Sub-programme Evaluation, combining three interlinked areas of focus (strategic relevance, sub-programme performance and factors affecting performance), five concentric units of analysis (UNEP as a whole, sub-programme, component, country and project) and a Theory of Change approach, allowed the evaluation team to cover the standard evaluation criteria in a comprehensive but concise manner, remaining strategic and without drowning in the details.

The ToC approach helped making a credible assessment of UNEP's contribution towards impact, sustainability and up-scaling, but did not allow this contribution to be quantified. In other words, the evaluation could not determine to what extent higher-level changes beyond stakeholder capacity (direct outcomes), such as changes in environmental management practices or greenhouse gas emissions, could be attributed to UNEP's efforts alone, and which changes might have happened anyway. In any case, a credible attribution of impact at the sub-programme or sub-programme component level would have been impossible without extensive impact assessments at the country or project level, which are currently not available in UNEP and could not have been realistically built into the sub-programme evaluation framework.

There appears to be a trade-off between the time that is invested in quality assurance and stakeholder involvement during the evaluation process, on the one hand, and the up-to-dateness of information provided and sustained stakeholder interest in the evaluation, on the other. Strong internal stakeholder involvement during the inception and data collection and analysis phases of the evaluation through interviews, discussions, surveys and commenting on intermediate products, boosted learning within UNEP during the evaluation process. However, the length of the evaluation process, due in part to the high quality standards applied by the Evaluation Office and the time required for receiving stakeholder comments on all evaluation products, created an important time lag between the data collection phase and the distribution of the draft main report. This had two consequences: information presented in the draft main report was more than 1 year old, and internal stakeholder interest for the main report, when it was finally shared within UNEP, appeared to be a lot less than it had been for the intermediate evaluation products.

The evaluation team decided to cover the cross-cutting Science and Outreach component as part of the three other components and not separately, as an acceptable way of dealing with the human resource and time constraints within the team. This was fine in principle, but as a result, some high visibility assessment products developed jointly by different units in UNEP under this component were not included in the project sample, and received therefore an only cursory treatment in the report. This undervalued some important UNEP-wide efforts and was also a missed opportunity in terms of learning lessons from cross-divisional collaboration. While it might not have been necessary to give the assessment of the Science and Outreach component the same level of depth as was given to the others, one or two projects from this component should have been included in the project sample.

As acknowledged in the evaluation report[7] under the section presenting the limitations of the evaluation, the size of the sample of the country case studies (six in total – or only one for most regions) was too small. Despite the logical and practical country selection criteria, this sample could not provide a representative and credible picture of UNEP's strategic relevance and performance at the country level. A larger sample size would, however, not have been possible within budget. An alternative approach could have been to base the country case studies on information collected from a country questionnaire sent over email, more in-depth desk review and interviews via Skype or video-link. A rough cost comparison with the actual approach suggests that about four additional country case studies could have been prepared using this alternative approach, bringing the sample to a more representative two case studies per region.

As also noted in the evaluation report, the evaluation would have benefited from more interviews with global partners and key informants outside UNEP with a good understanding of the global climate change arena. These would have increased diversity and credibility of views expressed in the evaluation and, possibly, generated more strategic recommendations. With hindsight, though some interesting views from partners were collected, the perception survey was not the most

---

[7]UNEP Evaluation Office 2014/2015, Evaluation of the UNEP Sub-programme on Climate Change: Final report. Web link: http://www.unep.org/eou/Portals/52/SPE%20Climate%20Change.pdf

appropriate tool to usefully explore these views and to tap partners' ideas on how UNEP's relevance and results could be enhanced. Alternatively, the evaluation team could have conducted a series of well-facilitated focus group discussions or a Delphi exercise with key resource persons. These could have yielded more credible findings but would have required additional resources.

Chapter 7
Enhancing the Joint Crediting Mechanism MRV to Contribute to Sustainable Development

Aryanie Amellina

Abstract This chapter looks at the initial progress of the JCM implementation in contributing to sustainable development in developing countries through facilitating diffusion of leading low-carbon technologies and implementation of mitigation actions. The current progress of the JCM in 16 partner countries looks promising with an established MRV system and efficient governance process. MRV methodologies are easy to use and benefits from standardized forms, default values, and practical monitoring system, but the methods in determining the reference emissions need to be strengthened. Rigorous project promotion is needed in underrepresented partner countries, especially least-developed countries, by supporting national programs and initiatives. The JCM should aim not only to complement, but also to improve preceding market mechanisms, by implementing a regulatory framework for evaluating its contributions to sustainable development. There is a need to clarify ways of credit allocation, arrange ways of credits accounting for national report and towards national pledge, and define the pathway of the JCM to a tradable crediting mechanism or retain its status quo of producing non-tradable credits.

Keywords Japan • JCM • Market mechanism • MRV • Technology transfer

7.1 Introduction

The Joint Crediting Mechanism (JCM) was initiated by the Government of Japan in pursuit of achieving global greenhouse gas emissions reduction/removals through facilitating diffusion of leading low carbon technologies, products, systems, services, and infrastructure as well as implementation of mitigation actions to contribute to sustainable development of developing countries. As of January 2016, 16 countries have signed bilateral agreement with Japan to implement the JCM;

A. Amellina (✉)
Climate and Energy Area, Institute for Global Environmental Studies, Hayama, Japan
e-mail: aryanie.amellina@gmail.com

© The Author(s) 2017
J.I. Uitto et al. (eds.), *Evaluating Climate Change Action for Sustainable Development*, DOI 10.1007/978-3-319-43702-6_7

Mongolia, Bangladesh, Ethiopia, Kenya, Maldives, Vietnam, Lao PDR, Indonesia, Costa Rica, Palau, Cambodia, Mexico, Chile, Saudi Arabia, Myanmar, and Thailand. The JCM promotes the use of advanced technologies and measure, report, and verify emissions reduced by the technologies.

7.2 The JCM Overview

The JCM was initially designed to complement the CDM. Some of its main differences with the Kyoto Protocol mechanism are its decentralized governance, simple and practical MRV system, and the credits its projects generate, up to the time of writing, are internationally non-tradable.

The JCM is 'decentralized' as it is implemented under bilateral cooperation between Japanese and partner countries government. The measurement, reporting, and verification (MRV) of the JCM are based on projects using the JCM MRV *methodologies* as the tool, which is developed under 'simplified' and 'practical' principles using clear technology-based eligibility criteria, list of default values, and ready-to-use monitoring templates. As depicted in Fig. 7.1, the Joint Committee between each partner country and Japan develops and approves the technology-based MRV methodologies to be used by projects to procure the greenhouse gas emission reductions/removals. Verified reductions/removals will be issued by each government as *JCM credits*. These credits are not financially valued and cannot be traded internationally. However, the JCM agreements do not rule out the possibility of domestic trade in line with partner country policy.

Instead of buying credits from partner countries, the Japanese government offers project developers upfront financial incentives for installing the advanced technologies. These incentives are expected to contribute to resolving the burden of high capital investments that have been hindering the development and utilization of advanced technologies in developing countries.[1] Currently, incentives to support projects implementation throughout their cycle are available from the Ministry of the Environment Japan (MOEJ), Ministry of Economy, Trade, and Industry Japan (METIJ), Asian Development Bank (ADB, through Japan Fund for JCM with contributions from MOEJ), and Japan International Cooperation Agency (JICA, in cooperation with MOEJ).

Technology installation is supported by either full grant (under long-term entrustment), partial subsidy (direct subsidy up to 50 % of project investment cost), loan, or loan interest subsidy. Development of methodology, project design document (PDD), monitoring, reporting, and verification (only the first time) are

---

[1] Mitchell, C., et al., in IPCC, 2011. *Special Report on Renewable Energy Sources and Climate Change Mitigation (SRREN)*. Cambridge: University Press, Cambridge, United Kingdom; Metz et al., in IPCC, 2000. *Methodological and Technological Issues in Technology Transfer*. Cambridge: Cambridge University Press, UK.

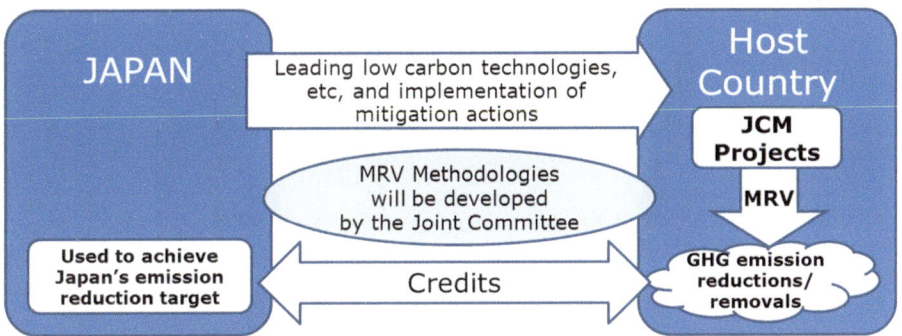

Fig. 7.1 The JCM scheme between Japan and partner country (Government of Japan, 2016)

also supported technically and financially. Feasibility study and capacity building are supported by full grant.

The interaction between the Japanese government, partner country, the JCM Joint Committee and other stakeholders which include project participants, third-party entities, and the Joint Committee secretariat is illustrated in Fig. 7.2.

The Japanese and each partner country government form a JCM Joint Committee with Co-Chairs appointed by each side. Co-Chair from Japanese side is an official of the Embassy of Japan in the partner country, and Co-Chair from the partner country government usually is a representative of the signatory or host ministry. Host ministry from partner countries is typically the Ministry of the Environment or related, except for Indonesia (Coordinating Minister of Economic Affairs), Saudi Arabia (Ministry of Petroleum and Mineral Resources), and Palau (Minister of Public Infrastructure, Industry and Commerce).[2] The role of Joint Committee is similar to those of Executive Board of the Clean Development Mechanism (CDM). It develops and approves rules and procedures, MRV methodologies, registers eligible projects, and approves request for credits issuance.

The secretariat serves the Joint Committee to support these roles. The Japanese government appoints a private company as its secretariat for all partner countries. Most partner countries appoint an office under its Co-Chairing host ministry, with mandates ranging from disaster management, natural resources management, climate change, sustainable development, to environmental conservation fund. In Indonesia and Mongolia, the Joint Committee established a dedicated entity to implement the JCM. Partner country secretariat collaborates with the Japanese secretariat in development of procedures and document reviews.

Project participants, who may be private companies, public organizations, foundations, or academic institutions from Japan and partner country establish a consortium, or a kind of joint venture, and propose projects jointly.[3] The Japanese

---

[2]The Joint Crediting Mechanism (JCM) official website, https://www.jcm.go.jp/

[3]Global Environment Center Foundation (GEC), secretariat for the Financing Programme for Joint Crediting Mechanism (JCM) Model Projects in FY2015. Financing Programme for JCM Model Projects Public Offering Guidelines (tentative translation). http://gec.jp/jcm/kobo/mp150907.html

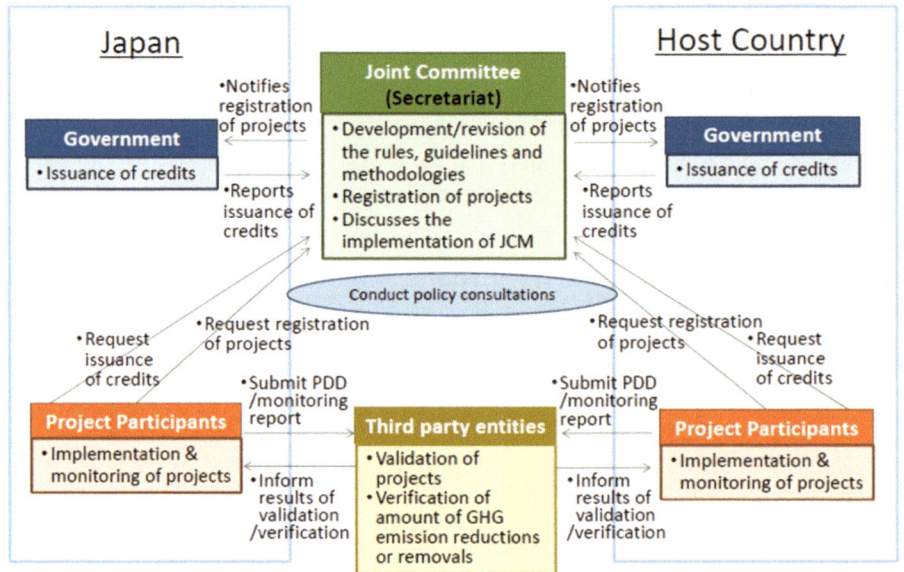

Fig. 7.2 The JCM stakeholders and their role

institution are required to represent the consortium and apply to the Japanese government for subsidy through open biddings. Financial support for selected projects are disbursed directly to participants or indirectly through intermediary organizations. For MOEJ Model Project financing program, the financing is limited only to those costs that can be verified as having been spent for implementation of eligible projects.[4] Once their project is registered, project participants start monitoring emissions reductions/removal by the project based on the relevant methodology, produce a monitoring report and ask third party entities to verify it.

Third party entities (TPEs) are ISO 14065 or CDM Designated Operational Entity (DOE)-certified organizations who are deemed eligible by the Joint Committee to conduct validation and verification activities in specific countries. Third party entities produce verification report on the project emissions reductions as reported by the project participants. This verification report is used by project participants to request credit issuance. Credits can be used to fulfil both Japan's and partner country's emissions reductions pledge.[5]

---

[4]MOEJ, 2015. JCM Financing Programme for JCM Model Projects Public Offering Guidelines (tentative translation) http://gec.jp/jcm/kobo/mp150907.html

[5]Written in bilateral agreement between Japan and partner countries, for example with Thailand (November 2015): (5) Both sides mutually recognize that verified reductions or removals from the mitigation projects under the JCM can be used as a part of their own internationally pledged greenhouse gases mitigation efforts.

The role of the JCM in supporting the Japanese government emissions reductions target post-2020 shows the level of its ambition.[6] Moreover, most of the JCM partner countries also implies their intention to report the reductions achieved from market-based mechanisms projects to fulfil their intended nationally determined mitigation contributions.[7] Therefore, the JCM needs a robust MRV system and policy arrangements to ensure the scheme's emissions reductions achievement fulfils the expectations, while contributing to sustainable development in the partner countries, as it aims to do.

7.3   Approach in Evaluating the JCM MRV

As one of the various approaches developed based on COP Decision 1/CP.18,[8] The overarching goal of the JCM MRV is to deliver *real, permanent, additional and verified mitigation outcomes, avoid double counting of effort and achieve a net decrease and/or avoidance of GHG emissions*.[9] Under the Paris Agreement, the use of market mechanisms is articulated in Article 6, which covers voluntary cooperative approaches resulting in *internationally transferred mitigation outcomes* that may be used towards nationally determined contribution. These cooperative approaches should promote sustainable development while upholding transparency and environmental integrity and avoiding double counting of outcomes. The JCM credits may be considered as these outcomes.

This chapter explores the early implementation of the JCM to find the possible answers to three questions:

1. What are the key enhancements needed for the JCM MRV to ensure real, permanent, additional and verified mitigation outcomes, avoid double counting of effort and achieve a net decrease and/or avoidance of GHG emissions?
2. What are the key challenges in the initial stage of the JCM and how can they be improved?
3. How can project contribution to sustainable development be properly evaluated through the JCM MRV?

Specific sources are referred to in this chapter. Assessment were mainly done to the publicly available information of official documents, publicly available

---

[6]In Japan's INDCs, the JCM is not included as a basis of the bottom-up calculation of Japan's emission reduction target, but the amount of emission reductions and removals acquired by Japan under the JCM will be appropriately counted as Japan's reduction. Apart from contributions achieved through private-sector based projects, accumulated emission reductions or removals by FY 2030 through governmental JCM programs to be undertaken within the government's annual budget are estimated to be ranging from 50 to 100 million t-CO_2.

[7]IETA INDC Tracker (2015), IGES INDCs And Market Mechanism Database (2016).

[8]Recent Development of The Joint Crediting Mechanism (JCM), November 2015.

[9]FCCC/CP/2011/9/Add.1 Decision 2/CP.17, para 79.

presentation materials, and databases. External reviews, either academic or institutional, are still very limited. Assessment on the JCM MRV was done by reviewing 19 approved methodologies (as of January 2016). On the JCM governance, project development, and capacity building, assessment were based on observation during author's work experience with the JCM partner countries such as Indonesia, Mongolia, Lao PDR, and Cambodia. Findings from Indonesia were taken from interviews conducted with Indonesia JCM Secretariat member, expert, and project developers from Indonesian side.

This chapter was developed under voluntary initiative. It is important to note the limitations to this assessment; first, examples provided in this chapter are taken only from partner countries in Asia and the Pacific, considering the current progress and experiences concentrated in this region; second, the limited number of interview; third, limited experience on some parts of the JCM MRV project cycle such as verification and credit issuance.

7.4 Enhancing the JCM Measurement, Reporting and Verification (MRV) Framework

Reductions/removals from the JCM projects are likely to be reported as national achievements to the international community. It is thus important to ensure that the JCM reports accountable emissions reductions from projects that contributes to sustainable development in partner country. Four aspects need to be strengthened to deliver this goal: *governance*, *MRV methodology development*, *project development and capacity building*, *and a framework to evaluate sustainable development contribution*.

7.4.1 Governance

The JCM Joint Committee typically consists of six to seven Japanese officials (the Embassy of Japan, Ministry of the Environment, Ministry of Economy, Trade, and Industry, Ministry of Foreign Affairs, and sometimes Forestry Agency) and eight to ten partner country officials related to the environment, foreign affairs, industry, trade, energy, agriculture, and economy and finance. Considering the nature of JCM support scheme which favours energy-related projects, it is important to involve ministries and agencies with a mandate in energy, industry, and infrastructure. It is also recommended to engage organizations with an established, strong relations with private companies such as investment bureaus, trade councils, and business councils.

As the main supporting entity for the Joint Committee, capacity of the JCM secretariat of the partner countries is a key aspect in enhancing the implementation of the JCM. Not only to support general management, the secretariat could push forward projects that meet the need of country stakeholders by gathering project proposals, review projects contribution to sustainable development, ensure additional features of projects, and enhance capacity building of local entities. Cooperation for capacity building between secretariats and other organizations including through international support not only from the Japanese government is very important. Despite its status as a bilateral partnership, acknowledgment of the JCM contribution in achieving the international goal is inevitable, not different with other regional or national market-based initiatives that are increasingly being developed. Furthermore, the sources or providers of advanced low-carbon technologies to be supported by the JCM are not limited. This opens the door for various countries, companies, and international organizations to be involved and cooperate. To encourage this, transparency and timely information should be enhanced especially on project approval process and financial aspects.

Another important aspect in governance is credit sharing. The existing rule that needs to be clarified is the general term of credits allocation between participants (or countries) to be based on consultations among participants 'on a pro rata basis'.[10] At the same time, the JCM Model Project are requested to deliver at least half of issued credits to the Japanese government, regardless of the finance rate.[11]

As the JCM clearly mentions, the credits need to be recorded in a registry system. An online JCM registry system (https://www.jcmregistry.go.jp/) is already operational, which also provides the space for partner countries to manage their registry. However, partner countries prefer to run its own registry system and thus the set of common specifications and rules of registry system were agreed beforehand. Both governments also agreed to request projects in Indonesia to allocate at least 10 % of issued credits to the Indonesian government. In May 2016, credits were issued from two projects in Indonesia to the registries of Japan and Indonesia. The information on issuance are available on JCM websites of Japan and Indonesia. From the first project, Japan side received 23 tCO credits (20 tCO allocated to the government and 3 tCO to the project participant) and Indonesia side received 6 tCO credits (3 tCO allocated to the government and the project participant each). From the second project, Japan side received 8 tCO credits (7 tCO allocated to the government and 1 tCO to the project participant) and Indonesia side received 3 tCO credits (2 tCO allocated to government and 1 tCO allocated to the project participant).

Towards and beyond 2020, it is crucial for the partner countries to consider how the JCM emissions reductions will be accounted in their national report to the

---

[10]Joint Crediting Mechanism Project Cycle Procedure in partner countries except Costa Rica (describes that 'part of the credits is allocated to the project participants from the developed country taking into consideration their contribution to GHG emission reductions or removals through the JCM project'), Chile, Myanmar, and Thailand (information not available).

[11]Recent Development of the Joint Crediting Mechanism (JCM), September 2015, GEC.

UNFCCC. There are indications that partner countries plan to report their achievement through their Biennial Update Report (BUR) under nationally appropriate mitigation actions (NAMA) umbrella or national registry system. In this regard, the way of accounting the JCM credits in these reports without double counting, double issuance, and double claiming, need to be arranged at the domestic and international level. Countries without an established inventory system may consider reporting outside the inventory for pre-2020 achievements. For post-2020, an international accounting rules and infrastructure for Internationally Transferable Mitigation Outcomes (ITMO) shall be in place.

7.4.2 MRV Methodology and System

As of January 2016, 19 MRV methodologies have been approved under the JCM, all for countries in Asia and Pacific region. More than 50 % (11 methodologies) are in the energy efficiency sector, more than 20 % (4 methodologies) in the energy industry (power generation by waste heat recovery, solar power), energy efficiency (energy-efficient chillers, refrigerators, LED) and the rest in energy distribution (improvement of electricity transmission and distribution grid), waste handling and disposal (anaerobic digestion of market waste for biogas), and transport (digital tachograph in vehicles). These methodologies are technology-based and applicable only in the country where they are approved, as shown in methodology ID number (VN_AM001 means the first approved methodology to be used in Vietnam). There are three key aspects of a JCM methodology: *ensuring net emissions reductions by conservative determination of reference emissions*, *eligibility criteria*, and *simple monitoring methods*.

Net emissions reductions are ensured by conservative measurements of reductions, by assuming the highest amount of emissions possible in the baseline scenario, to ensure the emissions reductions achieved by the projects are not overestimated. Baseline, or called "reference emissions" in the JCM, does not necessarily mean result of 'before project' emissions calculation. Instead, the calculation can be done in two ways.

The *first way* is to adjust "reference emissions" conservatively. They are set as high as possible while staying below business-as-usual emission, which represent plausible emissions in providing the same outputs or service level of the project. Emission reductions to be credited are defined as the difference between "reference emissions" and "project emissions". In this case, the reference emissions are assumed to be the highest plausible emissions. Most of the approved methodologies applied this approach.[12] *Second way* is to adjust "project emissions" conservatively. The methodology uses predefined default values instead of the actual values

---

[12]IGES JCM Database, January 2016 http://pub.iges.or.jp/modules/envirolib/view.php?docid=6185

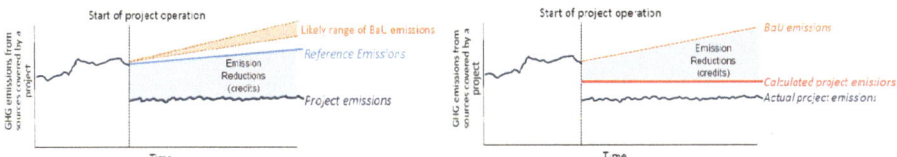

Fig. 7.3 Ways to realize net reduction (*left*: first way, *right*: second way)

measured and monitored from the project. This will result in "calculated project emissions" that are larger than "actual project emissions". In this case, emission reductions to be credited are defined as the difference between BaU emissions and "calculated project emissions" (Fig. 7.3).

Reference emissions can be determined by conducting a survey on the best available or most widely used technology in the partner country, from legal requirements, the current situation and performance or average historical performance at a relevant project site, and voluntary standards and/or targets, national or international. The use of national standards and regulations as methodology reference need to be supported[13] and The use of internationally-recognized default values and rules for equipment calibration need to be strengthened continuously.

Procedures for market surveys during methodology development also need to be strengthened as they are heavily used in determining reference emissions. The Joint Committees could consider setting a general standard for survey, data collection and renewal of reference. As reference condition are very likely to improve over time, reference emissions need to be adjusted periodically. A number of methodologies in Indonesia and Mongolia have set the requirement for the default values to be updated every 3 years. However, the responsible party for updating these methodologies needs to be clarified.

Eligibility criteria are developed in a concise manner to reduce the risk of project rejection. They provide specific requirements for each project and technology type, touching upon the concepts of 'additionality' and 'applicability' of projects under the CDM to a certain extent, through an easy-to-apply simple 'checklist' in each methodology. The number of project eligibility criteria defined by the 19 approved methodologies ranges from two to seven, with an average of four criteria in one methodology.[14] Highly used criteria are the specification and components of technology, capacity of service, and the types of eligible activity (for example, new installation or replacement of technologies). These criteria should be specific but also general enough to enable them for use by multiple projects, not only for the projects the methodologies are developed for. Ensuring this among countries will reduce the time and costs of project development.

---

[13]For example, methodology ID_AM005 refers to Indonesian national standard (SNI) for eligibility criteria on room illuminance and MN_AM001 refers to Mongolian national standard (MSN) for eligibility criteria on electricity transmission loss.

[14]IGES JCM Database, January 2016.

List of monitoring parameters is the shortlist of data for monitoring emissions to be collected by project participants, the approved sources, and default values from IPCC or other approved standards that are ready to use. These default values can be applied in the standardized Microsoft Excel-based *monitoring and calculation spreadsheets* in each methodology. The set of spreadsheets is arguably the most valuable aspect of the JCM MRV for the project participants, as they clearly indicate the options of data collection method, default values, and formulas to calculate reference emission, project emission, and emission reduction. In addition to that, project monitoring team is also described in the same spreadsheet. The basic requirements such as continuity of operation, production year, and availability of data are assumed as given, resulting in brief and easy-to-read documents. By using these sheets, project participants do not need to spend much time to justify the calculation formula and ways to acquire required data.

In the long run, quality of methodology and its development process need to be consistent between countries and improve inclusion of local programs or standards. It is also important to design more general methodologies to increase their applicability to similar projects. For example, as opposed to developing methodology for a specific 'air conditioners with inverters to public sector building' (VN_AM002), the methodology could be developed for air conditioners with inverters to any kind of building.

7.4.3 Processing Time

Strengthened method and time-efficient approval process will improve the efficacy of JCM MRV. So far approval process are being implemented in an efficient manner. Approving proposed methodologies take from 18 to 384 days with an average of only 107 days from the time of methodology proposed. Similar trend is observed in project registration process, which generally takes less than a month with an average of 10 days.[15] These effective processes benefit from close cooperation between Japanese and partner countries government, supporting agencies, project participants, and the JCM budget allocated for implementation support. This could be expected to continue in the long run, as long as institutional support and commitment from the involved countries at least remain.

As the number of JCM projects continue to increase, however, it may become more challenging to maintain such speed. Thus methodology reviewers such as JCM secretariat and the general public need to ensure time efficiency does not overrule quality. Capacity building, technical assistance, and resources are important factors in this aspect.

---

[15]Ibid.

TPE auditors that perform the future verification should be fully accountable for all their activities.[16] Since the JCM TPEs must be certified under CDM Designated National Authority or ISO 14064, their accountability should be high enough. Moreover, the Joint Committee and accreditation organizations have the authority to suspend or withdraw them in case of non-performance.

However, as next verification process is not financially supported, there is a need for a clear incentive for projects to verify emissions reductions and issue credits in the future, especially in 2021 and 2030, since the JCM credits are so far non-tradable and there has not been any indication of making it otherwise. Currently, the Japanese government supports the cost of first verification achieved at the end of first year of project. After the first request, participants may request issuance of the JCM credits for emission reductions achieved during several years in one time, but they shall request issuance of the JCM credits for emission reductions achieved by 2020 by the end of 2021.[17] The projects participants (especially from partner countries without any agreement with their government) who may not have enough budget for verification may be reluctant to continuously monitor and verify their achievements by the end of the project period, which is 13 years in average among registered projects.[18]

The pathway to transform the JCM into tradable scheme, if there is any plan, is important not only for the participants but also for partner countries. While Japan requires verification at the end of 2020 in the project financing rules, it is important for partner countries to take measures for the future use of credits owned by partner country participants.

7.4.4 Project Development and Capacity Building

Since the first batch of selected projects for funding in 2013, the JCM have registered eight projects, all in the Asia and Pacific region (Indonesia, Mongolia, Vietnam, and Palau). In total, 89 projects have been selected for funding and currently in the implementation pipeline as of January 2016.[19] They are concentrated in Indonesia and Vietnam and to a lesser extent in Bangladesh and Mongolia.

The "standard" project proposal procedure starts from submission by a project consortium to the Japanese government during call for request period, usually two to three times a year for 1 month. A consortium includes Japanese company and

---

[16]Stockholm Environment Institute (SEI). 2015. *Has Joint Implementation reduced GHG emissions? Lessons learned for the design of carbon market mechanisms.* Seattle: Stockholm Environment Institute.

[17]MOEJ, 2015. JCM Financing Programme for JCM Model Projects Public Offering Guidelines (tentative translation).

[18]IGES JCM Database, January 2016.

[19]IGES JCM Database, January 2016.

local company in partner countries. The application is to be made by Japanese institution as the main representative the consortium and showing an evidence of agreement. This procedure inadvertently leaves out local companies without a Japanese partner to submit on their own at times.

Due to this constraint, Indonesia created "Project Idea Note" procedure, learning from the CDM. Indonesian parties without Japanese partners can propose their technology needs to the Indonesian secretariat who will communicate it to the Japanese government. According to the Indonesian government, proposals from Japanese side are typically "normal" projects under previously established cooperation with Indonesian side, while proposals from Indonesian companies are usually more crucial for their own development and come with a guarantee that the project will face less non-technical burdens, but the Japanese technology may not always available.[20] This may be one of the reasons why only one out of six PINs were able to be followed up ("Power Generation by Waste Heat Recovery in Cement Industry").

This situation shows the need of strengthening project identification and development through close collaboration with partner countries. Some options for matchmaking could be innovated, for example (1) improving the use of available websites, also publishing a list of technology ideas and companies,[21] (2) promoting the involvement of state-owned companies, municipal governments and local companies, for example those experienced in the CDM, (3) working with the mass media and local company network.

Ideally, network of local companies and business-research institutions have the capacity to investigate the needs and potentials to support 'matching' between local and Japanese companies, as this needs to be done in business approaches. Effectiveness of processes and capacity building events must be enhanced through a long-term engagement with these organizations.

The INDCs from partner countries should also be promoted as key reference document for project development. Combining partner country emissions reductions potential and priority sectors under its INDCs shows a way to promote nationally-appropriate projects. In the future, the JCM also needs projects with bigger emissions reduction potential to increase cost-effectiveness. Some potential options are increasing partner country and local participants involvement, promote a 'program JCM' or group of projects, similar to CDM's Programme of Activities

---

[20]Manansang, Dr. Edwin, Head of Indonesia Joint Committee, Coordinating Ministry of Economic Affairs. *The JCM Development in Indonesia and Its Evolution Towards Sustainable Low Carbon Growth Cooperation.* Presented at International Forum for Sustainable Asia and the Pacific (ISAP) 2015, Parallel Session "Showcasing Successful Partnerships for Low Carbon Technology Transfer", Institute for Global Environmental Strategies.

[21]Asia Low-Carbon Development Collaboration Platform website, http://lowcarbon-asia.org/english/city.html

(PoA).[22] The Joint Committee can also consider identifying projects that were planned for the CDM but never could start. In any case, project approval must always uphold additional aspects of projects support, promote new activity, and more advanced technology than the prevailing technology in the country, and start after the earliest date decided by each partner country (most countries decided 1 January, 2013).

The functions of the JCM websites (so far, Indonesia, Vietnam, and Mongolia have established their own JCM websites in addition to the Japanese websites) should also be optimized to facilitate project developers showcasing their needs and availability, which can be followed up by the JCM secretariats and intermediary organizations. Transparency of information must always be ensured in communicating the JCM opportunities and approval process to eligible entities.

These information are crucial for private companies, whom are naturally attracted to the offered JCM subsidy of 'up to the half of investment cost', grant, and full ownership of the technologies. Although the price of advanced low-carbon technologies are generally higher than those available in the market, they are willing to invest because the subsidy helps pushing down payback period to a reasonable period of time. The emphasis on use of advanced technology should also be promoted. Current experience shows that the approved technologies have better efficiency and/or performance than those commonly used in partner countries. As a more specific criteria, MOEJ specifies cost effectiveness level and payback period as eligibility criteria for JCM Model Project support.

The role of state actors such as local government and their contribution to national emissions reductions efforts should also be encouraged. As the JCM has the additional value of realizing cost-intensive projects that may support local development plans, the global movement towards realizing sustainable cities and city-to-city cooperation programs can further encourage JCM projects. Schemes such as "sister city" and municipality international cooperation offices have been the playmakers for these cooperation. The City of Yokohama, for example, established a "Y-PORT" program in 2014, which aims to enhance collaboration between the government and private companies in the city to promote sustainable cities in other countries, utilizing, among others, the JCM. High social awareness and leadership on the environmental issues are helping to push these initiatives.

---

[22]Saito, Tetsuya. 2015. *Opportunities and challenges under the JCM scheme*. Presented at International Forum for Sustainable Asia and the Pacific (ISAP) 2015, Parallel Session "Showcasing Successful Partnerships for Low Carbon Technology Transfer", Institute for Global Environmental Strategies.

Case Study: PT Semen Indonesia Tuban

PT Semen Indonesia (Persero), Tbk. is the largest cement producing company group in the country. The company has experience in developing a CDM project in power generation by waste heat recovery in its Semen Padang factory. In 2013, PT Semen Indonesia proposed another "Power Generation by Waste Heat Recovery in Cement Industry" project, developed with their Japanese partner JFE Engineering.

The project introduces a waste heat recovery at a Semen Indonesia cement production plant in Tuban, East Java, Indonesia. The waste recovery system is designed to utilize waste heat emitted from the cement factory to generate electricity for own consumption, therefore reducing electricity import from the national electricity grid of approximately 165,000 MWh/year, which will lead to the reduction of fossil fuel combustion at grid-connected power plants. A power generation facility with 30.6 MW capacity was proposed.

After feasibility in fiscal year 2013, official Project Idea Note proposal was submitted in May 2014 and the project was selected for funding as JCM Model Project (expected to be registered in 2017). Benefiting from project participants' preparation, technical expertise, understanding and experience, as well as effective governmental consultation process, the approval of its MRV methodology (ID_AM001) took only 1 month.

PT Semen Indonesia sees three benefits from engaging in the JCM: environment, economic, and company image. Environmental benefits include CO_2 emission reduction (approximately 122,000 tCO_2/year, the largest selected project in Indonesia so far) and low stack gas temperature. Energy and water consumption reduction resulting from WHR process lead to both environmental and economic benefit. By utilizing about 30 MW electricity generated by the installed facility, more than 150 million kWH electricity per year can be saved, leading to more than 85 % cost saving. The company is also able to enhance company image, create jobs, and contribute to the community. The project investment costs around 50 million USD, and the JCM subsidy from the Ministry of Environment Japan, which accounts to around 18 % of this cost, decreases the project investment index (IDR/kW).

From their experience in the CDM, PT Semen Indonesia sees that the JCM offers a simpler, more reliable, and faster mechanism for the private sector and its MRV system preferable for its simplicity. The JCM encouragement for private sector active engagement is appreciated. PT Semen Indonesia suggests to address the following to further enhance the JCM implementation:

1. Improve appropriateness of MRV system by increasing the number of Indonesia TPE, ensure affordability of their services, and including MRV cost in the project budget

(continued)

2. Optimize capacity building for project host. For example, improve the role of project host to in equipment design and selection as well as supplier selection
3. Improve leadership of project host in the consortium, as the Japanese government requires Japanese entities to apply as head of consortium to apply for subsidy.

PT Semen Indonesia also addresses a concern on how to share the JCM credits among participants. To them, sharing the credits "on a pro rata basis" may relate to 'the benefit earned by each party along the MRV period'. On the other hand, under the JCM Model Project subsidy scheme, the Japanese government requires at least half of the credits to be delivered to its government. This concern needs to be clarified by the Joint Committee.

7.4.5   Sustainable Development Evaluation Framework

Environmental impact assessment, local stakeholder consultations, and capacity building as part of the JCM MRV are important aspects in ensuring contribution to sustainable development. Requirement of an environmental impact assessment for each project refers to the partner country regulation. To date, only two registered projects were required to conduct environmental impact assessment, both for installation of highly efficient heat-only boilers in Mongolia.[23]

The efforts by partner countries to enhance local stakeholders' engagement could be replicated. For example, Mongolia has been promoting local perspective in developing JCM projects by requiring project documents into Mongolian language in addition to Japanese. In Vietnam, a circular on the JCM implementation guidelines was distributed to governmental agencies.[24] To promote common standard and enhance the results of the JCM as a market mechanism,[25] these efforts and a guideline from the Joint Committee in conducting local stakeholders' consultation are useful. In the future, grievance mechanism could be established.

Public involvement in the public comment process also needs be further promoted, as the level of response for methodologies and projects are still low, receiving only up to three comments, with an average of two comments.[26]

---

[23]IGES Joint Crediting Mechanism (JCM) Database, January 2016.

[24]Ministry of Natural Resources and Environment. 2015. *Circular on regulations of development and implementation of JCM projects in the framework of cooperation between Vietnam and Japan.* http://en.jcmvietnam.vn/rules/circular-on-regulations-of-development-and-implementation-of-jcm-projects-in-the-framework-of-cooperation-between-vietnam-and-japan-a288.html

[25]Öko-Institut e.V. Institute for Applied Ecology. 2015. *Delivering Results-Based Funding Through Crediting Mechanisms: Assessment of Key Design Options.* Berlin: Öko-Institut e. V. Institute for Applied Ecology.

[26]Ibid.

Capacity building under the JCM is a must to ensure appropriate management and MRV implementation on the ground. Although capacity building is not regulated under the JCM, and surprisingly does not seem to be the main interest of project participants, it should be required from feasibility study phase to project implementation including on equipment use and maintenance training for end-users. The partner countries may also provide recommendations on project-level capacity building strategies. Attractive information media are always beneficial.

Ultimately, there is a need for an overarching framework for ensuring sustainable development contribution. At present, a guideline of Sustainable Development Criteria has been released by the Joint Committee between Japan and Indonesia for ex-ante and ex-post evaluation on the JCM projects contribution to sustainable development. The guideline assesses the benefits of each JCM project to the environment, economy, social conditions, and technological improvement. It consists of a Sustainable Development Implementation Plan (SDIP) and Sustainable Development Implementation Report (SDIR). Project participants are required to evaluate their own project using SDIP, which identifies potential impact of the project through a negative/exclusion checklist. After a chosen period, the project participant develop a SDIR to report the impact of their project, or outcomes of their SDIP, and review the contribution of their project to sustainable development, especially at the project area and its surrounding. The SDIP and SDIR are essential to integrate the JCM projects into the broader environmental, economic, and social impact management of the hosting entity, as well as to assess the co-benefits of projects.

The SDIP and SDIR are reviewed by the Joint Committee. The three registered projects were approved before adoption of these guidelines, but they may still be required to submit their plans. This procedure is applicable for all selected projects before they are registered by the Joint Committee. Availability of this procedure should be seen as an opportunity for the JCM to improve the sustainability assurance of projects in market-based mechanisms. There are benefits for the project participants, too, in preventing the project risks. These guidelines could be made as a mandatory part of submission. The application of similar guideline in the other partner countries and a procedure for grievance/complaints from stakeholders during project should also be considered as a follow up of SDIP and SDIR.

The collective evaluation on these project-based SDIPs and SDIRs by the Joint Committee or the Japanese government may be used to evaluate contribution to sustainable development at mechanisms-level for the JCM.[27] Negotiations under the Article 6 of the Paris Agreement could consider the development of

---

[27]Mapping the Indicators: An Analysis of Sustainable Development Requirements of Selected Market Mechanisms and Multilateral Institutions. Berlin: German Emissions Trading Authority (DEHSt) at the Federal Environment Agency.

mechanisms such as the JCM. Once the negotiations start discussing sustainable development the JCM approach to sustainable development evaluation can be an example.

7.5 Recommendations

The assessment could recommend several improvement opportunities as follows.

Governance: There is a need to discuss the ways of accounting and reporting emissions reductions by the JCM projects in national reports, and to discuss the pathway towards tradable scheme or to stay non-tradable.

MRV methodologies: Procedures for conducting market surveys in methodology development need to be strengthened as they are heavily used for setting reference emissions. In the long run, quality of methodologies and their development process as well as their applicability need to be consistent between countries and continue promoting local standards.

Project development and capacity building: Future capacity building activities should put more focus in project identification, development, and approval. The challenge of 'matchmaking' local companies in partner countries and Japanese companies may be tackled through business approaches with highlights on economic benefits, and increasing visibility of potential projects in NDCs. Enhancing capacities and promoting involvement of local entities as third-party entities and project participants are important to increase quality of projects and MRV.

Sustainable development evaluation framework: Focus is needed in the regulatory framework of the JCM sustainable development criteria evaluation, enhancing capacity building and local stakeholders engagement. Other ways to evaluate the JCM effectiveness should also be explored, for example by evaluating the variety of implemented sector and each sector's emission reduction to investigate how mitigation actions from multiple sector can be facilitated through the JCM.

Part II
Climate Change Mitigation

Chapter 8
Using Mixed Methods to Assess Trade-Offs Between Agricultural Decisions and Deforestation

Jyotsna Puri

Abstract Policies that target poverty reduction are often at odds with environmental sustainability. Assessing magnitudes of trade-offs between improved livelihoods on one side, and forest cover on the other, is important for designing win-win development policies that may help to mitigate climate change. I use a mix of panel data for 670 villages over a 10 year period, and combine it with historical land records and grey literature, to understand the drivers of deforestation *within reserved forests* of Thailand – an area where smallholder ethnic tribes are located. Given that reserved forests are the last bastions of forests in Thailand, examining what drives land clearing *within* these areas is important. I combine econometric findings with qualitative reports to infer that (i) it is important to measure the differential effects of policies on different crops, agricultural intensity and the agricultural frontier; and (ii) within forest reserves, policies that *encourage* cultivation overall may *not* be detrimental to forest cover after all. This has important implications for evaluators and policy makers.

Keywords Trade-offs • Poverty • Forests • Agriculture • Panel data • Thailand • Environment • Sustainability • Deforestation • Property rights • Evaluation

8.1 Background

Other than the ocean, standing forests constitute the most important carbon sinks in the world. Yet forests are being threatened and agricultural expansion is widely believed to be the main reason for deforestation in developing countries.[1] A study conducted by FAO (2001) of a stratified random sample of the world's tropical forests finds that 73 % of the world's forests are being converted to non-forest land due to agriculture. Barbier (2004) reports that cultivated area in the developing

---

[1]Barbier (2004) and Lambin et al. (2003).

J. Puri (✉)
International Initiative for Impact Evaluation (3ie), New Delhi, India
e-mail: jpuri@3ieimpact.org

© The Author(s) 2017   131
J.I. Uitto et al. (eds.), *Evaluating Climate Change Action for Sustainable Development*, DOI 10.1007/978-3-319-43702-6_8

world is expected to increase by more than 47 % by 2050, with two-thirds of the
new cultivated land coming from converting forests and wetland.[2] These figures
underscore the importance of examining factors affecting agricultural decisions
especially within forested areas, such as forest reserves.[3]

Using a mix of methods that includes an unbalanced panel dataset of 670 villages
located within Forest Reserves of Chiang Mai, Thailand, and a study of historical
accounts of the evolution of forest reserve legislation and land rights within forest
reserves, I examine the following questions: To what extent do policies that
encourage cultivation lead to deforestation? Is the forest frontier *always* adversely
affected by policies that encourage cultivation or is it possible to develop win-win
strategies? What is the net impact of policies that are otherwise expected to increase
agricultural profitability such as secure land rights, output prices and lower transportation costs, on the *forest frontier*?

Specifically I do two things: First I measure the effect of variables that can be
influenced by policy such as transportation costs, population and perceptions of
land rights on the agricultural frontier and cultivation intensity. Second, I combine
this data with reported land property records to understand and measure how
perceptions of land tenure security affect agricultural expansion and intensity. In
so doing I examine traditional assumptions about ethnic tribes that inhabit forest
reserves in Thailand. This analysis thus sheds light on the extent to which assumptions about land tenure security and particularly assesses claims that ethnic tribes
are significant drivers of deforestation within forest reserves.[4]

There are two main assumptions that are salient in this study. The first assumption is that population within Forest Reserves is exogenous to crop choice: during
the period of this study 1986–1996, population movement and size within reserved
areas of Thailand was controlled by administrative authorities who did not allow
mass migrations to occur.[5] Thus although during 1986–1996, the population of
Chiang Mai province rose by more than 15 %, population in villages that are located
within forest reserves (and are the subject of this study) grew at less than 1 % per
year. The second assumption is that access to markets is exogenous i.e. roads were
not built specifically to provide the ethnic tribes access to markets.[6,7] There is now
substantial evidence that road building in this region took place before the study
period and was undertaken primarily to provide military access to remote areas.

---

[2]Also see Fischer and Heilig (1997).

[3]See for example Alix-Garcia et al. (2011, 2014), Andamet al. (2007), Andersson et al. 2011, and
Bank and Sills (2014).

[4]See for example Delang (2002).

[5]Personal communication, Gershon Feder, The World Bank, 2004.

[6]There are some other agencies of the government and state, that construct roads for special
purposes, but their role is relatively minor.

[7]Road construction and investments related to improvements in access are undertaken by three
agencies in Thailand: The Department of Highways of the Ministry of Communications, the Office
of Accelerated Rural Development of the Ministry of Interior (ARD) and the Department of Land
Administration (DOLA).

Since road construction and road-quality related investments within study Forest Reserves took place for security reasons or to provide access to this area, this assumption is a plausible one.[8] I measure access to market using a composite variable – travel time to the market – which is a good proxy indicator for all three measures of access, and their combination – road presence, road quality and availability of transport.[9,10,11]

8.2 Reserved Forests in Thailand

Forest Reserves are the last bastions of forests in Thailand and more than one-fifth of the Thailand's villages are located within Forest Reserves. Until 1985, North Thailand, where the province of Chiang Mai is located, had the country's lowest population density and largest forested area, including large and critical watersheds. Before the study period in 1985–1993, Thailand as a whole lost 11 % of its forested area (Royal Forest Department 1994) and specifically the province of Chiang Mai lost almost 2000 square km of forest, which equals 10 % of its provincial land area.[12] Forest loss in the province has been attributed mainly to agricultural practices.[13,14]

8.2.1 Land Titles and Property Rights

Forest reserves in Thailand lie under the jurisdiction of the Royal Forest Department (RFD) that set boundaries, but unlike protected areas, do not strictly manage or patrol these. However this jurisdiction and indeed authority has not always been clear. Over the years, this ambiguity has led to frequent changes in legislation related to user rights, as well as, changes in boundaries of forest reserves themselves. Land rights for ethnic tribes living within forest reserves have frequently changed over the years (see Box 8.1). Boundaries of Forest Reserves in northern Thailand have changed leading to changes in the types of land titles especially on the edges of forest reserves which are most affected by boundary changes. Both

---

[8]Howe and Richards (1984) and Puri (2002a).

[9]Also, unlike other forms of investment, investments on roads occur in stages Puri (2002a).

[10]Puri (2002b) In addition, road-related investments are frequently assumed to be endogenous because the beneficiary communities can exert political pressure. To the extent that Forest Reserve villages are inhabited by minority communities, political pressure is not expected to have much sway on government investments.

[11]Howe and Richards (1984) and Puri (2002a).

[12]North Thailand lost approximately the same percentage of forest area. Forest area fell from $8,4126 \text{ km}^2$ in 1985 to $75,231 \text{ km}^2$ in 1993.

[13]Panayatou (1991) and Feeny (1988).

[14]Panayatou and Sungsuwan (1994) and Feeny (1988).

these changes have contributed to ambiguity about land rights for ethnic tribes living within forest reserves. Changes in legislations are summarized in Box 8.1. Arguably ambiguity in the type of land titles has had important implications for crop choice and agricultural decisions.

Box 8.1: Chronology of Important Events for Forest-Related Legislation in Thailand

(Note: Relevant important legislation are starred)

1874: Local Governor's Act of 1874 and Royal Order on Taxation of Teak and other logs. Central government/King becomes involved in managing logging concessions

1896: Royal Forest Department (RFD) founded

1897: Forest Preservation Order of 1897 regulates size of Teak to be logged

1901: Forest management completely under the control of the central government

1913: Forest Preservation Act controls species of Teak and others. Act legally defines a 'forest'. Gives a minister the authority to designate non-logging areas and issue orders to prohibit land clearing

1916: Draft of Forest Conservation Act. "First attempt" at introducing spatial conservation. Regional forest offices begin to select forests to conserve and designate as 'forest reserves'. Draft is not approved but temporary designations of 'forests' continue

1938: Forest Preservation and Conservation Act of 1938; Divides forests into two categories – 'Preserved Forests' and 'Forest Reserves'

*1941: Forest Act of 1941. Forest Reserves are promulgated

1952: Forest Ranger service for control and policing forests. However Rangers only monitor commercial logging concessions and are not assigned to particular Forest Reserves or Preserves

1953: Forest Preservation and Conservation Act is revises. Forest 'designating' committee must now contain a sub-district head as a member. Recognizing reality, temporary residence and use of forest start to be granted after investigation

1954: Forest Preservation and Conservation Act is made a ministerial order. 240 Preserved Forests and 8 Reserved Forests are counted in the country

1960: Forest Police founded as a department of the Police Department

1961: National Park Act passed. Fist NESDP (1961–1965) provides for 50 % of the country to be forested land. Forest rangers organized in 'forest protection units' are made responsible for forest protection

1963: Department of Land Development (DLD) established

*1964: National Forest Reserve Act of 1964 passed. The Act recognized that procedures for designating procedures are too time consuming. Therefore it omits the hitherto mandatory investigation of usufruct rights before designating an area a 'Reserve' or a 'Preserve'

1965: Rural Forest development Units established, to provide services additional to protection units such as extension services, while protecting forests

1966: Committee established to investigate local people's land use in National Forest Reserves

1967: RFD starts to designate 'project forests' for logging

1973: Ministry of Interior sponsors the 'Land distribution promotion project', conducted by RFD

*1975: Cabinet approves legislation for establishing 'Forest villages'

1979: The 'Cultivation Rights Project' in forest villages commences

Box 8.1 (continued)

1982: STKs start to be awarded
1993: Cultivation Rights Project ends
1989: All commercial logging is banned in Thailand
1991: Zoning of National Forest reserves starts (Zone A: Land suitable for agriculture; Zone C: Protected forest zone: Zone E: Economic Forest)
1992: Forest Protection Units transferred to provincial forest offices
1993: All degraded forest lands transferred to Agricultural Land Reform Office (ALRO), and excluded from National forest reserves. ALRO issues SPK4s to landless farmers
Sources: Various. Mainly Bugna and Rambaldi (2001), Fujita (2003), Thailand (2003), Buergin (2000), RFD (Various years), Wataru (2003)

Box 8.1 shows that the government of Thailand instituted many land titling programs before and during the period of study, that aimed to 'clarify' and 're-clarify' the status of property rights, often resulting in much confusion. Indeed village level data used in this study indicate that the modal perception of land title security did not remain constant over the 11 year study period (1986–1996). Table 8.1 shows that residents within Forest Villages changed their view of how secure their hold was over their land. We believe that understanding these perceptions of security are critical if we are to understand how residents within Reserved Forests made their cropping decisions.

All villages within the study dataset lay within forest reserves at least once during the 11 year period. Table 8.1 summarizes strongly held beliefs about land titles and shows that *perceptions* of land title (and therefore security of title) did not always match the type of land title households possessed. There were seven different types of land titles in the study region (see Table 8.1). Thus for example many residents within forest reserves believed that they could use their land as collateral. However forest legislation did not allow residents to have secure land titles or to use land as collateral. After discussing the implications of these land titles[15] and consulting literature around this, I differentiate between villages depending on whether they believe they have secure land rights or not. Villages that report possessing NS-4, NS3 and NS3-K are classified as possessing secure property rights. Nineteen percent of the villages in the study sample report that they had *secure* property rights even though *de jure* residents can possess only usufruct rights.[16] Another factor that contributed to this belief of secure ownership is that most residents pay property taxes. I discuss this more in the next section.

---

[15]Personal communication, Gershon Feder (2004).

[16]Feder et al. (1988a, b) and Gine (2004a, b) also document that residents of villages that have been in existence for a long period of time are likely to believe that they have secure property rights to the land that they cultivate, even if they do not possess land title papers. Feder et al. claim that despite the fact that land title documents are missing, there is an active land market in this part of the country, further underscoring this perception of secure land rights. Gine, when examining a sample of 191 villages in North East Thailand and Central Thailand, finds that 40 % of the households located in villages in Forest Reserve and Land Reform areas had titled land and only 20 % of the households were landless.

Table 8.1 Land titles and land use rights in Thailand (1954–1990)

Title type	Year introduced	Rights	Limits (as described by Feder et al. 1988a, b)
NS-4 (Chanod) Title Deed	1954	Most secure; full unrestricted ownership title	Issued only outside forest reserves
		Can be used as collateral and is fully tradable	
NS-3 (No-So-Sarm) Certificate of Use	1954	Very secure. Can be converted into NS-4	Issued only outside forest reserves, any transfer must be advertised for 30 days
		Tradable under certain conditions	
		Can be used as collateral	
NS-3K (No-So-Sarm-Kor) Exploitation Testimonial	1972	Very secure. Can be converted to NS-4	Issued only outside forest reserves. Ownership may be challenged if land lies fallow for more than 5 years
		Fully tradeable	
		Can be used as collateral	
NS-2 (Bai-Chong) pre-emptive certificate	1954	Authorizes temporary occupation of land. After a prescribed period may be converted to NS-3 or NS-4. Can be acquired only through inheritance. Cannot be used as collateral	Issued only outside forest reserves; validity of rights conditional on use within 6 months of issuance
SK-1 (So-Ko-Neung) Claim Certificate	1954	Particular to the period during which Thailand was adopting the Land code. Claim to ownership is based on possession before the enactment of land code	Issued only outside forest reserves
		Certificate tradeable only after transfer is advertised. Cannot be used as collateral	
STK (So-Tho-Ko) 1 and STK 2 Temporary cultivation rights	1982	Certificate of use only. Can be acquired only through inheritance and cannot be used as collateral. Cannot be converted into NS-3 or NS-4	Issued *inside* forest reserves; covers plots up to 15 rai. State reserves right to revoke usufruct rights if restrictions are violated
NK-3 (Nor-Kor-Sarm)		These are issued in specific areas under small official programs. They can usually be acquired through inheritance and usually cannot be used as collateral. These are usually usufruct rights and cannot be sold until 5 years after issue date	
Nk-2 (Nor-Kor-Som)			
Nk-1 (Nor-Kor-Neung)			
SPK (Sor-Por-Kor)			

Source: Adapted from Feder et al. Land Productivity and Farm Productivity in Thailand, 1988a

8.3 Study Area and Data Set and Study Area

The dataset used in this study was collected by Thammasat University for the province of Chiang Mai.[17] The data were collected for the National Economic and Social Development Board (NESDB). Data for the study were collected for six rounds, once every 2 years (biennially) starting in 1986 and then 1988, 1990, 1992 and 1996, for the province of Chiang Mai. Villages included in the study dataset all responded that they lay within Forest Reserves at least once during this 11 year study period. All villages in the dataset are registered with the Village Directory of the Department of Local Administration (DOLA). However because forest reserve boundaries changed a lot, all villages did not lie within Forest reserves during the study period. Inhabitants of Forest Reserve villages are mostly hill tribe people who are poor, and live in villages that are remote and have poor infrastructure.

Forest Reserve residents grew mainly three crops during this period – paddy rice, upland rice and soybeans. Thailand is among the largest growers of paddy rice and its biggest exporter. But rice is also a staple. Most villages in the study sample grew paddy rice. On average upland rice and soybean were grown by 25 % and 26 % of villages respectively.

The resulting panel dataset is unbalanced. Of the 670 villages that appear at least once in the dataset, 255 (38 %) are present for all six rounds in the panel; in contrast, 124 villages are present for only one round (Tables 8.2 and 8.3). Attrition in panel data is common: villages may choose to not participate in certain rounds or may not be asked to participate in certain rounds for several reasons (e.g. lack of resources with the survey agency). It is important to understand the cause of attrition or selection.[18] Villages that are surveyed and respond in all six rounds are the single largest group in the dataset (38 %). The second largest group is the villages that occur only once. These constitute 18.5 % of the villages.

In the survey conducted by Thammasat University, village communities were asked in every round of survey (there are a total of six rounds) if they had secure property rights (*'What land title did you have?'*) Using this information and the mapping above, from the type of land titles to the security of these land titles, I examine if these perceptions change over the different rounds. In Tables 8.2 and 8.3, I examine these responses for each round in the panel dataset. Table 8.2 shows that 62 % or 413 villages in the dataset never believed they possessed secure rights to village land. In contrast, only 36 villages claim to have secure rights during the entire study period (for all six rounds). For the remaining villages, the status of their land titles 'flips' from year to year. So for example Table 8.3 shows that 33 villages in the dataset were surveyed for all six rounds of data collection, in five of six of those rounds believed that their land title was insecure. They only report a secure

---

[17]The larger dataset consists of 784 villages.

[18]Missing observations in a panel data may not be randomly missing and, if so, estimators may be inconsistent. Ignoring attrition and using a balanced dataset, as is common practice, may lead to inconsistent estimates (See Heckman 1976; Nijman and Verbeek 1992).

Table 8.2 Security of land title cross-referenced with frequency of presence, forest reserve villages, Chiang Mai, 1986–1996

Number of years village is present in the panel dataset for – > perception about land title[a]	No. of villages	1 pt	2 pts	3 pts	4 pts	5 pts	6 pts
Never secure	413	89	29	60	78	23	134
Secure once	89	35	7	7	4	3	33
Secure twice	46		20	5	6	0	15
Secure three times	32			13	8	4	7
Secure four times	33				21	1	11
Secure five times	21					2	19
Secure six times	36						36
	670	124	56	85	117	33	255

Source: Data provided by Thammasat University
[a]Secure title to land implies, land can be used as collateral. These are responses from village headmen

Table 8.3 Frequency of occurrence of forest reserve villages cross-referenced with number of times villages are accounted for in the study dataset (1986–1996)

		Number of times a village is classified to be located within a forest reserve						
		Once	Two times	Three times	Four times	Five times	Six times	Total
Total number of times a village is present in the dataset	1	124						124
	2	17	39					56
	3	8	16	61				85
	4	1	19	24	73			117
	5				4	29		33
	6					91	164	255
Total no. of villages in 'forest reserves'		150	74	85	77	120	164	670

Source: Data provided by Thammasat University, Chiang Mai, Thailand

land title once. Similarly Table 8.2 shows that 117 villages were present in the dataset for four rounds. Of these, 73 said they were *within* Forest Reserves all four years; 24 said they were in Forest Reserves for only three out of the four years and 20 said that they were *within* Forest Reserves for at most two out of four years.

One difficulty with this dataset is that we don't know the location of villages. However we do know that all the villages were meant to be within forest reserves at least at some point so that they were included in this panel dataset. This provides one explanation for 'flipping' land titles which reflects changes in titles. As forest reserve boundaries change, it is likely that as a consequence land titles also change. We also hypothesize that it is more likely that villages located *just outside* forest reserves or along their boundaries, will witness more change in their boundaries than those in the interior. Box 8.1 shows the frequent change in legislation that led to changes in boundaries within these Forest Reserves. Villages that are located far in the interior of Forest Reserves are unlikely to see this change in boundary, and as a consequence their permitted land titles are unlikely to change.

The panel dataset for this study is thus divided into two types of villages: The first group of villages is a group of 257 villages that has 'ambiguous property rights' over the duration of the study period, caused in large part by changing forest legislation and by changing forest reserve boundaries. These villages witnessed frequent changes and had ambiguous property rights or APR villages and constitute 38 % of the study sample. The second group consists of villages that claim to have *no* secure rights *consistently* and are likely to be located deep inside Forest Reserves, where changing Forest boundaries create no ambiguity. The latter group of villages are called 'no secure property rights' villages NPR villages.

Two other features of the survey are that (i) village headmen provide responses to questions and, (ii) the biennially conducted survey records *modal* values of variables. Data are collected via questions such as: 'What is the mode of transport *most* (popularly) used by households in the village?' or, 'What is the method of sale for *most* households?' "*What is the most popularly grown short run (long run) crop this year?*" For crops other than paddy rice, crop area, the number of households growing the crop and other attributes are recorded only for the short-run or long-run crop that is 'most popular'. This means no crop is tracked for all years, other than paddy rice, unless that crop is 'popular' *every* year.[19] Furthermore, crops are tracked in groups i.e. 'short run crops' or 'long run crops'. Other challenges with the data including absence of price data and agricultural practices are discussed and addressed in Puri (2006). Main characteristics about villages included in this dataset are presented in Tables 8.4 and 8.5.

8.4   Characteristics of Data and Hypothesized Effects

In this section I discuss the hypothesized effects that different village level attributes are expected to have on two main agricultural variables: on agricultural area within a village and on average intensity of cultivation within a village.

The intensity of cultivation variable requires a brief discussion. Boserup (1965) in her classic exposition of factors governing agricultural expansion in developing countries, especially in Asia, defined agricultural intensification as "*. . .the gradual change towards patterns of land use which make it possible to crop a given area of land more frequently than before.*" (pp. 43). In this definition she thus departed from the definition of intensification that measured increased use of inputs per hectare of cropped area. In this study, I use this Boserup measure to understand intensity of cultivation: Intensity of cultivation is measured by a variable that is the response to the question "*What percentage of agricultural land is being used (for*

---

[19]Village headmen are also asked questions about "the *second* most important short (long) run crop" and the "*third* most important short (long) run crop". Data on these is scarcer.

Table 8.4 Basic characteristics of villages located in forest reserves, Chiang Mai (1986–1996), pooled dataset

Variable	All sample			APR villages			NPR villages		
	Obs	Mean	Std. dev	Obs	Mean	Std. dev	Obs	Mean	Std. dev
Population (no. of people)	2573	568.96	365.7	1067	597.8	354.7	1506	548.5	372
Village area (Rais)	2277	4354.2	6188.6	987	3561.5	5123.1	1290	4960.6	6833.6
Agricultural land (Rais)	2382	917.6	1085.4	999	945.15	917.1	1383	897.7	1192.2
Paddy rice area (Rais)	2563	279.6	292.4	1073	310.5	289.9	1490	257.2	292.3
Soybean area (Rais)	2654	41.9	143.8	1096	58.8	137.5	1558	29.9	147.1
Area devoted to upland rice (Rais)	2654	57.1	209.4	1096	23.3	98.7	1558	80.8	257.8
% villages with less than 10 % fallow land	2614	17.8		1081	24.6		1533	13.1	
% villages with more than 50 % fallow land	2614	27.1		1081	18.0		1533	33.6	
Assets									
Avg. no. of cows per household	2586	0.82	2.25	1073	0.46	1.53	1513	1.072	2.62
Avg. Proportion hhs owning small tractors	2586	0.07	0.10	1073	0.09	0.11	1513	0.054	0.095
Avg. Proportion hhs owning motor carts	2651	0.002	0.02	1095	0.003	0.02	1556	0.001	0.02
% of hhs working outside tambon	2037	3.6	6.2	994	4.54	7.6	1043	2.7	4.2
Access variables									
Avg. one way travel time to mkt. (min)	2408	70.6	66.4	1075	42.52	39.53	1333	93.21	74.5
Avg. one way travel time to district (min)	2509	83.2	90.4	1091	45.47	34.67	1418	112.23	107.8
Avg. proportion. Hhs owning motor bikes	2466	0.26	0.23	1064	0.38	0.24	1402	0.16	0.18
Proportion of literate population	2487	0.41	0.28	1058	0.54	0.23	1429	0.32	0.28
Avg. proportion of hhs with electricity	2586	0.49	0.43	1073	0.72	0.34	1530	0.32	0.41
Inputs									
Proportion of adults	2573	0.42	0.13	1067	0.45	0.12	1506	0.39	0.13
% villages using HYV rice	2654	71.6		1096	92.2		1558	57.1	
% villages w/sufficient (SR) water	2644	25.9	44	1094	32.5	47	1550	21.2	41
% villages w/sufficient (LR) water	2632	11.4	31.7	1087	9.6	29	1545	12.6	33
% villages that use BAAC credit	2650	43.7	49.6	1096	71	46	1554	25	43

Source: Data provided by Thammasat University, Thailand
APR Villages Villages with Ambiguous Property Rights, *NPR* Villages with No secure Property Rights

Table 8.5 Percentage of villages growing different crops, forest villages, Thailand, 1986–1996

Year	1986	1988	1990	1992	1994	1996
No crop at all	**0**	**6.6**	**6.8**	**5.1**	**0.8**	**6.7**
One crop	**78.8**	**57.1**	**45.2**	**41.6**	**44.4**	**48.1**
Only paddy	78.7	55.3	43.4	39.1	44.4	45.9
Only soybean	0	0.5	0.2	0.2	0.2	0
Only upland rice	0	1.3	1.6	2.3	0.2	2.1
Two crops only	**21.2**	**33.4**	**41.7**	**45.6**	**47.0**	**40**
Paddy rice and soy	21.0	24.7	21.4	19.0	18.5	16.0
Paddy rice and upland	0.3	8.6	20.3	26.6	28.5	23.5
Soy and upland rice	0	0	0	0	0	0.5
Three crops	**0**	**2.8**	**6.3**	**7.7**	**7.8**	**5.2**
Number of villages	367	392	429	469	477	520

Figures are for respondents who provide positive responses to the area question. Source: Data provided by Thammasat University

cultivation) in the village, in this year?" Implicit in this question is the understanding that the village has agricultural land that has been left fallow. Thus the percentage of land cultivated in time t, by village i, is assumed to be defined as:

% of land cultivated at time t in village i = [(Total land cleared and potentially fit for cultivation − Area left fallow at time t by village i)/Total land cleared and potentially fit for cultivation] × 100

I now discuss the hypothesized effect of village level variables on total agricultural area and on agricultural intensity.

Village Population   Village population is expected to have two types of effects on total village agricultural area and cultivation intensity. The first is a scale effect: A village with a larger number of households is expected to have a higher demand for agricultural land compared to one with a fewer households. The second effect is the 'food' (or subsistence) effect. A larger population also means larger subsistence requirements. The subsistence effect is likely to be stronger for food crops in villages located far from the market because it is not possible to buy food from the market. Both these effects are expected to be in the same direction.

Travel Time to Market   Travel time to the market is a proxy for the cost of transporting crops to the market and obtaining inputs from the market. I expect that farmers that are located far from the market are able to exercise less leverage in getting the best prices for their produce; are unable to spend much time searching for best bargains; are less willing to carry their produce back if a transaction does not go through; and, are likely to have limited access to information about markets.[20] Thus travel time is also a proxy for search costs, bargaining costs and, generally, costs of not being located in situ. Thus, for crops that are produced for the

---

[20]Minten and Kyle (1999).

market – such as soybean – travel time is likely to have a negative effect on the probability that they are produced and on the amount of land area devoted to them. To the extent that upland rice and paddy rice are grown for subsistence, this effect is expected to be insignificant. Moreover, if the only reason that the crop is grown is that it is a substitute for a staple that can be bought in the market, then the travel time coefficient is likely to be positive. The variable in the dataset measures the 'average time taken one way, in minutes, to reach the market, using the most popular mode of transport.' It thus takes into consideration mode of transportation and road quality.[21]

Proportion of Adult Population The proportion of adults in the village is expected to positively affect land brought under cultivation and the intensity with which it is cultivated. Adult labor is required to grow crops on virgin land that requires preparation.[22,23] The presence of more adults is likely to increase the amount of land cultivated and ameliorate labor scarcity. In this study, proportion of adult population is used as a proxy for available labor in the village and for the opportunity cost of labor.

Productivity of Land There are two variables that are used as a proxy to measure land productivity. These are water availability and a dummy for acidic soil.[24] (Please see below.) Additionally I also use a time invariant binary variable to indicate whether the village grew high yielding varieties (HYV) of rice at any time during the study period. (So HYV rice dummy =1 if the village *ever* grew HYV rice during the study period, and =0 otherwise). I expect this variable to have two impacts on productivity. The first is on paddy rice area: HYV rice is more productive than non-HYV rice. I expect it to have a positive effect on area devoted to paddy rice. The other effect this variable is likely to be a proxy for is the presence (or absence) of 'attention' from local authorities. To the extent that growing HYV rice requires additional knowledge and training provided by field officers and that

---

[21] See for example Dawson and Barwell (1993).

[22] It would be useful to gauge the different impacts of adult males and adult females.

[23] See Godoy et al. (1997) for a similar argument.

[24] Another possible variable is yield per acre but there are problems with measuring the variable since it is measured only when crop data are available. It is also potentially endogenous. For example for upland rice yield/hectare is available only for 541 observations, or 248 villages for at least one point in time. For the subset of variables for which data are available: For soybean, there is a positive time trend when the log of productivity is regressed on year, while controlling for other variables (~3 %). when we regress this variable for soybean on time dummies, the time dummies are insignificant (and indeed in the first two years, negative, compared to 1986. They are positive in the next 2 years but insignificant. Only in 1996 is the time dummy significant and positive – when an average increase of almost 30 % occurs). Similarly for upland rice, the time trend is not significant or large (although it is positive). This indicates that there were not very many productivity increases among farmers located in Forest Reserve villages of Chiang Mai, during 1986–1996, although some may have taken place in the last year of the study period, for soybean. Witnessing an increase in area despite there being an increase in productivity, further strengthen my results.

the government has been encouraging the cultivation of HYV rice, mostly via the BAAC (Bank of Agriculture and Agricultural Credit), the dummy is expected to be positively correlated with BAAC presence.[25]

Water Presence Scarcity of water is an important resource constraint in this region. Walker (2002) in a detailed study of the Mae Uam catchment area of the Mae Chaem district of Chiang Mai, finds that even cultivation of dry-season varieties of soybean, which requires relatively less water, has reached its hydrological limit. Dry season varieties of soybean (typically grown in the region) and upland rice are crops that require little water.[26] On the other hand paddy rice requires a lot of water to grow. Availability of water is used as a proxy for productivity of land. In this analysis, presence of water is measured by the response to the question *"Did this village have sufficient water to grow short run (long run) crops?"* The dummy variable is equal to 1 if there is sufficient water and is zero otherwise. Irrigation is usually provided by rain and, to a lesser extent, by small man-made weirs and canals.[27]

Acidic Soil The other variable used to measure the productivity of land is acidity of soil. Acidity of soil is an undesirable quality. The variable is expected to have a negative effect on agricultural area and intensity of cultivation. In this dataset it is recorded as 1 if soil within a village suffers from high acidity, and 0 otherwise.

Perceptions of Land Ownership Secure land titles are defined as titles that allow land to be used as collateral or sold. I expect that farmers who have secure land titles will be more willing to invest in land and grow cash crops. I use a dummy variable, which is equal to 1 if the village headman responds that *"secure land titles were held by most farmers in the village"*.

Credit Use Credit use is expected to increase the intensity of cultivation. The BAAC is the lender of first resort in most of these villages since it provides relatively low interest credit. Credit obtained from the BAAC is assumed to be mainly for agriculture, unlike credit provided by private money lenders (because of the conditions that BAAC imposes). Clearly, credit use is endogenous.[28] The variable used to indicate use of credit in this study is *"Do villagers use credit from the BAAC"*. This variable equals 1 if people in the village use credit from the Bank of Agriculture and Agricultural Credit, and 0 otherwise.

---

[25]Thus the variable is used as an instrument in the BAAC credit use equations.

[26]Although upland rice requires rainfall, it does not require standing water like paddy rice does.

[27]Palm et al. (2004).

[28]One reviewer suggests the use of a BAAC credit dummy which is =1 for the year that a village starts using BAAC credit and then, irrespective of response, is coded =1, for all years thereafter. The object here is to measure the use of credit and not so much the availability of credit. The endogeneity of credit use is not discussed more in this paper but is discussed in detail in Puri (2006).

Using this dataset and the relationships hypothesized above, I estimate two estimation models for total village agricultural area and cultivation intensity:[29]

$Log(\text{Agricultural Area})_{jit} = a_{i0} + a_{i1}Log(\text{Population})_{jt}$
$\quad + a_{i2}\,Log(\text{Travel time to market})_{jt} + a_{i3}(\text{Water availability dummy})_{jit}$
$\quad + a_{i4}(\text{Acid soil dummy})_{jit} + a_{i5}(\text{Property rights dummy})_{jt}$
$\quad + a_{i6}(\text{BAAC use dummy}) + a_{i7}(\text{Proportion of adult population})_{jt}a_{i7}\text{Time trend}$
$\quad + u^{*}_{ji} + \varepsilon_{jit}$

$\text{Intensity of cultivation}_{jit} = a_{i0} + a_{i1}Log(\text{Population})_{jt}$
$\quad + a_{i2}\,Log(\text{Travel time to market})_{jt} + a_{i3}(\text{Water availability dummy})_{jit}$
$\quad + a_{i4}(\text{Acid soil dummy})_{jit} + a_{i5}(\text{Property rights dummy})_{jt}$
$\quad + a_{i6}(\text{BAAC use dummy}) + a_{i7}(\text{Proportion of adult population})_{jt}a_{i7}\text{Time trend}$
$\quad + u^{*}_{ji} + \varepsilon_{jit}$

8.5   Results

Results are analyzed in two ways. First, I examine the effect of different crops on total agricultural area. Tables 8.5 and 8.6 discuss results from these equations. Second, I examine how policy variables affect overall agricultural area and intensity of cultivation.

I use random effects models in Tables 8.6 and 8.7, to estimate the effect of these variables on agricultural area and intensity of cultivation:

Table 8.6 shows that an *increase* in village agricultural land is associated with an *increase* in area devoted to paddy rice (coefficient $= 0.46$; $z = 7.8$) and upland rice (coefficient $= 0.21$; $z = 2.36$). On the other hand, an increase in area devoted to soybean is *not*: Villages that grow Soybean are likely to be those that have little agricultural land, and can only cultivate intensively. Speaking with agriculturalists, this is expected: Soybean is an input intensive cash crop and is usually cultivated on land that is fertilizer rich and input rich. Table 8.7 shows that an increase in *intensity of cultivation* is associated with an increase in area devoted to Soybean (0.01399; $z = 3.76$) and Paddy rice (0.0037; $z = 1.95$). Upland rice area does not contribute significantly to increasing cultivation intensity (measured by the number of crops grown on a plot of land in a year). This too is expected. Observational data and conversations with folks at the university reveal that upland rice is grown on forest frontiers, and typically on land with low fertility that is vulnerable to erosion.

---

[29]Where u^{*}_{ji} is distributed normally and is the unobserved influence of the village on repeated observations. ε_{jit} is the unobserved error term also distributed normally with mean 0 and variance σ^{2}_{ε}. For each of these equations, to account for BAAC credit use being endogenous, I estimate a first stage random effects equation to get the predicted value for BAAC credit use. To model BAAC credit use, for each of the equations above, I estimate the following random effects equation, which includes all exogenous variables in the system, including the three identifying instruments.

Table 8.6 Linear random effects regression results for land devoted to agriculture, forest reserve Villages, Chiang Mai (1986–1996)

Dep. variable: village agricultural area	Coefficient	Std. dev	Z	P > Z
Year	15.35**	3.15	4.87	0
Area devoted to paddy rice	0.46**	0.06	7.8	0
Area devoted to upland rice	0.21*	0.09	2.36	0.018
Area devoted to soybean	−0.19+	0.11	−1.69	0.091
Constant	−641.17*	294.84	−2.17	0.03
Sigma-u	1054.16			
Sigma-e	375.89			
Rho	0.89			
Observations	1979			
R-square within	0.042			
Groups	622			
R-square between	0.054			
R-square overall	0.056			
Gaussian wald statistic (chi2, 4df)	85.5			
Prob > Chi2	0			

Source: Data provided by Thammasat University, Thailand
** denotes significance at the 1 % level; * at the 5 % level, and + at the 10 % level

Table 8.7 Random effects interval regression results for intensity of cultivation, forest reserve Villages, Chiang Mai (1986–1996)

Dep. variable: intensity of cultivation	Coefficient	Std. error	Z	P > z
Year	0.0379	0.125	0.3	0.762
Area devoted to soybean	0.01399**	0.0037	3.76	0
Area devoted to upland rice	0.0011	0.0034	0.33	0.738
Area devoted to paddy rice	0.0037*	0.0019	1.95	0.051
Constant	63.314**	11.5450	5.48	0
Sigma-u	14.0611	0.6781	20.74	0
Sigma-e	16.392	0.3547	46.21	0
Rho	0.4239	0.0273	0.3712	0.478
Observations	2174			
Groups	629			
Gaussian wald statistic (chi2, 4 df)	20.03			
Prob > Chi2	0.0005			

Source: Data provided by Thammasat University, Thailand
** denotes significance at the 1 % level; * at the 5 % level, and + at the 10 % level

Furthermore land devoted to upland rice does not require much preparation. On the other hand soybean and paddy rice require large amount of inputs and preparation. They are usually grown on land that is agriculturally fertile and productive. They are usually cultivated on fertile and flat river beds and in watershed areas, and this land is much more likely to have other crops grown on it, once soybean and paddy rice have been harvested.

I now discuss the effect of variables that can be affected by policy on agricultural land and intensity of cultivation. Results are presented in column 1, Table 8.8 [30,31] Results in Table 8.8 show that a 1 % decrease in travel time to market increases the percentage of agricultural land cultivated by 2.9 % points. Population has no effect on the intensity of cultivation for either group of villages. Short run crop water availability increases the percentage of area cultivated by almost 6 percentage points. This may be occurring if short run crops such as soybean and mung bean are grown on intra-marginal lands.

Results show that the effects of explanatory variables are different for villages that have no secure property rights (NPR villages). On average NPR villages cultivate land less intensively than APR villages by 71 percentage points. Additionally in NPR villages, there is almost no effect of a change in travel time to market (travel time estimate for NPR villages = 0.343 (which is coefficient for log (travel time estimate) = -2.868 + coefficient (NPR = 1*Log(travel time estimate)) = 3.212) = 0.343; z = 0.42; Prob > Chi-square =0.67). Short run water availability also has no effect on intensity of cultivation in NPR villages (the short run water coefficient in NPR villages = 5.716–4.11 = 1.6; Z-statistic = 1.04; Prob > Z = 0.30).

To investigate land expansion as measured by village agricultural land, the same variables are used to explain the equation as used for agricultural intensity. This is because variables that affect intensity of cultivation should also affect land expansion. Results are presented in column 2, Table 8.8.[32]

Results in column (2) show that a 1 % increase in village population leads to a 0.4 % increase in area devoted to agricultural land in the villages in the estimation sample. BAAC credit use increases agricultural land by 1.1 % in these villages. A 1 % increase in travel time to the market increases the area under cultivation in APR

---

[30]Since the intensity of cultivation is measured as a categorical variable, with each value representing an interval, I estimate the equations for intensity of cultivation using a random effects interval regression model. Similar to the procedure followed for the crop area equations, I estimate a reduced form equation where BAAC credit use is endogenous. The results I discuss here use a two-step variant of the interval regression model in which the first step estimates a reduced form model for BAAC credit use, using a random effects probit model. Column (5) is a two-step variant of the random effects interval regression, where the first stage uses a random effects probit equation to estimate the model for BAAC credit use. Results from the first stage are reported in Table 5.17.

[31]The different specifications and sensitivity analyses are presented in Puri 2006.

[32]I estimate a random effects equation via generalized two stage least squares to estimate the model for agricultural land. The dependent variable is in logs. In Table 5.16 I present only one specification. BAAC credit use instrumented for, by using three identifying instruments. These are proportion of population with compulsory education, travel time to the district and HYV rice dummy. The results from the first stage random effects equation for BAAC credit use are not shown here.

Table 8.8 Random effects reduced form interval regression for intensity of cultivation and log (village area) in forest reserve villages, Chiang Mai, Thailand 1986–1996

	Random effects interval regression instrumental variables (intensity of cultivation)	Instrumental variables random effects (agricultural area)
	(1)	(2)
NPR dummy =1	−71.091	−0.651
	(2.15)*	(−0.35)
Year	−0.269	−0.022
	(−1.26)	(1.76)+
(NPR =1)*year	0.485	0.018
	(−1.43)	(0.86)
Log (Village Population)	0.328	0.427
	(0.2)	(5.33)**
(NPR =1)*Log (Village Popn)	0.083	0.021
	(0.04)	(0.21)
Log(Travel time to market)	−2.868	0.158
	(2.91)**	(4.04)**
(NPR =1)*Log (Tr time to mkt.)	3.212	−0.126
	(2.52)*	(2.59)**
Short run water dummy	5.716	−0.055
	(3.87)**	(−1.12)
(NPR =1)*SR water dummy	−4.11	0.005
	(1.92)+	(0.08)
LR water dummy	−3.06	−0.041
	(−1.37)	(−0.62)
(NPR =1)*LR water dummy	3.343	0.145
	(1.1)	(1.52)
Proportion of adults	−7.157	−0.056
	(−1.16)	(−0.31)
(NPR =1) *Propn of adults	28.112	0.227
	(3.25)**	(0.83)
Acidic soil dummy	−7.973	−0.28
	(3.82)**	(4.13)**
(NPR =1) *Acidic soil dummy	−0.85	0.314
	(−0.3)	(3.22)**
BAAC credit use dummy	4.93	1.11
	(1.09)	(2.41)*
(NPR =1) *BAAC credit dummy	−1.904	−0.83
	(−0.28)	(−1.32)
Constant	104.437	4.339
	(4.88)**	(3.94)**
Observations	2204	1989
Number of ID	628	622

Source: Data provided by Thammasat University, Thailand
NPR = 1 if villages have no secure property rights; = 0 otherwise. ** significant at 1 % level; *significant at 5 % level; + significant at 10 % level

villages by 0.16 %. Presence of acidic soil in APR villages reduces agricultural land by 0.3 %.

The effects of travel time to market and acidic soil disappear in NPR villages: the travel time coefficient for NPR villages = 0.032; z = 1.12; P > Z = 0.26) and, acidic soil dummy for NPR villages = 0.033; z = 0.48; P > z = 0.63). Presence of sufficient water for long run crops increases the total agricultural land in a village by 0.14 %.

8.6 Discussion of Main Results

In this paper I explain the direction and magnitude of impacts on agricultural intensity and extensive frontier using random effects equations for village agricultural land and intensity of cultivation. I discuss findings below:

8.6.1 Effect of Population

The study finds that a 10 % increase in population leads to a 4.3 % increase in agricultural land. This is consistent with the findings in Cropper et al. (1999), who report that a 10 % increase agricultural household density in North Thailand increases agricultural land by 4 %. However, it is higher than the elasticity of cleared land with respect to population reported in a spatially explicit study of the effects of population and transportation costs in Cropper et al. (2001): In that study, a 10 % increase in population leads to a 1.5 % increase in cleared land in the forested areas of North Thailand.[33] It is lower than the elasticity reported by Panayatou (1991) for Northeast Thailand. That study reports that a 10 % increase in population leads to a 15 % decrease in forest cover.

The effects of population do not differ across the two sets of villages explored in this paper i.e. villages with ambiguous property rights and villages with no secure property rights. There is some evidence of a significant difference in direction of impact for soybean cultivation, but the magnitudes of impact are very small.

8.6.2 Effect of Travel Costs

I find that transportation cost has a quantitatively modest impact on agricultural decisions in the study area. For total agricultural land in a village, the effects of

---

[33]In this study the elasticity for cleared land with population is smaller.

travel time remain small. A 10 % increase in travel time to the market increases agricultural land by 1.6 %.

This finding that travel time has modest effects on agricultural decisions in Forest Reserves of Chiang Mai is consistent with other studies of the region: Cropper et al. (1999) find that a 10 % increase in road density leads to a 2 % decrease in forest cover in North Thailand. Cropper et al. (2001) find that a 10 % increase in travel time to the market leads to a 2.4 % decrease in forested area in the forest areas of North Thailand. Similarly in North-east Thailand, Panayatou (1991) finds that changes in road density have an insignificant impact.

One policy conclusion from this is that road building may not have a deleterious effect on forest cover in this area. This is different from what has been found in other parts of the world. To the extent that roads provide increased access to services and markets, improving access within Forest Reserves might help to alleviate poverty without affecting forests. However this result should also be treated with caution.[34]

8.6.3 Property Rights

In this study, I make a distinction between NPR villages and APR villages. It is important to make this distinction: villages with no secure property rights are likely to be more remote and poorer than villages that have ambiguous property rights.

An important effect in the study is that villages with no property rights are likely to likely to cultivate their land less intensively (being in an NPR village reduces intensity of cultivation by 71 percentage points). However magnitudes of impact of the two main variables – travel time and population – on cropping decisions are not very different for the two groups of villages. Particularly, travel time to market has a negligible effect on upland rice cultivation and agricultural land in NPR villages. The mixed evidence is explained by the fact that the distinction between the two groups with respect to their property rights is not sharp. Villages with no property rights (NPR villages) are located in the same region as those with ambiguous property rights and are likely to behave similarly. Feder et al. (1988a, b) in their study of Forest Reserves in Northeast Thailand show that villages without secure property rights are less likely to invest in land. This may help to explain the significantly lower intensity of cultivation in NPR villages. They also conclude that secure property rights allow better access to credit. In this study the distinction

---

[34]The random effects estimators in the study reflect primarily cross-sectional variation in the data. Differences in effects of transportation costs could thus be picking up differences between location of villages.

between the two groups may also be muted because residents may have different perceptions about their claims to land they occupy according to their length of residence (see for example Lanjouw and Levy (2002)).

8.7   Overall Discussion

Anecdotal evidence in Thailand shows that North Thailand witnessed a large increase in deforested area during the years 1986–1996. One of main reasons for this is claimed to be agricultural expansion. ASB (2004) reports that during the same period, area devoted to upland rice area grew rapidly as well. To the extent that both these occurred concomitantly, and that upland rice cannot be grown on land devoted to other crops, the study suggests that it may be important to do a more detailed analysis of the factors affecting upland rice cultivation especially since it is seen as being detrimental to the environment. Upland rice is grown on mountain slopes with thin soil and low fertility, i.e. on land that is otherwise agriculturally marginal and undisturbed. Upland rice also has a much larger effect on the surrounding ecosystem compared to paddy rice and soybean. On the other hand, paddy rice and soybean can be intercropped and are usually grown on agriculturally important land while upland rice is usually not grown with other crops (in these contexts). Specifically speaking upland rice is grown on lands which is deserted after two or three crops have been planted and harvested.

This study suggests that a reduction in travel time to market reduces the area devoted to upland rice. It also suggests that *while not affecting forest cover, a reduction in travel time to market may also help to reduce the incentive to adopt and cultivate upland rice.* One policy implication from this study is to encourage crops that allow multiple rotation in the lowlands, and thus reduce pressures that push the agricultural frontier to mountain slopes that are prone to erosion. Understanding the magnitudes of impacts on crop adoption and acreage of population and roads can also help understand certain trade-offs. If for example, road building is being considered as a policy option in a region, but there is evidence that it affects crop adoption and acreage, then understanding which crops are affected most, can help to understand otherwise unintended repercussions of this policy.

References

Alix-Garcia, J., et al. (2011). *The ecological footprint of poverty alleviation: Evidence from Mexico's Oportunidades Program*.

Alix-Garcia, J., Aronson, G., Radeloff, V., Ramirez-Reyes, C., Shapiro, E., Sims, K., & Yañez-Pagans, P. (2014). *Environmental and socioeconomic Impacts of Mexico's payments for ecosystem services program*.

Andam, K. S., Ferraro, P. J., Pfaff, A. S. P., & Sanchez-Azofeifa, G. A. (2007). *Protected areas and avoided deforestation: A statistical evaluation* (Final report). Washington, DC: Global Environment Facility Evaluation Office.

Andersson, C., Mekonnen, A., & Stage, J. (2011). Impacts of the productive safety net program in Ethiopia on livestock and tree holdings of rural households. *Journal of Development Economics, 94*(1), 119–126. Available at: http://dx.doi.org/10.1016/j.jdeveco.2009.12.002.

Bank, I. D., & Sills, E. O. (2014). *Have we managed to integrate conservation and development? ICDP impacts in the Brazilian Amazon.* World Development. Available at: http://dx.doi.org/10.1016/j.worlddev.2014.03.009.

Barbier, E. B. (2004). Explaining agricultural land expansion and deforestation in developing countries. *American Journal of Agricultural Economics, 86*(5), 1347–1353.

Buergin, R. (2000, July). *'Hill Tribes' and forests: Minority policies and resource conflicts in Thailand.* Socio-Economics of Forest Use in the Tropics and Subtropics (SEFUT) Working Paper No. 7, Freiburg.

Bugna, S., & Rambaldi, G. (2001). *A review of the protected area system of Thailand.* Asean Biodiversity, Special Report, July–September, 2001

Cropper, M., Griffiths, C., & Mani, M. (1999). Roads, population pressures, and deforestation in Thailand, 1976–1989. *Land Economics, 75*(1), 58–73.

Cropper, M., Puri, J., & Griffiths, C. (2001). Predicting the location of deforestation: The role of roads and protected areas in North Thailand. *Land Economics, 77*(2), 172–186.

Dawson, J., & Barwell, I. (1993). *Roads are not enough: New perspectives on rural transport planning in developing countries.* London: Intermediate Technology Publications.

Delang, C. O. (2002). Deforestation in northern Thailand: The result of Hmong farming practices or Thai development strategies. *Society and Natural Resources, 15*, 483–501. https://www.ncsu.edu/project/amazonia/Delang.pdf.

Feder, G., Onchan, T., Chalamwong, Y., & Hongladarom, C. (1988a). *Land productivity and farm productivity in Thailand.* Baltimore: John Hopkins University Press.

Feder, G., Onchan, T., & Chalamwong, Y. (1988b). Land policies and farm performance in Thailand's forest reserve areas. *Economic Development and Cultural Change, 36*(3), 483–501. Fujita 2003.

Feeny, D. (1988). Agricultural expansion and forest depletion in Thailand, 1900–1975. In J. Richards & R. P. Tucker (Eds.), *World deforestation in the twentieth century.* Durham: Duke University Press.

Fischer, G., & Heilig, G. K. (1997). Population momentum and the demand for land and water resources. *Philosophical Transactions of the Royal Society Series B, 352*, 869–889.

Gine, X. (2004a). *Cultivate or rent out? Land security in rural Thailand.* World Bank Working Paper, Development Economics Research Group, The World Bank, Washington, DC.

Gine, X. (2004b). *Access to capital in rural Thailand: An estimated model of formal vs. informal credit.* Policy Research Working Paper, The World Bank. Godoy et al. (1997).

Godoy, R., O'Neill, K., Groff, S., Kostishack, P., Cubas, A., Demmer, J., McSweeny, K., Overman, J., Wilkie, D., Brokaw, N., & Martinez, M. (1997). Household determinants of deforestation by Amerindians in Honduras. *World Development, 25*(6), 977–987.

Heckman. (1976). The common structure of statistical models of truncation, sample selection, and limited dependent variables and a simple estimator for such models. *Annals of Economic and Social Measurement, 5*, 1976.

Howe, J., & Richards, P. (1984). *Rural roads and poverty alleviation.* London: Intermediate Technology Publications.

Lambin, E. F., Geist, H. J., & Lepers, E. (2003). Dynamics of land-use and land cover change in tropical regions. *Annual Review of Environment and Resources, 28*, 205–241.

Minten, B., & Kyle, S. (1999). The effect of distance and road quality on food collection, marketing margin and traders' wages: Evidence from the former Zaire. *Journal of Development Economics, 60*(2), 467–495.

Nijman, T., & Verbeek, M. (1992). Nonresponse in panel data: The impact on estimates of a life cycle consumption function. *Journal of Applied Econometrics, 7*, 243–257.

Palm, C., Vosti, S., Sanchez, P., & Ericksen, P. J. (Eds.). (2004). *Slash and-burn agriculture, the search for alternatives.* New York: Columbia University Press.

Panayatou, T., & Sungsuwan, S. (1994). An econometric study of the causes of tropical deforestation: The case of Northeast Thailand. In K. Brown & D. Pearce (Eds.), *The causes of tropical deforestation: The economic and statistical analysis of factors giving rise to the loss of the tropical forests*. London: UCL Press.

Panayatou, T. (1991). *Population change and land use in developing countries: The case of Thailand*. Paper prepared for the Workshop on Population Change and Land Use in Developing Countries, organized by the Committee on Population, National Academy of Sciences, Washington, DC, December 1991.

Puri, J. (2002a). *Guatemala: A transportation strategy to alleviate poverty approach and concept paper*. Study prepared for the Guatemala Poverty Assessment, The World Bank, Washington, DC.

Puri, J. (2002b). *Are roads enough? A limited impact analysis of road works in Guatemala*. Study prepared for the Guatemala Poverty Assessment, The World Bank, Washington, DC.

Puri, J. (2006). *Factors affecting agricultural expansion in forest reserves of Thailand: The role of population and roads*. Ph.D dissertation, University of Maryland, Department of Agriculture and Resource Economics.

Thailand: National report on protected areas and development – review of protected areas and development in the lower Mekong Region, ICEM, Indoroopilly, Queensland, November 2003.

United Nations, Food and Agricultural Organization (FAO). (2001). *Forest resources assessment 2000: Main report* (FAO forestry paper 140). Rome: FAO.

United Nations, Food and Agricultural Organization (FAO). (2003). *State of the world's forests 2003*. Rome: FAO.

Walker, A. (2002). *Agricultural transformation and the politics of hydrology in Northern Thailand: A case study of water supply and demand* (Resource Management in Asia-Pacific Working paper no. 40). Canberra: The Australian National University.

Wataru, F. (2003). Dealing with contradictions: Examining national forest reserves in Thailand. *Southeast Asian Studies, 41*(2), 206–238.

Chapter 9
Methodological Approach of the GEF IEO's Climate Change Mitigation Impact Evaluation: Assessing Progress in Market Change for Reduction of CO$_2$ Emissions

Aaron Zazueta and Neeraj Kumar Negi

Abstract This chapter presents the methodological approach adopted in the evaluation of GEF support to market change for climate change mitigation in four emerging markets: China, India, Mexico and Russia. The evaluation was completed in October 2013. This evaluation included 18 completed and fully evaluated GEF mitigation projects covering various sectors with opportunities for renewable energy, energy efficiency and methane emission reduction. A theory of change approach was used to undertake a comparative analysis across projects aiming to tease out changes across diverse markets or markets segments in different countries as a consequence of GEF support. While attention was given to the extent to which projects resulted in actual greenhouse gas (GHG) emission reductions, more emphasis was placed on understanding the extent and forms by which GEF projects contributed to long term market changes resulting in GHG emission reductions and assessing the added value of GEF support in the context of multiple factors affecting market change.

Keywords Impact evaluation • Climate change mitigation • Sustainable development • Mixed methods • Theory of change • Market barriers • Complex systems

A. Zazueta (✉)
Independent Consultant, New York, USA
e-mail: a@zazuetagroup.com

N.K. Negi
Global Environment Facility, Independent Evaluation Office, Washington, DC, USA
e-mail: nnegi1@thegef.org

9.1   Introduction

The Global Environment Facility (GEF) is a partnership for international cooperation to address global environmental issues related to biodiversity, climate change, international waters, land degradation, and chemicals and waste.[1] Since its inception in 1991 GEF has provided more than US 14.5 billion dollars for addressing these concerns, of which at least $ 4 billion has been provided to support activities that directly address climate change mitigation.[2] Within the GEF partnership, The GEF Independent Evaluation Office (GEF IEO) has the central role of ensuring the independent evaluation function.[3]

The OECD DAC 'Glossary of Key Terms in Evaluation and Results Based Management' (OECD 2002) defines impact as *"Positive and negative, primary and secondary long-term effects produced by a development intervention, directly or indirectly, intended or unintended."* The OECD DAC's Principles for Evaluation of Development Assistance' (OECD 1991) defines evaluation as *"an assessment, as systematic and objective as possible, of an on-going or completed project, programme or policy, its design, implementation and results."* Thus, impact evaluations may be understood as systematic and objective assessment of the long-term effects of a development intervention. The impact evaluations undertaken by GEF IEO seek to gauge the long term effects of GEF support, how these were achieved and how GEF's effectiveness in achieving them may be improved. These evaluations have a strong focus on learning.

The GEF IEO undertook *"Climate Change Mitigation Impact Evaluation*[4] to assess impact and learn lessons from GEF supported climate change mitigation projects. This paper discusses the methodological approach adopted for the evaluation, the challenges faced and choices made in developing and implementing the evaluation, which was carried out by the GEF Independent Evaluation Office in four emerging economies: China India, Mexico and Russia.[5] The evaluation was implemented from 2012 to 2013.

---

[1]Instrument for the Establishment of the Restructured Global Environment Facility, March 2015. GEF docs.

[2]Accessed on November 30th 2015. https://www.thegef.org/gef/whatisgef

[3]The GEF Monitoring and Evaluation Policy, 2010. GEF Docs. https://www.thegef.org/gef/sites/thegef.org/files/documents/ME_Policy_2010.pdf

[4]https://www.thegef.org/gef/sites/thegef.org/files/documents/Impact%20-%20Climate%20Change%20Mitigation%20IE.pdf Under Publication.

[5]Within the GEF partnership, The GEF Independent Evaluation Office (GEF IEO) has the central role of ensuring the independent evaluation function. The impact evaluations undertaken by the GEF IEO seek to determine the long term effects of GEF support, how these were achieved and how GEF's effectiveness in achieving them may be improved. These evaluations have a strong focus on learning.

The purpose of the evaluation was to promote accountability and learning about GEF's mitigation programme and across GEF overall. It assesses the extent and ways in which GEF support contributes to market change to reduce CO2 emissions and mitigates climate change, and derives lessons to improve the effectiveness of future GEF support.

The evaluation concluded that GEF projects achieved significant direct GEF emission reduction, although indirect emission reduction – which is difficult to measure – may account for much larger reduction. The evaluation found that of the 18 projects covered, in 17 cases there was broader adoption of promoted technologies, approaches and strategies, beyond the direct scope of the project. It found that the projects that demonstrated high progress towards long term impact were those that had adopted comprehensive approaches to address market barriers and specifically targeted supportive policy frameworks. The evaluation found that the methodologies being used by project teams to measure GHG emissions and to calculate ex-post emissions reduction at project completion were inconsistent and contained uncertainties.

The experience gained through conducting the evaluation made methodological challenges in evaluating GEF support salient. It was challenging to draw conclusions and lessons from a large diversity of projects that GEF finances and the wide range of sectors that it covers. Another challenge is the assessment of GEF contributions to change when multiple actors, factors and conditions affect outcomes. Similarly, inconsistency and inaccuracies in measurement pose difficulties. This paper presents methodology adopted to evaluate the contributions of GEF support to initiatives seeking to reduce climate change emissions.

9.2 Utility as a Guiding Factor to Define What Needs to Be Evaluated

The initial step was to identify the overall topic and the key questions that the evaluation would address. The key criteria were the extent to which the evaluation could provide useful information to inform future GEF support on climate change and the extent to which there were sufficient completed projects to carry out an impact evaluation. The climate change mitigation strategies and programs supported by the GEF were reviewed. Given that most climate change mitigation projects supported by the GEF aim at transforming markets for reducing greenhouse gas emissions, this emphasis became a starting point for developing the evaluation questions. Evaluation questions were developed in consultation with the GEF Secretariat staff especially those from the climate change mitigation program, and GEF Partner Agencies that were responsible for implementing these projects on the ground. This process led to three overall evaluative questions.

These are: (1) What have been the GEF contributions to GHG emission reduction and avoidance? (2) What has been the progress made by GEF supported activities towards transforming markets for climate change mitigation? And (3) What are the impact pathways and factors affecting further progress towards market transformation?

The composition and trends of the GEF mitigation portfolio were analyzed to identify the types of projects that GEF has been supporting and where this support was more concentrated. The next step was to identify a set of projects from which relevant lessons could be derived in addressing climate change mitigation while simultaneously assessing the results of GEF support. Based on consultations with the stakeholders, a decision was made to focus on the major emerging economies based on their respective share within the GEF climate change mitigation portfolio, the potential for climate change mitigation, and their continued importance for future GEF support in this area. Due to budget and time considerations, it was difficult to cover more than four countries. Selection of these countries was based on an iterative process of portfolio analysis and consultations with key stakeholders. Firstly, the GEF climate change mitigation portfolio in all the emerging economies was compared. Based on criteria of overall size of the climate change mitigation portfolio, share in the climate change mitigation portfolio approved before 2002, share in the technology transfer portfolio; and, share in STAR[6] allocation for climate change mitigation, six countries were identified for further consideration. These were Brazil, China, India, Mexico, Russia and South Africa (Table 9.1). Further analysis showed that GEF climate change mitigation portfolios in China, India and Mexico stand out both in terms of total cumulative GEF funding and total GEF funding in projects that were approved before 2002. Among the remainder, GEF IEO had completed a Country Portfolio Evaluation[7] in Brazil when the preparation for the Climate Change Mitigation (CCM) impact evaluation started. Therefore, to avoid evaluation fatigue, Brazil was dropped. In South Africa the GEF climate change mitigation portfolio was relatively small compared to other major emerging economies both in terms of completed projects and projects that were under implementation. Therefore, it too was dropped. Russia, where the portfolio of completed projects was also relatively small, was selected because a sizable amount of investment was under implementation and it also had the third largest allocation for climate change mitigation for GEF-5 (2010–2014) period.

---

[6]System for Transparent Allocation of Resources (STAR) is GEF's performance based allocation framework for the recipient countries.

[7]Country Portfolio Evaluations analyze the totality of GEF support across GEF Agencies, projects, and programs in a given country, with the aim of reviewing the performance and results of GEF-supported activities and assessing how those activities align with country strategies and priorities as well as with GEF's priorities for global environmental benefits. https://www.thegef.org/gef/CPE accessed on March 10th 2016.

Table 9.1 GEF CCM portfolio in the countries considered in this evaluation (in US $ million)

Country	Small grants programme[a]	Enabling activities[b]	Medium-size projects[c]	Full-size projects[d]	All modalities
Brazil	0.0	5.7	0	78.0 (9)	83.8
China	0.0	8.6 (2)	1.8 (2)	502.1 (38)	512.5
India	1.8	3.5 (2)	3.8 (5)	199.4 (20)	208.5
Mexico	0.2	0.3 (1)	1.0 (1)	159.0 (14)	160.5
Russia	0.0	0.0 (0)	2.7 (3)	111.5 (13)	114.2
South Africa	0.2	0.3 (1)	3.8 (5)	27.2 (5)	31.5

Number of projects in *parentheses*, except for Small Grant Programme (SGP). Note: the assessment was conducted in 2012 and it takes into account data up to August 2011. Source: Climate Change Mitigation Impact Evaluation, GEF IEO. The number in the parantheses signify the number of projects
[a]The GEF Small Grants Programme (GEF SGP) is a corporate programme of the GEF. The Programme provides financial and technical support to communities and civil society organizations (CSOs) to address environmental concerns including climate change mitigation through community-based initiatives and actions
[b]Enabling Activities are short duration projects that generally receive up to US $ 1.0 million in GEF grant. These are means of fulfilling essential communication requirements to Conventions, provide a basic level of information to enable policy and strategic decisions to be made, or assisting planning that identifies priority activities within a country
[c]Medium Size Projects (MSPs) are projects with up to US $2 million in GEF funding. Expedited procedures are followed for approval of MSPs so that they can be designed and executed more quickly and efficiently
[d]Full Size Projects (FSPs) are projects that involve GEF funding of more than US $ 2 million. An overwhelming majority of GEF funding is provided through FSPs

9.3 Defining the Scope of the Evaluation

Among the four selected countries the extent of coverage was based on substantive and operational considerations. Completed projects that addressed concerns that were still relevant for GEF and likely to receive funding in future were considered (Table 9.2). In India and Mexico all completed full size projects were included. In Russia two of the three completed projects were included.[8] However, in China where the GEF climate change mitigation portfolio was the largest, only some of the completed full size projects were selected so as to keep the cost of the evaluation manageable. In selecting projects in China, it was ensured that the major targeted

---

[8]The project that was excluded from the coverage through the evaluation was "Removing Barriers to Coal Mine Methane Recovery and Utilization" (GEF ID 1162, UNDP, Russia). The GEF IEO assessed the project to have been completed after "satisfactory" achievement of its expected outcomes. The project was excluded because it pertain to coal bed methane recovery, a line of investment that had been discontinued. Nonetheless, a coal bed methane recovery project was covered in India, where all the completed projects were already being covered as part of a Country Portfolio Evaluation being undertaken concurrently by the GEF IEO. Although the evaluation team could have excluded the Coal Bed Methane project (GEF ID 325, India) as well from the evaluation, it chose to include is because it found that the overall findings of the evaluation were not sensitive to inclusion or exclusion of this project.

Table 9.2 Technologies/Markets addressed by projects covered by the evaluation

Targeted Market	China	India	Mexico	Russia	Total
Renewables/wind	2	1	1	0	4
Renewables/biomass or methane	0	2	1	0	3
Renewables/solar	2	1	1	0	4
Renewables/hydro	0	2	0	0	2
Energy efficiency/all – mixed	0	1	0	1	2
Energy efficiency/industry	1	0	0	0	1
Energy efficiency/lighting	0	0	1	0	1
Energy efficiency/buildings	0	0	0	2	2
Transportation	2	0	1	0	3
Total number of projects	5	6	5	2	18

Some projects addressed more than one technology, so columns may add up to more than the total number of projects. Rows would add up as none of the project covered two or more of the four countries simultaneously
Source: Climate Change Mitigation Impact Evaluation, GEF IEO

markets such as wind energy, solar energy, and transportation, which were also covered in at least one of the other three countries, were represented. While implementation success was a criteria in selection of the projects within China, the eventual outcomes and long term impacts were not considered for selection. Thus, the completed projects covered as part of this evaluation are representative of GEF's support to climate change mitigation in the countries covered by the evaluation.

In all 18 completed climate change mitigation projects were covered. These projects account for more than US $ 180 million in GEF funding and more than US $ 680 million in total financing. GEF requires that the project proponents also seek co-financing from other sources so that GEF funds only the incremental costs of implementing the projects. Total GEF funding for covered projects ranged from US $ 1.0 million to $ 40 million, whereas total financing ranged from US $ 3.0 million to $ 284 million.

The date for start of implementation of the projects covered through the evaluation ranged from 1992 to 2007 and their completion dates ranged from 1997 to 2012. Although the projects were designed to be implemented for a duration of 3–6 years, during implementation several projects needed extension. As a result, the actual duration of project implementation ranged from about 4 years to 12 years. Inclusion of projects that ended at different points in time meant that at the time the evaluation was conducted, different time duration had elapsed post project completion. That this difference is a factor was established when comparison was made between the observed progress to impact at the point of project completion and at the time evaluation was conducted.

9.4 Assessing Impacts of GEF Support

An intervention theory of change is meant to explain how inputs and activities will lead to outputs and impacts and to make explicit the key assumptions about how impacts will be achieved. Many publications discuss the use of theories of change in evaluation (Chen 1990; van den Berg and Todd 2011; Weiss 1972). These approaches are particularly well suited to evaluate specific projects or programs. The climate change evaluation included a wide variety of projects covering diverse technologies and markets that are affected by different factors and conditions including policy instruments, institutions, and interactions among producers, suppliers and consumers. Thus, this specific challenge required an overall framework that allowed systematic comparison among such different interventions. Since its inception, GEF has been supporting generation of global environmental benefits in different focal areas. However, for a long period there was no consistent overall conceptual framework that was applicable across its different focal areas to assess how GEF intends to achieve the global environmental benefits. The GEF IEO has, however, found that having such an explicit framework is important for its impact evaluation work. GEF IEO has prepared a generic framework for the development of theories of change (TOC framework) to facilitate these comparative analysis (GEF IEO 2012, 2013, 2014). This general framework was used as a basis to assess impact of GEF climate change mitigation activities (Fig. 9.1).

The generic TOC framework shows that the GEF support seeks to change behavior and institutions by focusing on three broad realms of intervention: generation and sharing of knowledge and information; development of institutional

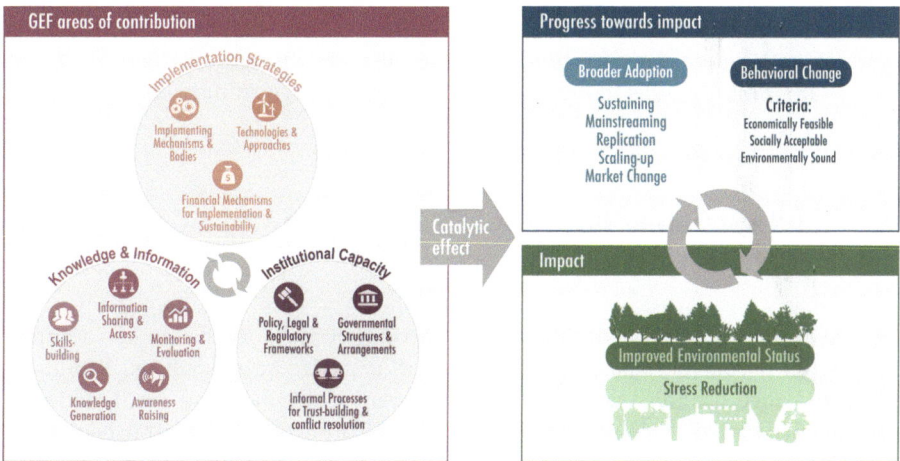

Fig. 9.1 General framework for GEF theory of change (Source: GEF IEO 2014)

capacities; and, testing implementation strategies for behavior change. Through its support to one or more of these realms, GEF support aims to bring about conditions and behavior that if broadly adopted can result in transformations in the long run. The framework identifies the following as pathways for broader adoption: sustainability, mainstreaming, replication, upscaling and market change. Depending on the intervention, one or more of these pathways can be at play. Thus while some carbon emission reduction can take place over the short run, emission reduction at scale is assumed to take place gradually over a longer period of time as behavior changes and systems transform. It is also assumed that the extent and trajectory of change is likely to be affected by multiple factors, some of which may have been addressed by the project, while others may not have been envisioned during project preparation and/or addressed through project design. In time, the spread of tested approaches and behavior that reduce environmental stress (carbon emission reductions) result in changes leading to improved environmental status and human wellbeing. This process is assumed to be unpredictable, non-linearand affected by multiple confounding factors, thus requiring constant attention and adaptation to emerging conditions (Zazueta and Garcia 2014). In case of GEF support for climate change mitigation activities the expected long term impact includes reduction in GHG emissions through the transformation of the structure and function of targeted markets (Fig. 9.1).

The TOC framework was used to develop theory of change for specific project clusters that were covered in the evaluation. The specific theories relevant to the projects were used in this evaluation to develop instruments that would ensure comparability of the information gathered.

Information gathered on the specific contributions of GEF support to conditions (knowledge and information, institutional capacities and effective implementation strategies) to reduce CO_2 emissions as well as expected impact pathways, along with information on the rival hypothesis on observed changes, formed a basis to assess GEF contributions to observed changes in the targeted markets.

9.5 Understanding the System Targeted by the Intervention

The definition of the system which the intervention seeks to change has a strong bearing on the factors that the evaluation will consider in its analysis. While the climate change phenomena take place at different scales including local, national, regional and global, to determine the specific evaluation questions that needed to be asked, in addition to the underlying project's theory of change for the given intervention, the evaluation also focused on understanding the system that the GEF supported activity was trying to influence, including system boundaries, system components, interactions among components and emergent properties characteristic of each system (Holling 2001; Wörlen and Consult n.d.). While acknowledging that processes affecting climate change take place in various interlinked scales, it is important to identify the relevant system boundaries pertinent to the evaluation. For this evaluation the targeted market/technology was the unit of the analysis and the system boundaries were set at the national scale. Attention was given to identifying various components, and to the interactions among these components or segments of the market targeted by the project. Special attention was given to identifying market barriers to change and how these barriers affect the functioning of the system and the systems likely emergent properties. Wörlen (2014) and Wörlen and Consults (n.d.), have analyzed changes in the status of market barriers addressed by climate change mitigation projects, including those supported by the GEF. The evaluation built on their work to assess changes in targeted market barriers and factors contributing to change.

Subsequent steps focused on assessing the extent and the way in which specific elements of the intervention's theory of change interacted with elements of the system (Mayne 2008). The focus of enquiry was on how the intervention became part of the system and the changes (intended or unintended) which this brought about (Garcia and Zazueta 2015). This perspective seeks to emphasize the interconnectedness of the intervention and elements of the system unlike other contextual perspectives that emphasize the effects of context on the intervention (Pawson et al. 2004; Blamey and Mackenzie 2007). This approach was woven into the instruments, which were designed to gather information on the GEF activities and on the links of these activities to support provided by other actors that were relevant to the targeted market. The instruments also took into account the extent the activities undertaken by the other actors were influenced by the GEF support and vice-versa.

The decision to restrict the system boundaries at the national scale was influenced by the fact that flow of information and learning is easier, and policy framework more consistent, within the national boundaries. Similarly, barriers related to suppliers, finance and expertise are more consistent within a country than among countries. GEF projects too are generally geared towards influencing the targeted markets at the national scale (Eberhard and Tokle 2004). This, however, does not mean that the systems at the national scale are insular and may not be affected by factors that have origins in other countries. The evaluation itself documented three instances – ILUMEX (GEF 575), BRT (GEF 1155) and Landfill (GEF 784) in Mexico which had been replicated/scaled up in other Latin-American countries.

9.6   Measurement of Emission Reduction Benefits

The direct and indirect tons of CO_2 emission reductions for each project, were small when compared to global emissions needed to have any effect on climate change mitigation. However, this analysis is important to assess the extent to which GEF supported approaches work and to determine if there is a potential for wider application. To determine the level of GHG emissions, assumptions made by the project proponents on the benefit stream of the technologies promoted by a given project including the estimated duration of the benefit stream were recorded; and, the expected GHG emission reduction – including the changes in the measurements of the underlying indicators for calculating emission reduction – expected at the project start, realized during project implementation period, and revised estimates at the point the evaluation was conducted, were noted. This information, along with information provided in the terminal evaluations of the completed projects and that was gathered through interviews and documents accessed during field verification, formed a basis to prepare revised estimates of the GHG emission benefits.

The evaluation found that although most of the GEF projects covered by the evaluation tracked direct and indirect emission reduction and/or avoidance, in most instances regular monitoring of the emissions related benefits stopped at project completion. Moreover, the information on the indicators specified in the project M&E plan was not being gathered and analyzed regularly. Methodological approaches used by different project proponents to track emission reduction and/or avoidance were often inconsistent. Table 9.3 lists the type of errors that were encountered. To address these errors, the evaluation team recalculated the emission reduction benefits using the available information. Although results for individual projects differed from what had been calculated by the project proponents, the overall figure at the portfolio level were similar.

The evaluation found that of 18 projects, 16 resulted in direct GHG emission reduction. Aggregate direct emission reduction is estimated to be about 6 million tons of CO_2 equivalent per year. However, of the 16 projects that were assessed to have had direct GHG emission reduction impact, for two projects the extent of GHG emission reduction could not be ascertained. Of the 16, for three projects actual GHG emission reduction exceeded expectations at the start of the project. For the remainder actual achievement was lower than the expectations. Among the projects, the China TVE II (GEF 622) alone contributed a third of the direct emission reductions achieved by the 18 projects covered by the evaluation. It was found that the key determinants of the scale of the direct GHG emission reduction achieved included market size, maturity of the promoted technology, and the emission factor for the country, which were positively correlated to the scale of direct emission reduction achieved. Projects that tend to address the prevalent market barriers more comprehensively tended to achieve emission reduction at a higher scale. Overly optimistic projection of the expected benefits – which probably also makes project more attractive during appraisal – was also a reason why several projects had lower than expected direct emission reduction benefits.

Of the 18 projects, 14 led to indirect GHG emission reduction. Of these, in 11 instances quantitative assessment of the indirect GHG emission reduction was possible – for the other three projects, the information required to carry out this analysis was not available. Overall, the indirect emission reduction was assessed to be ten times more than direct reductions.

Table 9.3 Types of Errors encountered in GHG Calculations among projects covered by the evaluation

GHG methodology concern	Type of error	Examples
Installed capacity	Over or under estimation	China RESP (GEF #943): sometimes 28 MW small hydro, sometimes 24 MW small hydro
Capacity factor (power that can be generated from a MW of installed capacity)	Over or under estimation	China RE: assume average capacity factor of solar PV systems of 35–14 % would be more realistic
	Over or under estimation	China REDP (GEF #446) and RESP (GEF #943): assume average of 2,500 h of full load operations of wind systems – 29 % is more realistic
	Over or under estimation	Full load hours within the same project for small hydro power varies from 2,000 to 8,100 full load hours
Operating hours	Calculation errors	Mexico Agriculture (GEF #643): pumps would have to be on average over 70 kW if they are under operation 3,000 h/a
System size	Digits	Mexico Agriculture (GEF #643): Typical irrigation pumps are <10 kW
Emission factors: CO_2 emission reduced per unit of fuel/electricity	Using marginal or Average emission factors	Marginal: can, e.g., be coal with 1,000 g/kWh or gas CHP with 350 g/kWh vs. average can be anywhere lower or higher
	Using outdated emission factors	Emission factor of India and China reduced from 2003 to 2012. The change was not factored in
Benefit period	Inconsistent with methodology or comparison between technologies	India Energy Efficiency (GEF #404): 20 years for all promoted technologies

Source: Climate Change Mitigation Impact Evaluation, GEF IEO

Some of the projects provided relevant actors an opportunity to learn about new technologies and approaches, whereas others were geared towards providing support to the locally nurtured initiatives on climate change mitigation. Of the 14 projects where indirect GHG emission reductions were reported, 9 projects were part of the ongoing process within the country for addressing barriers related to the targeted market/technology. In five instances the GEF supported project supported the first application of the promoted technology in the country. Project design and delineation of project boundary were assessed to be a major factor on whether GHG are counted as direct or indirect result of the project.

9.7   Assessing Market Change

More important than the carbon emission reductions is the extent to which projects contributed to change that in the long term will result in the needed market, technological and behavioral transformations. To bring about these changes, the projects covered through the evaluation addressed barriers related to different sectors and markets. These projects promoted technologies and removal of barriers to markets/technologies on wind energy, biomass energy, methane, hydro power, solar energy, industrial efficiency, efficient lighting, building efficiency, and, transportation. The instruments developed for the evaluation tracked barriers and changes in these markets/technologies in four spheres: Consumers/users; supply chain and infrastructure; financing; and, policy environment. Based on applicability, market barriers considered in the analysis for a sphere included: information gap on promoted technology or approaches, lack of interest or motivation to adopt, lack of relevant expertise, lack of access to relevant mitigation technologies, lack of cost effectiveness, and lack of a viable model. The instrument developed for analysis also captured the intensity with which the given project targeted each of the barriers prevalent in the given market, the specific activities that the GEF implemented to target each barrier, efforts by other actors in addressing the given barrier, and the extent to which the change evident in the status of the barrier could be attributed to the given project. Information on these indicators was gathered through desk review of available information, field verification, interviews and information from independent sources.

Of the 18 projects covered through the evaluation, for 14 projects market change was observed. The observed changes in the targeted markets may be classified into four categories: adoption of higher quality product/technology in the market (8 projects); reduction of production/technology cost (7 projects); availability of more and/or better suppliers (12 projects); and greater demand for promoted product/technology (7 projects). Generally, achievement of improvement in availability and quality of suppliers and improvements in products were linked to each other and often due to a requirement to meet a predetermined quality standard or to achieve a certification.

9.8   Establishing Causality and Accounting for Alternative Hypothesis

Determining the causal variables was more demanding than determining the observed change. The log frames and other logic models that articulate a project's theory of change identify its expected effects. The theories also sketch the expected pathways through which the project outputs and outcomes would lead to its expected long term impacts. While a project's theory of change provides a useful tool to understand its rationale, there are two main limitations in relying on it as a basis for an impact evaluation. Firstly, although the necessary conditions predicted by a project's theory of change may have been met the observed change may have been due to factors that were independent of the project. Second, there may be some uncertainty involved in how and when the intended impacts manifest, particularly in the cases in which important causal relationships may be non-linear in nature. Focusing entirely on the project's theory of change has a risk of overstating GEF's role in effecting the observed change or may lead to neglect of conditions that are impeding future progress. Furthermore, exclusive focus on the causal links proposed by a project's theory of change can function as blinders that constraints an evaluation in recording and assessing the unintended impacts of the project (Garcia and Zazueta 2015). Therefore, in addition to taking into account the given project's theory of change, the evaluation also addressed other factors that may have a causal relationship with the observed change but were independent of the GEF project. The evaluation also searched for alternative explanations that could explain the observed change and assessed their merit in contributing to the observed change vis-à-vis a given GEF project. During the visits to the field the evaluation team gathered information on this issue from different stakeholders such as project implementers, beneficiaries, other agencies that were not involved in project implementation but were familiar with the project, and government officials.

Despite limitations of the theory of change approach, for the most part it remains a useful basis for tracking a given project's provable impact pathways. As the general framework of GEF's theory of change suggests, for any observed change to be attributable to GEF project, the behaviors promoted by the project should be adopted by the targeted actors within a market. This broader adoption in turn creates a basis to assess the progress towards the expected long term transformations. To assess this progress, the evaluation tracked the intensity, the scale, and the processes through which it was taking place.

Of the 18 projects in 17 instances there was evidence that broader adoption was taking place through one or more of the following processes: sustaining project supported activities; mainstreaming; replication; scaling up; and, market change (Table 9.4). For each of these processes, the manner in which it was happening and the extent to which it was linked with the GEF project was determined. In 14 cases the evaluation was able to establish causal links between the project activities and the progress made. This involved linking specific activities or components of the GEF supported projects with the intended observed outcomes based on the information gathered through terminal evaluations and interviews conducted and documents accessed during field verifications. The next stage was to also assess the effects of the other actors and factors that could account for the observed results. Based on the qualitative assessment of the information gathered, in ten cases the evaluation was able to discard rival theories and establish primacy of the GEF supported project in effecting the observed change. For example, in India the technologies and approaches promoted by projects on Photo Voltaic Systems (PVMTI GEF 112) and Hilly Hydel (GEF 386) were scaled up at the national level with significant link established with the underlying project. For four other projects (India Alternative Energy, PVMTI and Energy Efficiency; Mexico

Table 9.4 Progress to impact and causal links to GEF projects

Country	GEF ID	Short name	Sustaining	Mainstreaming	Replication	Scale-up	Market change	Causal link to GEF projects
India	76	Alternate Energy	X	X	X	X	X	Significant
India	112	PVMTI	X	X	X	X	X	Moderate
Russia	292	Boilers				x (?)		Low to negligible
India	325	CBM	X	X	X		X	Significant
India	370	Biomethanation	X	X	X	X	X	Significant
India	386	Hilly Hydel	X	X	X	X	X	Significant
India	404	Energy Efficiency	x	X	X	x	X	Moderate
China	446	CREDP	X	X	X	X	X	High
Mexico	575	Ilumex	X		X	X	X	High
China	622	TVE II	X	X	X	X	X	High
Mexico	643	Agriculture	X	X	X		X	Moderate
Mexico	784	Landfill gas	X	X		X	X	Significant
China	941	FCB I			x			Unable to assess
China	943	CRESP	X	X	X	X	X	High
Mexico	1155	BRT	X	x	X	X	X	High
Russia	1646	Education					X	Moderate
Mexico	1284	Wind	X	x	X			Low/negligible
China	2257	FCB II						Unable to assess

Source: Climate Change Mitigation Impact Evaluation

The symbol (x) within parentheses means broader adoption took place but no causal link or negligible link to the project could be established. The symbol X? Means that change took place but the evaluation did not have sufficient information to determine if there were causal links to GEF support (The ratings were provided on a four point scale using the following criteria: High – Stress reduction occurring at a large-scale (i.e. across the targeted market); Significant – Substantial stress reduction occurring but at a local scale (i.e. in specific or disconnected areas or sub-markets); Moderate – Some stress reduction occurring but at a local scale (i.e. in specific or disconnected areas or sub-markets); and, Low/Negligible – No positive environmental impact observed, or negative impact observed)

Agriculture) although causal links were established for some of the changes, these could not be established for others and rival theories were also difficult to discard.

In the remaining four cases, for two projects no link or very tenuous link could be established between the GEF project and the changes observed. In Russia Boilers project (GEF ID 292) although there was some evidence of scale-up it was not linked to the activities supported by the project. Similarly in the case of the Wind project in Mexico, at project end important regulatory changes had be undertaken by the Mexican government, while the project design included such reforms as an important intended outcome, the evaluation found that other factors accounted for such reforms, and that the contributions to these changes by GEF supported activities were marginal. In the two remaining cases (china FCB I and II) the evaluation did not have enough evidence to assess the causal links of the project with the changes observed.

9.9   Assessing What Would Have Happened If GEF Support Had Not Taken Place

The work presented so far in this paper assesses the extent to which market change took place, whether there is a causal link between GEF support and the changes and whether there are alternative explanations for the observed changes. However, projects take place through partnerships which include governments, other donors and civil society organizations. While linked to GEF support, changes can also be a result of other factors and conditions, some which might not be readily apparent. To assess GEF contribution more fully, understanding GEF role within the change process is also important. Thus, the evaluation also needs to assess the extent to which the given project (or a comparable activity) would have taken place without GEF support. For each of the 18 projects the evaluation carried out an inquiry to assess the extent to which other factors (projects, activities, events) could have bring about or contributed to the observed changes. This was done through interviews with key informants, including people whom had been part of the process and other third parties in the countries, as well as through analysis of publications, gray literature and other relevant reports. The findings are summarized in Table 9.5.

The analysis shows that of the 18 projects, 8 were assessed to be very unlikely or not likely to have taken place without GEF support and 9 projects were very likely or likely to have taken place without GEF support. However, the likelihood that a project would have taken place without GEF support does not mean that the support did not bring additional value.

Of the nine projects that were very likely or likely to have taken place without GEF support, in seven instances the GEF support was assessed to have accelerated the process of the project (or comparable activity) being implemented. In two instances it was assessed that the GEF support to the project allowed its design and implementation to be of a higher quality than would have otherwise been

Table 9.5 Added value of GEF financing

Question	Classification based on assessment
How likely is it that the project (or comparable activity) would have taken place without the GEF support?	**Very unlikely or not likely: 8 projects**
	Very likely or likely: 9 projects;
	Unable to assess: 1 project
For 9 projects that were assessed to be "very likely or likely" to take place without GEF support: • If the project would have taken place any way, what was the added value of GEF financing?	• **Would have taken place more slowly: 6 projects (6/9)** (enhanced speed)
	• **Would have not been implemented as per international standards: 1 project (1/9)** (enhanced quality)
	• **Would have taken place more slowly and would have not been implemented as per the international standards: 1 project (2/9)** (enhanced speed and quality)
	• Added value difficult to determine: 1 project (1/9)

Source: Climate Change Mitigation Impact Evaluation, GEF IEO

possible. Overall it was assessed that GEF support did add value in addressing concerns related to eight of these targeted markets. In one of these nine cases the evaluation was not able to determine if GEF support added value. In summary in 16 of the cases GEF support contributed to the desired change.

9.10 The Critical Role of Indicators in Impact Evaluation

Determining the extent to which expected long term impacts have taken place requires that relevant impact indicators are being tracked. Often there are severe gaps in base line information (Tokle and Uitto 2009), the results of the project may not be monitored consistently – especially after project completion – and/or the methods used to track changes may not be appropriate. The evaluation found that projects were generally tracking too many indicators and often not tracking them well. Even though projections of GHG emission reduction benefits extended over the post project completion period, often projects made no provisions for gathering data after project completion. Furthermore there are inconsistencies in calculating GHG emission reduction benefits and calculations of these varied considerably across projects. A major recommendation of the evaluation was that GEF should improve its methodology of GHG emission reduction calculations, which was accepted by the GEF Council in its November 2013 meeting. As a follow up to the Council decision, GEF Secretariat has developed new (and updated) methodologies for GHG emission reduction calculation.

9.11 Conclusion

The project's theory of change are an important resource to track the extent that projects realized their intended results. Nevertheless there is a need to go beyond a project's theory of change for a fuller understanding of "why" and "how" the observed change took place. While the theory of change is useful, it by itself is not sufficient because other factors may turn out to be more influential in effecting the observed change. Under these conditions dependence on the project's theory of change runs the risk of functioning as a blinder hindering assessment of unexpected factors and making it difficult to get a handle over the unintended consequence and alternative explanations to the results realized. This risk can be mitigated going beyond a mere assessment the extent to which intended results of projects were achieved and the causal links between project activities and results and carry out an analysis of the forms in which the project intervention interacts with other components of the system. In other words to fully assess the contributions of an intervention vis-à-vis other interventions and other factors, the evaluator must assess project interventions as part of the system that is targeted. This requires the evaluator to develop a good understanding of the system that the project has targeted, including the system boundaries, components, interactions among components and the unexpected changed resulting from these interactions.

References

Blamey, A., & Mackenzie, M. (2007). Theories of change and realistic evaluation peas in a pod or apples and oranges? *Evaluation, 13*, 439–455.

Chen, H. T. (1990). *Theory-driven evaluations*. Newbury Park: Sage.

Eberhard, A. A., & Tokle, S. E. (2004). GEF climate change program study. Washington, DC: Global Environment Facility.

Garcia, J. R., & Zazueta, A. (2015). Going beyond mixed methods to mixed approaches: A systems perspective for asking the right questions. *IDS Bulletin, 46*, 30–43.

GEF IEO. (2013). *OPS 5 technical document #12: Progress towards impact*. https://www.thegef.org/gef/sites/thegef.org/files/EO/TD12_Progress%20toward%20Impact.pdf

GEF IEO. (2012, October). *GEF support in the South China Sea and adjacent areas*. https://www.thegef.org/gef/Impact%20Evaluation%3A%20South%20China%20Sea%20and%20Adjacent%20Areas

GEF IEO. (2014). *Fifth overall performance study of the global environment facility* (Final report: At the crossroads for higher impact). https://www.thegef.org/gef/sites/thegef.org/files/documents/OPS5-Final-Report-EN.pdf

Holling, C. S. (2001). Understanding the complexity of economic, ecological, and social systems. *Ecosystems, 4*, 390–405.

Impact Evaluation: GEF International Waters Support to the South China Sea and Adjacent Areas | Global Environment Facility [WWW Document] (n.d.). https://www.thegef.org/gef/Impact%20Evaluation%3A%20South%20China%20Sea%20and%20Adjacent%20Areas. Accessed 14 July, 2015.

Mayne, J. (2008). *Contribution analysis: An approach to exploring cause and effect*. (ILAC Brief Number 16).

OECD. (1991). The *DAC principles for the evaluation of development assistance*. Paris: OECD.

OECD. (2002). *Glossary of key terms in evaluation and results based management*. Paris: OECD.

Pawson, R., Greenhalgh, T., Harvey, G., & Walshe, K. (2004). *Realist synthesis: An introduction*. Manchester: ESRC Research Methods Programme University of Manchester.

Tokle, S., & Uitto, J. (2009). Overview of climate change mitigation evaluation: What do we know. In *Evaluating climate change and development* (World Bank series on development, Vol. 8). New Brunswick: Transaction Publishers.

van den Berg, R. D., & Todd, D. (2011). The full road to impact: The experience of the Global Environment Facility fourth overall performance study. *Journal of Development Effectiveness, 3*(3), 389–413.

Weiss, C. H. (1972). *Evaluation research: Methods of assessing program effectiveness*. Englewood Cliffs: Prentice-H.

Woerlen, C. (2014). Meta-evaluation of climate mitigation evaluations. *Evaluating Environment in International* Development, 87.

Wörlen, C., Consult, A. (n.d.). Guidelines for climate mitigation evaluations.

Zazueta, A. & Garcia, J. R. (2014). Multiple actors and confounding factors: Evaluating impact in complex social-ecological systems. In J.I. Uitto (Ed.), *Evaluating Environment in International Development* (pp. 194–207). Oxon and New York: Routledge.

Chapter 10
Integrating Avoided Emissions in Climate Change Evaluation Policies for LDC: The Case of Passive Solar Houses in Afghanistan

Yann François and Marina Gavaldão

Abstract In many Least Developed Countries, the minimum level for basic services like energy access is not reached. In the cases of long-term investment in carbon intensive technologies, the expansion of basic services is likely to carry with it a significant increase in GHG emissions. This chapter discuss the importance of accounting for these avoided emissions through the case study of the Passive Solar Houses (PSH) in Afghanistan.

In Kabul winters are cold and 48 % of households cannot afford enough fuel to heat their house. To reduce fuels expenses and improve living conditions, the NGO GERES is supporting local artisans to disseminate a PSH model made of a veranda built on the south-facing part of the house to conserve the sun energy captured and stored in the walls. During the 2013–2014 winter, the fuel consumption and indoor temperature of PSH and control houses were monitored to assess the impact of the technology.

The results show an energy saving of 23 % resulting in annual greenhouse gases emission reduction of 0.37 tCO_2e/year as well as an average indoor temperature increase of 1.43 °C to reach 18.22 °C. Then, a regression model was developed to estimate the emissions that would have occurred if the control group had reached the same indoor temperature than the PSH and, in a second scenario, the minimum indoor temperature of 18 °C recommended by the WHO. For both scenario, the avoided emission represent approximately half of the total climate change mitigation impact with 0.40 tCO_2e/year and 0.34 tCO_2e/year respectively.

Keywords Climate change mitigation • Suppressed demand • Avoided emission • Passive solar houses • GERES • Afghanistan

Y. François (✉) • M. Gavaldão
GERES - Group for the Environment, Renewable Energy and Solidarity, Aubagne, France
e-mail: y.francois@geres.eu; m.gavaldao@geres.eu

© The Author(s) 2017
J.I. Uitto et al. (eds.), *Evaluating Climate Change Action for Sustainable Development*, DOI 10.1007/978-3-319-43702-6_10

171

10.1   Introduction

Globally, the International Energy Agency (IEA) estimates that 1.3 billion people still live without access to electricity and that 2.6 billon are still reliant on biomass for cooking.[1] There is a staggering inequality in access to services and in the quality of services between rich and poor societies – the poorer three quarters of the world's population use only 10 % of global energy.[2] Lack of access to minimum levels of basic services is a serious barrier to socio-economic development and progress toward the Sustainable Development Goals (SDG).[3] While imperative, cost-effective expansion of minimum services throughout the developing world is likely to carry with it a significant increase in greenhouse gas (GHG) emissions. In the cases of the development of infrastructures, institutions and cultural practices based on carbon intensive development, the achievement of the SDG could come with a "carbon lock-in" inhibiting a future switch towards low-carbon technologies.[4]

Therefore the integration of the expansion of basic services into the climate change evaluations is crucial. The latent demand for basic services that is "suppressed" due to barriers such as low income, weak infrastructure and inadequate access to technology have to be measured and accounted in the decision making process. Pure mitigation instruments that only focus on reducing emissions and not on avoiding emissions are therefore likely to have minimal impact in the long-term for developing countries and offer no incentives for alternative "cleaner" development pathways to the poorest.

In many Least Developed Countries (LDC), the low level of historic emissions means that there is little CO_2 emissions to reduce, rendering the gains from result-based finance mechanisms like the Clean Development Mechanism (CDM) marginal or negligible. To foster the achievement of the sustainable development objectives of the CDM, the concept of "suppressed demand" has been integrated in some of the methodologies allowing to account for minimum services for the baseline GHG emission levels. In this case, instead of the historical emissions, the baseline scenario would account for the GHG emissions if the minimum level for basic services such as energy access, clean water or sanitation was reached. The baseline emissions may be calculated using the baseline technologies in a case where the barrier to meet the minimum level is the cost of operation of the technology like fuel consumption, or new technologies corresponding to the minimum level of services in contexts where the service is available. The determination

---

[1] WEO, *Would Energy Outlook 2012* (Paris, France: International Energy Agency, n.d.), http://www.iea.org/publications/freepublications/publication/WEO2012_free.pdf.

[2] Summary conclusions of the Vienna Energy Forum, June 2011

[3] United Nations, "Resolution Adopted by the General Assembly on 25 September 2015 – 70/1. Transforming Our World: The 2030 Agenda for Sustainable Development.," 2015, http://www.un.org/ga/search/view_doc.asp?symbol=A/RES/70/1&Lang=E.

[4] Gregory C. Unruh and Javier Carrillo-Hermosilla, "Globalizing Carbon Lock-In," *Energy Policy* 34, no. 10 (July 2006): 1185–97, doi:10.1016/j.enpol.2004.10.013.

of the minimum levels (minimum energy consumption, daily amount of clean water per capita etc.) and the corresponding GHG emissions level can be challenging, this is the case for the housing sector with the indoor temperature.

According to IPCC's latest report,[5] in 2010, buildings accounted for 19 % of the GHG emissions. This same report highlighted the need for scaling up low-energy demand housing systems in LDCs. Low-energy buildings aim to achieve minimum service level without relying on energy-intensive equipment for heating or cooling to avoid a locking in carbon-intensive buildings for several decades. In order to support climate change policies, the evaluation of housing projects in LDCs should integrate this potential carbon-locking and therefore assess the impact of low-energy housing systems using the suppressed demand. But the minimum level for indoor temperature is difficult to estimate as it is highly context specific. Globally, the minimum indoor temperature recommended by the WHO is 18 °C with up to 20–21 °C for more vulnerable groups, such as older people and young children.[6]

Achieving sustainable development in the housing sector of cold regions like Afghanistan requires important improvement of the indoor temperature while mitigating the emissions of the business-as-usual technologies and practices. Therefore the evaluation of the climate change impact of projects in the housing sector of cold regions should also account for the avoided emissions of the intervention compare to the business-as-usual development pathway in addition of the actual emission reduction. This study presents an application of the suppressed demand approach for the housing sector in a difficult context through the case study of the Passive Solar Houses project in Afghanistan. This case study presents the importance of accounting for the avoided emissions when minimum service level are not reached due to incomes barriers but also the methodological challenges of estimating the emissions to reach the same level of service using baseline technologies.

10.2   Approach

All over Afghanistan, winters are severe and access to sufficient fuel is a challenge. Most of the households rely on biomass fuels like wood, sawdust or cow dung or mineral coal for heating. In Kabul, energy expenses represent roughly 20 % of households' annual expenses with 6 % only for heating. These fuel expenses are

---

[5]O. Lucon et al., "Buildings," in *Climate Change 2014*: *Mitigation of Climate Change. Contribution of Working Gourp III to the Fifth Assessment Report of the Intergovernmental Panel on Climate Change* (United Kingdom, New York, USA: Cambridge University Press, 2014), 671–738, https://www.ipcc.ch/pdf/assessment-report/ar5/wg3/ipcc_wg3_ar5_chapter9.pdf.

[6]WHO, "Health Impact of Low Indoor Temperatures" (Copenhagen: Would Health Organization – Regional Office for Europe, 1985), http://www.theclaymoreproject.com/uploads/associate/365/file/Health%20Documents/WHO%20-%20health%20impact%20of%20low%20indoor%20temperatures%20%28WHO,%201985%29.pdf.

particularly important during the winter when incomes are at the lowest and goods prices like food or gas for cooking are at the highest. According to GERES survey, 15 % of households contract debts partly or totally to purchase fuel and 48 % of the household report difficulties to meet their energy needs. Thermal comfort during winter months remains very problematic as the current levels of indoor temperature do not reach the WHO recommended threshold of 18 °C minimum service level.

In order to improve the thermal comfort during the winters while contributing to climate change mitigation, GERES – Group for the Environment, Renewable Energy and Solidarity, French NGO working in Afghanistan since 2002, has developed and transferred to local entrepreneurs the Passive Solar Housing technology. Passive Solar Housing construction design rely on collecting, storing and distributing solar energy during the winter without any mechanical or electrical equipment. The GERES housing innovation is comprised of a veranda with a wooden frame and plastic sheeting added to the south-facing part of the house. The air inside the veranda is heated during the day by the sun's radiation. By keeping an enlarged window open between the veranda and the house, the warm air is transmitted to the room. At night, the window is closed and curtains are drawn in order to keep the heat inside the room. In addition, the veranda also provide an extra warm room during the day for housework and social events for a very affordable cost (Fig. 10.1).

In 2012, GERES started a 3 year project with funding from the *Agence Française de Développement* (AFD) and *Fondation Abbé Pierre* to support local artisans for the wide dissemination of PSH in Kabul. During the winter 2012–2013, a Socio-Economic Assessment of Domestic Energy Practices (SEADEP) survey was conducted to assess the socio-economic and energy consumption profile of households of Kabul. During the following winter, between 2013 and 2014, a monitoring campaign was conducted in Kabul, with the objective of assessing the impacts of PSH technologies on livelihoods and GHG emissions. Two groups of houses each (non-PSH and PSH) were monitored during 8 weeks, indoor and outdoor temperatures were measured using data loggers and fuel consumptions were recorded daily.

Using these data, the impact of the PSH technology in terms of indoor temperature and energy consumption has been assessed to determine the energy efficiency of the PSH compare to the non PSH. Then, using the suppressed demand approach a regression model has been built to assess the GHG emissions that would have occurred if the same indoor temperature was reached using technologies in non PSH. This case study illustrate the importance as well as the limitations of applying the suppressed demand in the housing sector in LDC.

10.2.1 Sampling and Data Collection

The houses selected for the study are located in three police districts in the southern part of Kabul and spread from central part of the city to its outskirts, including semi-

Fig. 10.1 The Passive Solar House (PSH) technology in Kabul

rural areas with agricultural activities. No significant differences appear between the districts that are all characterized by internal heterogeneity: planned and unplanned areas, individual and vertical housing, rich and poor areas. Most residential areas of these districts are occupied by houses built according to the traditional Afghan pattern in mud or cooked bricks, with flat roofs, one to three living rooms, a yard, and the house facing south whenever possible. Therefore, 75 % of the houses in these three districts match GERES' technical requirements for the construction of verandas (South-oriented houses, no direct obstruction and shadow, more than 3 m in front of the house).

The winter monitoring lasted for an overall period of 8 weeks, from 5 December 2013 to 5 February 2014.

Two groups of houses are classified by type:

- Type 1: Control group – Houses not equipped with the veranda PSH system
- Type 2: Treatment group – Houses equipped with veranda PSH system

To assess the impact of the PSH technology, 13 houses of Type 2 equipped with the PSH are compared with 13 houses of Type 1 selected as a control group. The house were strictly selected using the SEADEP database according the number of heated family room (only houses with one family room were selected), its size and orientation, the household socio-economic profile and energy consumption practices criteria. The house construction plans were survey by GERES technician to insure an unbiased comparison between the two groups of houses. Both PSH and non PSH used traditional heating devises ("bukhari") along with wood, coal, sawdust or other fuels like cow dung, shells or cardboard.

The main data collected during the study were the fuel consumption (collected once per day, five times a week) and the indoor and outdoor temperature collected using thermometers with data-loggers.

The thermometers were positioned based on the following criteria:

- Thermometers measuring indoor temperature

  - Room: main family room (heated room)
  - High: 50 cm from the ceiling
  - Opposite to the heating system (at least 3 m from the heating system)
  - Protected from direct sunrays
  - Not close to the windows or doors (at least 1.5 m from windows and doors)
  - Protected to any activity to not be disturbed or damaged

- Thermometers measuring outdoor temperature

  - On outer north face of houses
  - Not easily accessible (Height: at least 2 m)
  - Protected from snow and rain (below roof overhang)
    Only the fuel consumption of the devises situated in the main family room was recorded. Fuel was not provided to households as it can promote over-consumption, household were using their own fuels and a stock was made close to the heating devise that was monitored. The remaining stock of fuel was monitored every day before refilling the stock to insure that the stock was sufficient to support the household energy needs.

10.2.2   Data Analysis

Once the data were collected, the fuel consumption and temperature records were cleaned and treated.

10.2.3   Fuel Consumption and Temperature Data Treatment

The daily fuel consumption monitored in kilogrammes has been transformed in kWh and the consumption of all fuels was summed to get an average energy consumption per week.

Equation 10.1: Calculation of the Weekly Energy Consumption

$$Weekly\,Consumption\;(kWh) = \sum_{d=day}\left(\sum_{n=fuel} Daily\,Consumption(kg)_{n,d} \times NCV_n\right)$$

The outdoor temperature data collected by the thermometers and data loggers situated outside the house has been transformed into Heating Degree Day (HDD) and summed for each week of measurement. The HDD is a measurement designed to reflect the demand for energy needed to heat a building. It is calculated by counting the missing degrees to reach a comfort temperature. The comfort temperature has been determined at 18 °C and the HDD18 (explain abbreviation) is calculated as follow:

Equation 10.2: Calculation of the Weekly Heating Degree Day Value

$$Weekly\,HDD18 = \sum_{d=day} 18 - \overline{Daily\,Temperature_d}$$

Similarly the indoor temperature data collected by the thermometers and data loggers are used to calculate the weekly average indoor temperature.

Based on these data first analysis of the differences in energy consumption and HDD18 are available between PSH and non-PSH.

10.2.4   Greenhouse Gas Calculation

The GHG emissions were calculated from the energy consumption using the Gold Standard GHG calculation methodology "*Technologies and practices to displace decentralized thermal energy consumption*[7]" and IPCC emission factors.[8] For biomass fuels, the calculation of the fraction of non-renewable biomass (fNRB) was based on the Gold Standard-approved methodology using FAO data.

The overall GHG reductions achieved by the project activity are then calculated as follows:

$$ER = \sum BE - \sum PE - \sum LE$$

Where:

ER = Emission reduction (tCO$_2$e/year)
BE = Baseline emissions for the non PSH (tCO$_2$e/year)

---

[7]Available on http://www.goldstandard.org/wp-content/uploads/2011/10/GS_110411_TPDDTEC_Methodology.pdf
[8]IPCC, "Energy," in *2006 IPCC Guidelines for National Greenhouse Gas Inventories*, vol. 2, 5 vols. (Intergovernmental Panel on Climate Change, 2006).

PE = Project emissions for the PSH (tCO$_2$e/year)
LE = Leakage (tCO$_2$e/year)

The baseline and project emissions are calculated as follows:

$$E_y = FC_y \times \left(\left(f_{NRB,y} \times EF_{fuel,CO2} \right) + EF_{fuel,nonCO2} \right) \times NCV_{fuel}$$

Where:

E = Emissions for baseline/project situation in tCO$_2$e
FC = Quantity of fuel consumed for baseline/project situation in tonne
f_{NRB} = Fraction of non-renewable biomass
NCV_{fuel} = Net calorific value of the fuel that is substituted or reduced
$EF_{fuel,CO2}$ = CO$_2$ emission factor of the fuel that is substituted or reduced
$EF_{fuel,nonCO2}$ = Non-CO$_2$ emission factor of the fuel that is substituted or reduced

Then, the GHG avoided emission are calculated using the suppressed demand to assess for the impact of a higher comfort in PSH. This requires to build a model linking the non PSH indoor temperature to the level of greenhouse gas emissions and the outdoor HDD18.

To account for the different emission factors of the different used, an Ordinary Least Square regression is developed to link the GHG emission to the indoor temperature for the same outdoor temperature. This model is finally used to estimate the extra GHG emission that would have occurred in non PSH to reach the same indoor temperature level than the PSH as well as to reach the WHO recommended minimum indoor temperature of 18 °C.

10.3   Results

10.3.1   Energy Efficiency

10.3.1.1   Heating Degrees Day Required to Be at 18 °C (Outside Temperature)

In order to compare the average indoor temperature or energy consumption between PSH and non PSH it is necessary to validate that the test conditions are similar, i.e. that there is no significant difference of the cumulative Heating Degree Day necessary to obtain a weekly indoor temperature of 18 °C between PSH and non PSH (Table 10.1).

Student's t-test shows that there is no significant difference between the Heating Degree Day required to be at 18 °C for PSH and non PSH which indicates that the energy requirement to reach the minimum level of service are the same for the PSH and non PSH groups.

Table 10.1 Heating Degrees Day for PSH and non PSH

	Non-PSH	PSH
Mean (°C)	109.69 °C	109.19 °C
Standard deviation	15.36	14.95
Coefficient of variation	0.14	0.14
Standard error	2.13	2.53
Uncertainty of the mean *(90 % confidence interval)*	0.03	0.04
Conclusion	The level of precision (±10 %) for a 90 % confidence interval is met	

10.3.1.2 Energy Savings

For the two groups, firewood was the main fuel consumed, representing approximately the two-thirds of the energy consumption for heating purpose. The remaining energy consumed was a mix of coal, sawdust, cow dung, husk and cardboard. This fuel mix is representative of the energy consumption patterns of Kabul according to the SEADEP results (Fig. 10.2).

The results of the analysis of the energy consumption of the monitored room over the 8 weeks monitoring period are presented in the Table 10.2.

The PSH energy consumption was 23 % lower than non-PSH houses, with a net energy consumption decrease of 60 kWh per week in average (Fig. 10.3). Extrapolated to 1 year considering 110 heating days that represents a saving of 938 kWh per year for each house equipped with PSH.

10.3.1.3 Indoor Temperature

The analysis of the temperature recorded in the main heated room, attached to the veranda for the PSH, during the monitoring period shows the following results (Table 10.3).

Over the monitoring period, the PSH reached an average weekly indoor temperature of 18.22 °C. The PSH average temperature is 8 % higher than non-PSH houses, with a net increase of 1.43 °C of the weekly indoor average temperature. The variation of the PSH indoor temperature was also much lower than the non PSH (Fig. 10.4).

This difference show that, in addition of reducing fuel consumption, the PSH group reached the WHO recommended minimum indoor temperature of 18 °C.

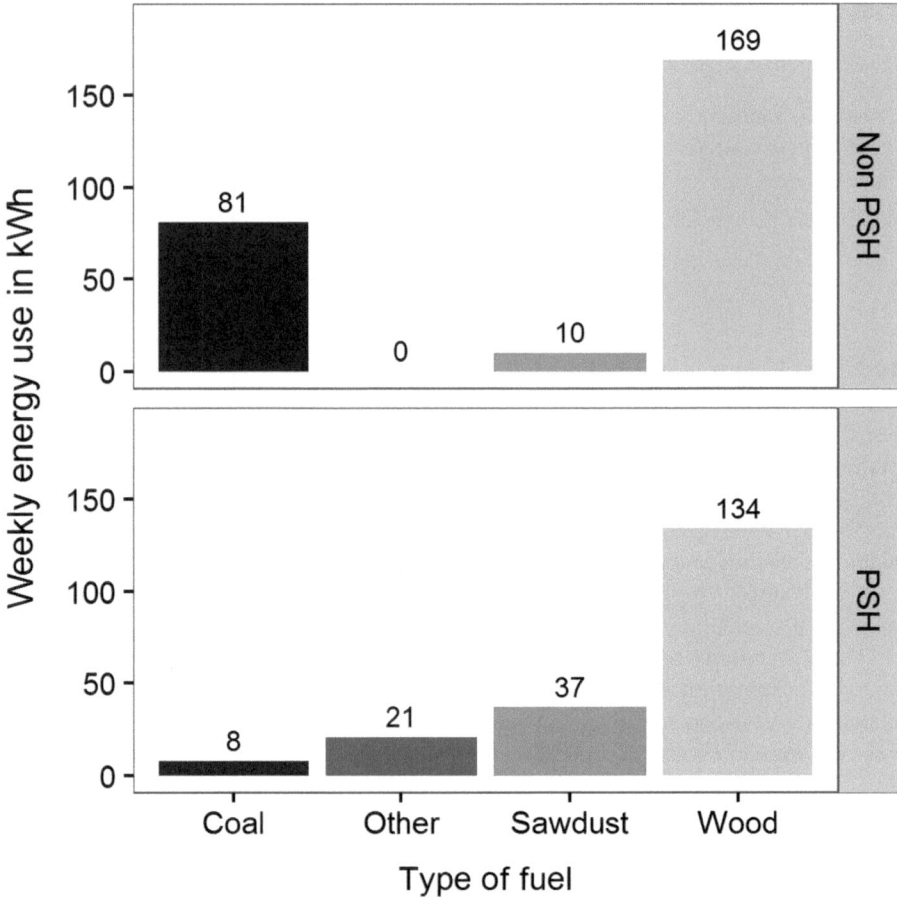

Fig. 10.2 Fuel mix of the PSH and non PSH groups

Table 10.2 Energy consumption of PSH and non PSH

	Non-PSH	PSH
Mean (kWh/week)	259.76	200.05
Standard deviation	120.73	56.47
Coefficient of variation	0.46	0.27
Standard error	16.74	9.55
Uncertainty of the mean *(90 % confidence interval)*	0.11	0.10
Conclusion	The level of precision ($\pm 10\,\%$) for a 90 % confidence interval is met for the PSH group	

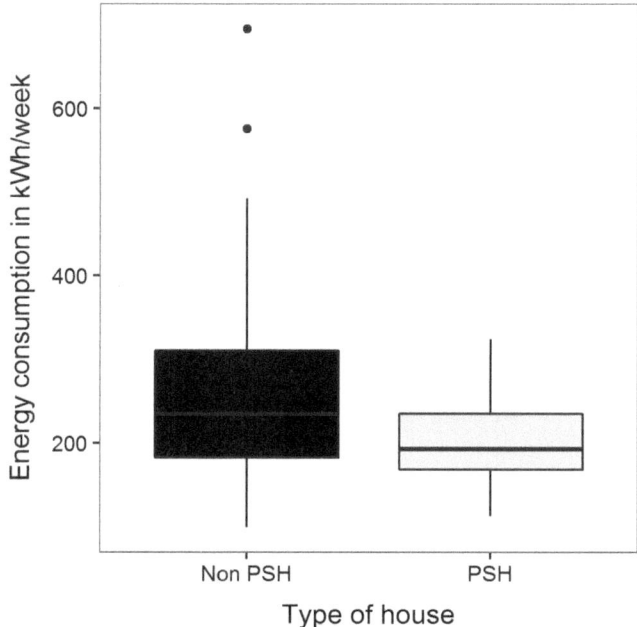

Fig. 10.3 Comparison of the energy consumption between PSH and non-PSH houses

Table 10.3 Indoor temperature of PSH and non PSH

	Non-PSH	PSH
Mean	16.78 °C	18.22 °C
Standard deviation	3.54	1.38
Coefficient of variation	0.21	0.08
Standard error	0.49	0.23
Uncertainty of the mean *(90 % confidence interval)*	0.05	0.02
Conclusion	The level of precision (±10 %) for a 90 % confidence interval is met	

10.3.2 Greenhouse Gas Emission Reduction and Avoided Emissions

The emission reduction are calculated based on the recorded actual energy consumption for both the non PSH and PSH. The results extrapolated for the whole winter season considering 110 days of heating are presented in the following table (Table 10.4).

These results show an emission reduction potential from the PSH technology of 0.366 tCO_2e/year. However this does not take into account the higher indoor temperature in the PSH compare to non PSH and the emissions avoided using the PSH technology.

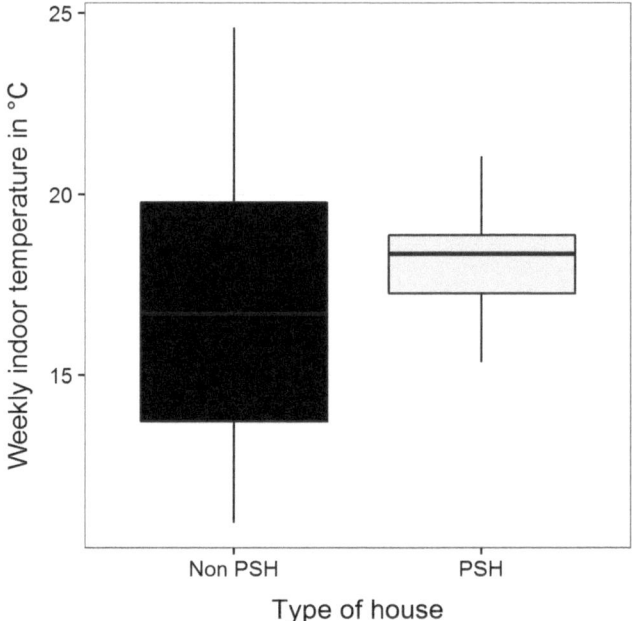

Fig. 10.4 Comparison of the weekly indoor temperature between PSH and non-PSH

Table 10.4 GHG emission per type of house and sampling in tCO_2e

Non PSH	PSH	Emission reduction
1.414	1.049	0.366

To assess the avoided emissions of the PSH it is necessary to assess what would have been the additional GHG emissions to reach the same indoor temperature in non PSH. For that, an OLS regression model is developed to assess the relation between the indoor temperature and the GHG emission considering a similar outdoor temperature between non PSH and PSH. For the purpose of this study a simplified model is developed linking the GHG emission and the outdoor temperature to the indoor temperature (Fig. 10.5).

Based on the model developed the avoided emission from the PSH technology if the non PSH household were reaching the same indoor temperature as well as the total climate change mitigation impact are calculated for the whole winter season and presented in the Table 10.5.

The difference between the emission reduction and the avoided emission is significant. This is explained by the fact that PSH indoor temperature was significantly higher than non PSH despite the lower energy consumption. In that case, accounting to avoided emissions led to an impact 110 % higher than accounting for actual emission reduction only.

Another option for calculating the avoided emission is to take the WHO recommended minimum indoor temperature as the minimum service level instead

Regression formula				
Indoor average temperature = f(GHG emissions + HDD Outside)				
Coefficients				
Parameter	Estimate	Standard Error	t value	p value
(Intercept)	21.66	2.34	9.27	2.36e-12
GHG	0.06	0.007	7.735	4.87e-10
HDDOut	-0.09	0.02	-4.499	4.20e-05

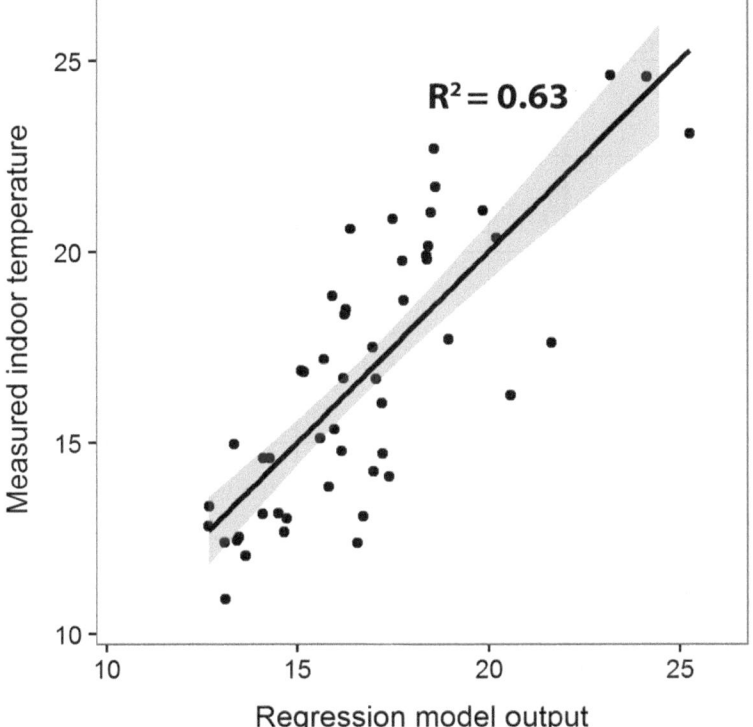

Fig. 10.5 Results of the suppressed demand OLS regression model

Table 10.5 Emission reduction and avoided emission to reach the PSH indoor temperature

	tCO_2e emissions
Emission reduction	0.366
Avoided emission	0.402
Total impact	0.768

of the PSH indoor temperature. In this case, considering a minimum service level of 18 °C instead of the PSH indoor temperature, the avoided emission are slightly lower with 0.344 tCO_2e, but still significant with an increase of the emission reduction by 94 % as describe in Table 10.6.

Table 10.6 Emission reduction and avoided emission to reach the WHO recommended minimum indoor temperature

	tCO$_2$e emissions
Emission reduction	0.366
Avoided emission	0.344
Total impact	0.709

Overall, using both minimum service level, the avoided emissions represent approximately half of the total climate change mitigation impact of the PSH technology.

10.4 Implication for Policy Makers and Development Practitioners

This study shows that accounting for the suppressed demand can have a significant impact on the climate change mitigation potential of a project with strong social improvement components. With a fuel savings of 23 %, the PSH technology contributes to reduce the households vulnerability during winter period when the source of income is the most irregular and the expenses are at the highest level. Equally important, the veranda increases significantly the indoor temperature in the main family room during the cold winter bringing health and confort benefits. To assess the full benefits of this technology it is therefore necessary to account for avoided emissions in addition of emission reductions.

The avoided emissions from increase in indoor temperature with the PSH represented approximately the same amount than actual emission reductions from reduced fuel consumption. This could indicate that the household balance the reduction in fuel expenses when their financial vulnerability is the highest and the increase in indoor temperature during the coldest period of the year. The barrier to reach the minimum service level being financial, it reinforces the relevance of applying a suppressed demand approach to account for household preferences and therefore comprehensively assess the climate change impact of a project.

In this case study, two minimum service levels have been considered: the indoor temperature reached using the PSH technology, as well as the WHO recommended minimum indoor temperature. With an average weekly indoor temperature of 18.22 °C, considering the PSH indoor temperature or the WHO recommended value of 18 °C as minimum service level leads to marginal differences in avoided emissions. However in other contexts, differences between recommended values and real achievement might be much higher. In such cases the level to be accounted for the suppressed demand should be, whenever possible, the level achieved using the technology or practices introduced, as long as it doesn't exceed the agreed minimum service level.

Linking differences in services levels to greenhouse gases emission may represent the major challenge to include the suppressed demand in climate change mitigation evaluations. This evaluation used a simple linear regression model to assess the suppressed demand. With an important work on the sampling, the sources of variation have been minimized which allowed to achieve relatively good coefficient of determination considering the simplicity of the model. But this model suffer from strong limitations and very low external validity. In the future, considering the important limits of the OLS to model the indoor temperature, the development of context specific dynamic thermal models appears as a relevant option to account for the suppressed demand. However, housing conditions are heterogenious in characteristics and therefore the model should be adaptable to a wide number of houses using parameters that could be easily collected. Alternatively, default factors of energy consumption could be developed to assess the baseline emission levels for different levels of indoor temperature.

Applying the suppressed demand also requires important data collection and significant equipment in the case of temperature monitoring. In the PSH case study, the potential factors of variation between the houses are very important which require a very carefull sampling. This study rely on an important household survey as well as highly trained surveyors to assess the houses characteristics and select the samples. In difficult contexts like Afghanistan undertaking long house intrusive surveys comes with numerous challenges in terms of social acceptation and cost in comparison of the budget for project implementation. To insure a high quality of analysis, fuel consumption measurements had to be done 5 days a week during 8 weeks. In the Afghan context, the fact that a man cannot enter in a house when a woman is alone and the security context of Kabul have strongly affected the study. This led to numerous visits to the same household to gather data and in some cases can lead to withdrawing some houses from the study.

Despite a limited immediate climate change mitigation potential, the investment in energy efficient housing is crutial to achieve a low-carbon pathway and avoid a critical carbon locking considering the time frame of housing investment. Future research should focus on the development of suppressed demand models that could be adaptable to different contexts. Emission default factors accounting for both emission reductions and avoided emissions could help decision-making on climate change mitigation policies by highlighting the significance of the emissions that would result from long-term investments in carbon-intensive technologies. In LDCs, successful climate change mitigation action requires to anticipate the socio-economic development that will lead to investment in carbon-intensive infrastructures or practices. The generalization of the suppressed demand in the project screening procedures for climate financing and project evaluation methodologies can contribute significantly to meet both climate change mitigation objectives and Sustainable Development Goals.

Chapter 11
Sustainable Development, Climate Change, and Renewable Energy in Rural Central America

Debora Ley

Abstract Decentralized renewable energy (DRE) projects have the potential to contribute to climate change mitigation, climate change adaptation, and sustainable development objectives. DRE systems are considered for emissions reduction or poverty alleviation purposes while their role for climate change adaptation has hardly been analysed. In terms of adaptation, DRE provides electricity that can be used both to prepare for and recover from disasters, and to provide additional income and livelihood opportunities, thus reducing dependency on natural resources. For example, DRE can power early warning systems, telecommunication systems, health clinics and potable water systems. Although it might be said that climate change adaptation applications of DRE systems have already been implemented, the vulnerability of these systems towards climate impacts, and the robustness of these systems to climatic impacts are oftentimes not even considered.

The assessment of 15 community-owned renewable energy projects in Guatemala and Nicaragua show that, under certain conditions, renewable energy projects can simultaneously meet the triple objective of sustainable development and climate change mitigation and adaptation. Research also points to specific drivers which can facilitate or hinder projects meeting their own stated objectives and, consequently, the triple objective, and their long-term functioning. These drivers include the specific background of the beneficiary community, the financing and implementing entities and the local governance structures in place.

Keywords Renewable energy • Climate change mitigation • Climate change adaptation • Central America • Sustainable development • Rural electrification • Evaluation

D. Ley (✉)
Environmental Change Institute, School of Geography and the Environment, University of Oxford, Oxford, UK
e-mail: debbieannley@yahoo.com

© The Author(s) 2017   187
J.I. Uitto et al. (eds.), *Evaluating Climate Change Action for Sustainable Development*, DOI 10.1007/978-3-319-43702-6_11

11.1   Introduction

Renewable energy technologies can provide energy to rural populations to which it is technically or economically infeasible to extend the electricity grid. Electricity can be used for applications ranging from lighting to a wide array of productive uses to energy services supporting health, education, and sanitation. Current research has mainly focused on the impacts and case studies of DRE on poverty alleviation and sustainable development.

Climate change adaptation is necessary due to the adverse impacts of increasingly frequent extreme weather events. The poorest and most vulnerable populations within developing countries suffer the worst effects of extreme weather events, especially populations in which natural resource bases are fundamental for their livelihoods (Adger et al. 2003; Thomas and Twyman 2005). The United Nations Development Programme (UNDP) 2007/2008 Human Development Report (HDR) first emphasized the importance of adaptation integrated with development since 'adaptation is about development for all' (UNDP 2007). Therefore, failure to address adaptation will deter developing countries from growing economically and alleviating poverty (UNDP 2007).

Adaptation literature has focused on specific topics that include crop diversification (Bradshaw et al. 2004; Naylor et al. 2007), insurance (Crichton 2007, Linnerooth-Bayer and Mechler 2006; Mills 2007; Moser et al. 2007; Romilly 2007; Johnson et al. 2007), the ski industry, and flood risk management (Johnson et al. 2007; Tol et al. 2003). However, there is scant literature on the use of renewable energy to increase adaptive capacity. Eriksen and O'Brien (2007) and Venema and Rehman (2007) hypothesize DRE may be one strategy to meet the triple objective, although they don't provide in-depth details on how this will happen.

The role of renewable energy systems to meet climate change mitigation goals has been well documented (CEPAL 2007a,b). Market-based policy instruments have been created to mitigate climate change without sustainable development objectives always being met. For example, ocal, small-scale renewable energy projects, which have a larger development component, haven't been main participants within CDM project portfolios, while they have figured more prominently under Voluntary Carbon Offset (VCO) initiatives. As such, VCO projects include have a greater focus on development objectives than the CDM. Even though rural development projects have been included within the CDM, there is a need to create a clear set of guidelines to effectively incorporate sustainable development objectives into the projects.

The use of DRE is the only cost-effective and environmentally sound option to provide access to electricity to many rural populations. Only recently has energy access been viewed as a necessary, though not sufficient, enabler for development, including the achievement of the Millennium Development Goals (MDGs) and now of the Sustainable Development Goals (SDGs). The lack of basic infrastructure, including energy, has prevented some countries from achieving the MDG's in rural areas, while meeting them in the urban sector. As in the case of the MDG's, energy serves as an enabler for the achievement of other goals under the SDG's. Additionally, SDG 7 addresses the energy sector specifically by 'ensuring access to affordable, reliable, sustainable and modern energy for all'. The linkage with climate change comes in SDG 13, which calls to 'take urgent action to combat climate change and its impacts'.

My research examines the relationships between sustainable development, climate change and renewable energy in rural Central America. The main research question I answer is '**Can rural renewable energy projects simultaneously meet the multiple goals of sustainable development, climate change mitigation and climate change adaptation? If so, under what conditions**?' and I use three guiding questions:

1. How well are RE projects meeting their goals of sustainable development, climate change mitigation and climate change adaptation?
2. What are the relative roles of local historical background and physical characteristics, type of community governance, and funding source and project implementation process in the success of projects in meeting adaptation, mitigation and development goals?
3. What are the challenges in integrating development and climate change adaptation policies in rural Central America? How might the evolving international climate regime contribute to this integration?

I also look at how the climate change mitigation, climate change adaptation and sustainable development *mainstreaming* and *integration* can take place. For this research, and as defined by Sperling, **mainstreaming** indicates that climate issues are being used for planning and budgeting decision making while **integration** is used when specific adaptation measures are added to design and implementation strategies (Sperling 2003). That is, mainstreaming includes climate change considerations, that go beyond adaptation, from the outset during project planning.

11.2   Approach

I used the political ecology approach to assess the importance of, and relationships between, political economy, social and community structures, local historical backgrounds and the use of natural resources. The approach provides a useful framework for evaluating rural renewable energy projects, focusing on institutions (such as common property resources), markets, local response to development interventions and to the material effects of development on the physical environment (for example, water, soil, and carbon).

Political ecology studies of Latin America are mostly related to the relationship between poverty and environmental degradation: poverty and conservation efforts in protected areas, development, land degradation, wildlife and livelihoods, land use change, land use and food security, shrimp mariculture and fisheries, and irrigation and water resources. The energy sector has also been an area of study for political ecology and political economy, including the use of wood fuel, the wind turbine industry and U.S. energy policy; however, other RE systems haven't been analysed. The existence and type of local governance structures, the level of poverty, and population displacement due to civil wars are among the considerations important to the 'surrounding causes, experiences, and management of environmental problems' (Blaikie and Brookfield 1987) that will contribute to the debate around mainstreaming development with climate change mitigation and adaptation.

Key ways that political ecology influences research design are through the attention to material carbon reductions, climate impacts, renewable energy in the structures of markets and policies and their and responses to changes.

Common Property Resources (CPR) were analyzed as an institution under political ecology, since all the development projects evaluated were community owned. I also used the Pressure and Release (PAR) model for the analysis of renewable energy systems meeting climate change adaptation goals.

Research on CPR has covered topics surrounding natural resources and their uses, including aquaculture, trade, forestry, neoliberalism, ecotourism and coastal livelihoods. Energy use, including renewable energy, has also been studied through a CPR approach, mainly focusing on the optimal use of finite sources.

Ostrom designed principles to determine the failure or success of CPR. As part of the research design, I analysed whether the 'design principles for common property resources' identified by Ostrom (2002) also apply to community-owned renewable energy systems (Table 11.1).

CPR appears as a major set of institutions for managing resources. However, agency (actions of individuals) does influence CPR's when the CPR rules are changed by the people/community. Political ecology has had very few studies of renewable energy in relation to climate governance, local communities and the actions of individuals (agency). Figure 11.1 shows the relationship between CPR and PE.

Based on Political Ecology and CPR, I would expect that the success of projects would be explained by:

1. Political and economic structures that secure property rights; access to resources; equitable benefits; communal ownership and local management of the renewable energy system; taking into account the role and impact of local institutions and the influence of government and foreign and international donor agencies.
2. The agency of individuals in a community and project managers who seek the success of a project and work towards it.
3. Constraints and opportunities afforded by the physical environment, historical background, and cultural and religious diversity.
4. Relationship with Ostrom's rules for successful CPR management, and defined rules, sanctions and incentives.

The evaluation for potential for adaptive capacity and adaptation to climate hazards was carried out using the Pressure And Release (PAR) model. The PAR

Table 11.1 Ostrom's common property resources design principles

1. Resource system characteristics
a. Well-defined boundaries
2. Group characteristics
a. Well-defined boundaries
(1 and 2) Relationship between resource system characteristics and group characteristics
3. Institutional arrangements
a. Locally devised access and management rules
b. Ease in enforcement of rules
c. Graduated sanctions
d. Availability of low-cost adjudication
e. Accountability of monitors and other officials to users
(1 and 3) Relationship between resource system and institutional arrangements
a. Match restrictions on harvests to regeneration of resources
4. External environment
a. Technology
b. State
i. Central governments should not undermine local authority
ii. Nested levels of appropriation, provision, enforcement, governance

Ostrom (2002)

Fig. 11.1 Relationship between the theories of common property resources and political ecology

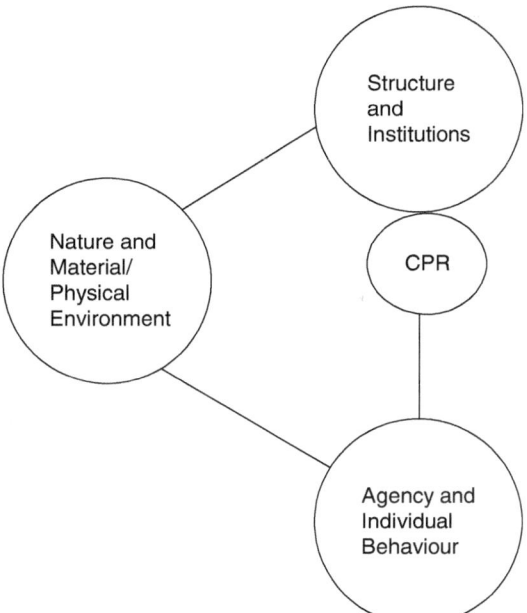

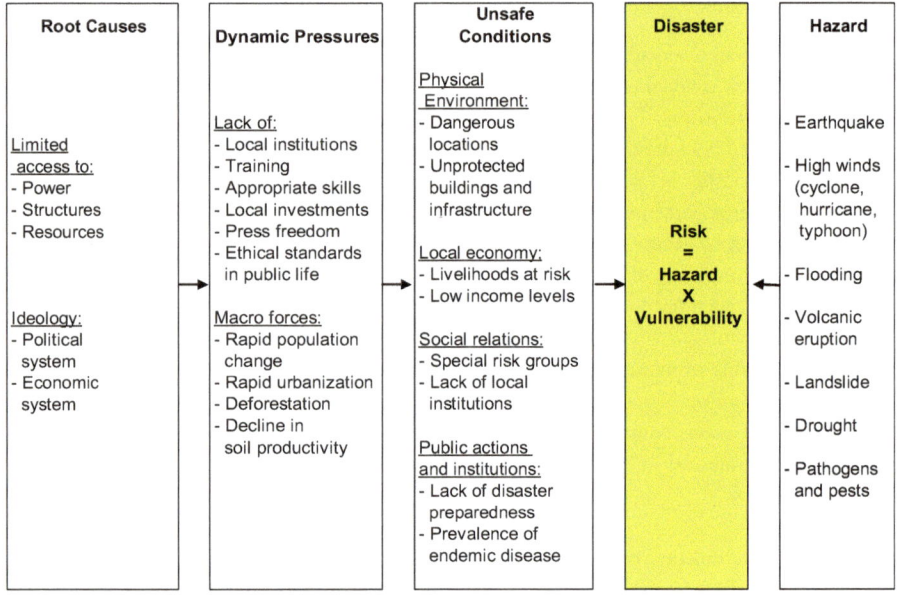

Fig. 11.2 Pressure and release model (Wisner and Blaikie 2004)

model, and for this research using a political ecology lens, examines the relationships between political and economic structures, the physical environment, and communities, to understand 'processes that generate vulnerability' (Wisner and Blaikie 2004) and explain differences in exposure, impacts and ability to cope with previous or future hazards (Eakin and Luers 2006).

The PAR Model explains disasters as the 'intersection of the natural hazard and the processes that generate vulnerability' (Wisner and Blaikie 2004; Blaikie and Brookfield 1987; Birkmann 2006). These processes, explained in part by political ecology, are categorized as root causes, dynamic pressures and unsafe conditions, as shown in Fig. 11.2, and are based on physical, political, economic and social environments and variables.

For this research, the analysis included the role of DRE systems in improving the dynamic pressures and unsafe conditions that decrease vulnerability as well as how DRE systems can be more robust in order to decrease t impact of the hazards on them, and reducing the overall risk of the disaster.

11.3 Methodology

I assessed 15 community-owned renewable energy projects in Guatemala and Nicaragua, which were selected based on general and project type specific criteria.

General criteria included projects:

1. Small-scale (less than 5 MW)
2. Renewable energy (solar photovoltaic, wind energy, run of the river hydroelectric, biogas)
3. Located in a rural community

Following are the criteria for development project selection:

1. Productive-use (income-creating or enhancing) application
2. Implemented for at least 2 years and still working
3. Community owned

The criteria for climate change mitigation project selection follow:

1. A Clean Development Mechanism (CDM), a Voluntary Carbon Offset (VCO) or an Early Warning System (EWS) project

Disaster Relief projects were chosen following these criteria:

1. Developed as part of a relief or reconstruction program.

And adaptation related criteria?

I added two projects because their governance structures provided useful answers to the research questions although they did not fit the criteria of being community owned and of a productive-use application. These two separate projects consisted of individual home lighting solar photovoltaic systems; one of them a loan program implemented by a government Ministry in communities which would benefit from the national electric grid extension in the short to medium terms and one implemented by a national NGO in isolated communities that would never benefit from grid extension. Table 11.2 categorizes the case studies by country, type and renewable energy resource.

Figures 11.3 and 11.4 below show the geographical distribution of the projects. As mentioned above, in the cases where I evaluated programs, the star indicates where the cluster of projects is located.

The projects were evaluated on economic, developmental and climate change indicators, which included indicators focusing on sustainable development, poverty alleviation, emissions reductions, and climate vulnerability. I examined how the type of common property governance, local historical and environmental background and project implementation process influenced the project success in meeting multiple objectives of climate adaptation, mitigation and development. Data collection methods included participatory poverty assessment techniques, semi-structured interviews, stakeholder analysis, and a combination of rapid and participatory methods. The analysis of sustainable development and vulnerability used Sustainable Livelihoods Approach methodologies and emissions reductions were calculated using carbon reduction methodologies of the IPCC.

Figures 11.5 and 11.6 portray the logical flowcharts from which the indicators for this research were derived for each of the two main research questions. Tables 11.3 and 11.4 list the specific indicators used.

Table 11.2 Case study projects

Country	Type	Renewable energy source	Name	Capacity
Guatemala	Development	Hydroelectric	Nueva Alianza	16 kW
		Biodiesel		48 gal/48 h
		Biogas		N/A
Guatemala	CDM	Hydroelectric	San Isidro	3.92 MW
Guatemala	VCO	Hydroelectric	Chel	165 kW
Guatemala	Disaster relief	PV	Cahabón Post-Mitch reconstruction[a]	40 W
Guatemala	Development	PV	Chapín Abajo women's coop	60 W
Guatemala	Development	PV	Cancuén Archaeo-logical site	105 W (in 3 different locations)
Guatemala	Early-warning systems	PV	Early warning systems[a]	35 W
Guatemala	Development	PV	Ministry of energy and mines loan[a]	45 W5
Guatemala	Development	PV	ADIM Quiché[a]	12–65 W
Nicaragua	Development	PV battery charging station	Francia Sirpi and Awastingni	2.4 kW in 3 arrays
Nicaragua	Development	PV water pumping	El Trapiche	600 W
Nicaragua	CDM	Hydro	El Bote	930 kW
Nicaragua	Development/in process of CDM	Hydro	Río Bravo	180 kW
Nicaragua	Development	PV	Solar women of Totogalpa	95 W
Nicaragua	Early warning systems	PV	Early warning systems[a]	35 W

[a]Although these are referred to as projects, these constituted programs with installations in diverse communities

Different methodologies were applied to each one of the sub-research questions as explained:

1. *How well are projects meeting their goals of sustainable development, climate change mitigation and climate change adaptation?*
   Based on development literature, the main variables that are used to measure sustainable development include economic feasibility, social acceptance and environmental responsibility (Najam et al. 2003; Olsen 2007; Swart et al. 2003). The inspection protocol for the photovoltaic systems included the following:

   1. System status and history:
      (a) Previous technical inspections
      (b) Previous and current failures
      (c) Equipment replaced

Fig. 11.3 Location of systems in Guatemala (Source: CIA World Factbook)

2. Photovoltaic array

   (a) Array technical specifications
   (b) Mounting structure, orientation, inclination
   (c) Damaged, shaded, dirty modules
   (d) Status of cables, connectors, grounding system and lightning and surge
       protection

Fig. 11.4 Location of systems in Nicaragua (Source: CIA World Factbook)

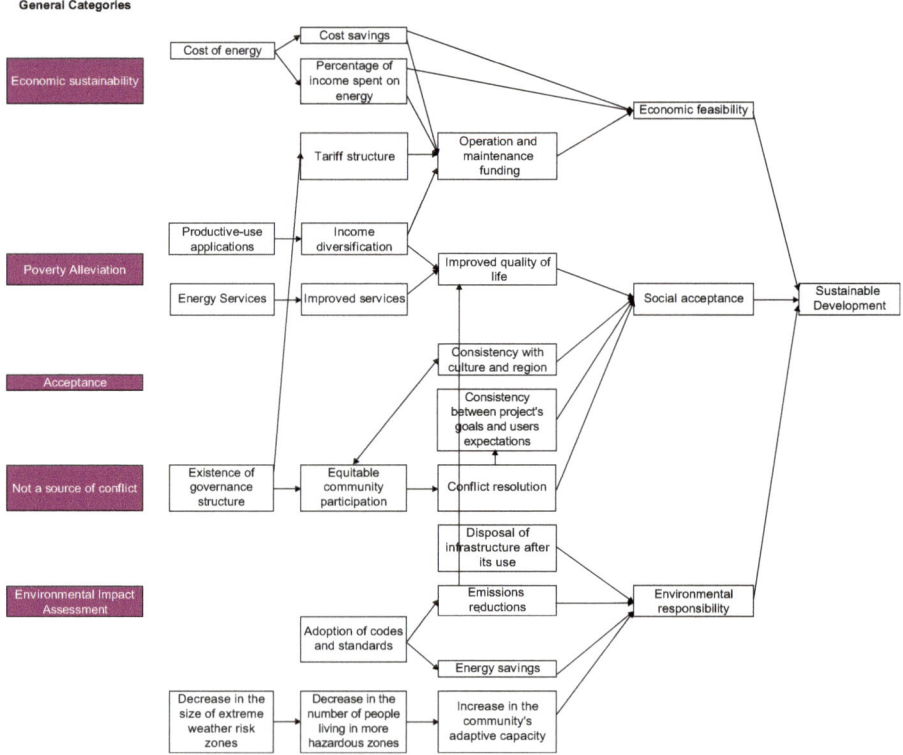

Fig. 11.5 Sustainable development indicators

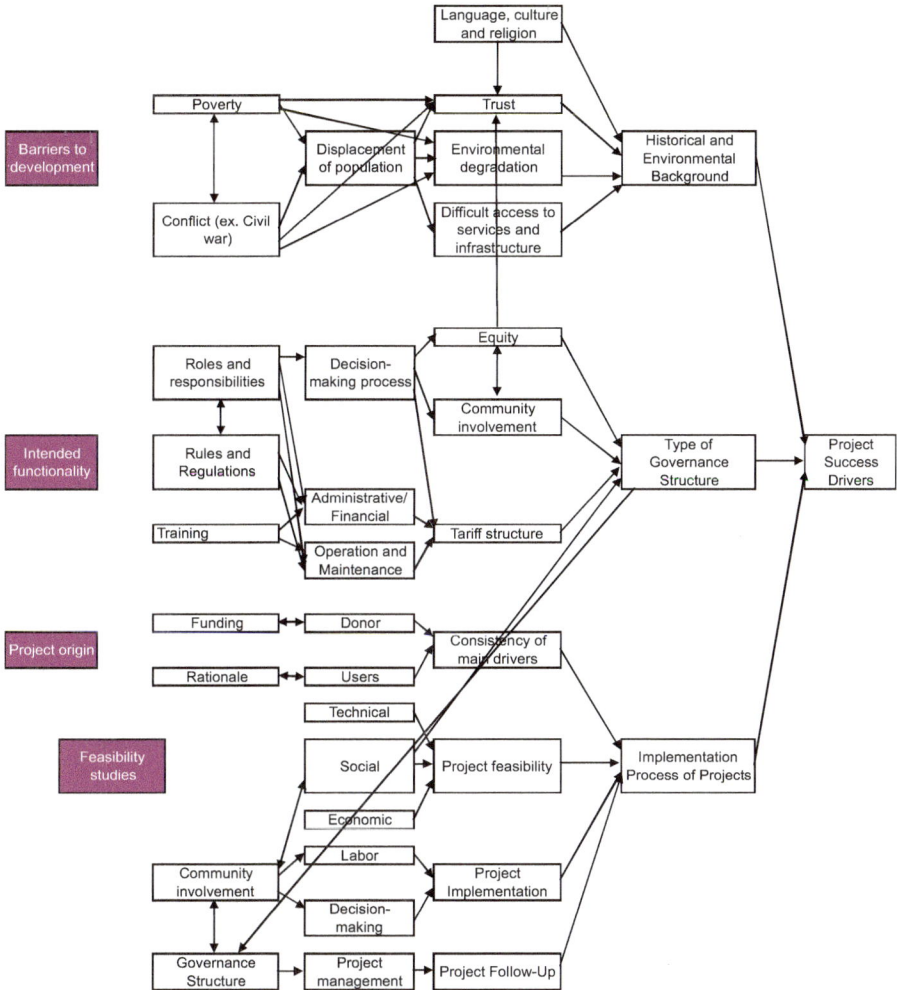

Fig. 11.6 Hypothesized project success drivers

3. Battery bank

   (a) Bank technical specifications
   (b) Battery protections
   (c) Status of connectors, terminals, electrolyte level

4. Lights and other loads

   (a) Technical specifications
   (b) Status of electrical connections and indicators

Table 11.3 Indicators and methodologies

Indicators	Data gathering and analysis methodology
Sustainable development:	
Policy objectives achieved (including National Action Plans Adopted). These policy objectives should have explicitly defined goals that can be measured	Survey question to government officials and donor program managers. Official publications from governments and donor institutions
Policy objectives maintained since project inception	Survey question to government officials, donor program managers, and community members
Local capacity developed: institutions	Survey question to government officials, donor program managers and community members
Local capacity developed: technical skills	Survey question to government officials, donor program managers and community members
People with increased access to energy services	Survey question to project implementer and direct observation
Homes adopting improved cooking/heating/lighting techniques	Survey question to project implementer and direct observation
Number of people with reduced exposure to combustion pollutants indoors	Survey question to project implementer and direct observation
Number of governance structures created and their functioning	Survey question to project implementer and community members. Focus groups
Improvement in livelihoods (natural, physical, financial, social and human capitals)	Survey questions on the five assets to community members. Survey questions to community members, and different levels of government and project implementers about the 'Transforming Structures and Processes'
Poverty alleviation:	
Reduction in the cost of energy	Survey question to community members and direct calculations
Reduction in the percentage of income spent on energy sources	Direct calculations
Increase in productive time	Survey question to community members
Diversification of income sources	Survey question to community members
Increase in number of microenterprises generated	Survey question to community members
Improvement in health and education infrastructure	Survey question to community members, project implementers and government
Improvement in health and education services	Survey question to community members, project implementers and government
Formalization of land rights	Survey question to community members, project implementers and government
Economic feasibility:	
Existence of tariff or fee for electricity use	Survey question to community members
Existence of a bank account or other form of tariff management	Survey question to community members

(continued)

Table 11.3 (continued)

Indicators	Data gathering and analysis methodology
Use of the tariff or fee to cover operation and maintenance (O&M) costs (this includes preventive, short term and long term maintenance)	Survey question to community members
Social acceptability:	
Cultural and religious acceptance	Survey question to community members
Consistency between project goals and user expectations	Survey question to community members
Additional benefits (for example, spending more time with family)	Survey question to community members
Source of conflict (for example, misuse of tariff)	Survey question to community members
Environmental responsibility:	
Existence and implementation of an environmental impact assessment	Survey question to project implementers and government agencies
Consideration for disposal of used components	Survey question to community members, project implementers and government
Emissions reductions:	
Increase or decrease in CO2 emissions (tons of carbon)	Simplified estimate based on IPCC methodologies
Energy savings (tons of oil equivalent)	Simplified estimate based on IPCC methodologies
Cost savings	Calculation
Standards adopted and implemented	Inspection of renewable energy systems following standard protocols and survey question
Adaptive capacity:	
Change in the number of people living in more hazardous zones	Direct observation, survey question to community members and government
Decrease in the size of extreme weather risk zones	Direct observation, survey question to community members and government
Increase in the community's adaptive capacity (creation of social networks or increased knowledge of technologies that can help cope with disaster, through the use of renewable energy systems)	Direct observation, survey question to community members and government

5. Charge controller and inverter

   (a) Number and capacity of each light and appliance
   (b) Indication of functionality of each light and appliance

2. *What are the relative roles of local historical background and physical characteristics, type of community governance, and funding source and project implementation process in the success of projects in meeting adaptation, mitigation and development goals?*

The background of a location can give more insight into its current poverty, development and climate vulnerability status and how the project can be designed.

Table 11.4 Indicators and methodologies

Historical and environmental background of locations:	
Previous conflict (for example, civil war)	Literature review and survey question (to who?)
History of extreme poverty/poverty	Literature review
Displaced populations	Literature review
Disenfranchisement due to language barriers	Literature review
Governance structure:	
Type of governance structure	Focus group and survey question to community members
Existence of other community governance structures	Focus group and survey question to community members
Functionality and effectiveness	Focus group and survey question to community members
Existence of internal rules and regulations	Focus group and survey question to community members
Equitable access	Focus group and survey question to community members
Funding sources and implementation process of project:	
Funding sources of the project	Project implementer, project documents
Existence of pre-feasibility and feasibility studies	Project implementer, project documents
Community socialization and training process	Project implementer, project documents
Existence of O&M plan	Project implementer, project documents
Monitoring and evaluation	Project implementer, project documents

For example, a background that includes previous conflict and displacement of populations can have an impact on trust, environmental degradation, and access to resources, which also shape projects and need to be considered during project planning and implementation. Environmental degradation can impact the renewable energy system design and dictate other activities that the users will need to carry out for the system to continue working properly. A common example is reforestation activities in the upper watersheds of small-scale hydro systems. The implementation process of projects is another hypothesized driver. Pre-feasibility studies must be conducted to determine if projects are technically and economically feasible, and to highlight relevant social concerns. Having proper operation and maintenance (O&M) plans will ensure that the system will continue working and providing benefits after the donor and implementers are gone.

11.4   Analysis

11.4.1   Meeting the Triple Objectives

The results show that, *under certain circumstances and design considerations*, renewable energy projects can simultaneously meet these three objectives, and thus that responses to climate change mitigation and adaptation can be integrated with poverty alleviation and sustainable development. Small scale hydroelectric and solar systems can reduce emissions, enable adaptation and help local liveli-hoods although there are numerous problems that limit the success of projects including poor design, inequitable distribution of benefits, negative user percep-tions, and poorly designed or non-existent governance and maintenance structures.

Although the design of some case study projects did not allow for the triple objective to be currently met, this does not preclude the projects from meeting it in the future. In some projects, a proper PAR analysis wasn't carried out as there were no extreme weather events reported or any other emergency that showed the usefulness, robustness or vulnerability of the technology or of the population thanks to the infrastructure. Indeed, some DRE projects can be more robust, and some have already been rebuilt after specific extreme weather events. Some users indicated that their systems could still be working had the local donor or implementing NGO been more aware and visited more often and not disappeared. This points to the need for greater and better monitoring as well as evaluation, which hadn't been carried out in some of the projects visited, despite their being implemented for more than 5 years.

Table 11.5 gives a summary of the results of all the case studies, according to each major category of indicators.

The notes below explain in greater detail the concept of each column.

11.5 Renewable Energy and Climate Adaptation

As hypothesized, DRE systems have been seen to both increase and decrease vulnerability to extreme climate events. To date, the potential response to extreme weather events of DRE systems has hardly been considered and it has been seen that they are vulnerable to extreme weather events which can harm users and hamper their stated goals such as in Nicaragua. For example, a woman in Guatemala had a nervous breakdown when her solar PV system had a short circuit inside her house during a particular storm. In this case, the fault was due to improper system installation which wasn't reported earlier as this was the first external visit to the system and household. On the other hand, one case study, Nueva Alianza, used their biodiesel system after only 1 day of being installed, and it was robust enough to withstand the force of Hurricane Stan. In general, the case studies helped identify the main vulnerabilities of DRE systems to extreme climate events. The case studies also showed that communities in which adaptation goals are being met are communities in which, more often than not, development goals are also being achieved. Actions that enable adaptation also enable development, such as com-munications, alternate income sources and more community unity. However, research results also indicate that in most cases, adaptation to natural disasters is better in communities where there is a good governance structure and where the renewable energy system is commonly owned. Although this result might have

Table 11.5 Results summary

Country	Project	Adaptation (Chap. 4)			Mitigation (Chap. 5)		Sustainable development (Chap. 6)					Governance, exists, works[k]	Governance, exists, doesn't work[l]	Governance, doesn't exist[m]
		Good[a]	Bad[b]	Neutral[c]	Energy savings[d]	Emissions reductions[e]	Domestic uses[f]	Productive uses[g]	Communal uses[h]	Cost savings[i]	Social acceptance[j]			
Guatemala	Nueva Alianza	✓			✓	✓	✓	✓	✓	✓	✓	✓		
	San Jerónimo			✓	✓	✓				Grid connected	?			✓
	Chel			✓	✓	✓	✓	✓	✓	✓	✓	✓		
	ADIM PV			✓	✓	✓	✓	✓	✓	✓	✓	✓		
	Cancuén	✓			✓	✓		✓		Project exists because of RE system			✓	
	Chapín Abajo			✓	✓	✓	✓	✓		NGO maintains system	✓	✓		
	MEM PV			✓	✓	✓	✓	✓		✓	✓		✓	
	CONRED PVEWS	✓			✓	✓			✓	NA	✓	X		
Nicaragua	PVBCS		✓		✓	✓	✓	✓		?		✓		
	SINAPRED PVEWS		✓		✓	✓	✓		✓	NA	?			
	El Trapiche			✓	✓	✓			✓	Donor maintains system			✓	
	Río Bravo			✓	✓	✓	✓		✓	Cost of energy higher than national utility, but lower than	✓	✓		

							diesel generators or traditional biomass
El Bote	✓	✓	✓	✓	✓	✓	Cost of energy higher than national utility, but lower than diesel generators or traditional biomass ✓
Women's solar center	✓	✓	✓	✓	✓		Project exists because of RE system ✓

Grid connected means that the project is interconnected to the national grid; therefore, users are subject to the tariffs set by the national regulatory agency

Where an external source maintains the system, it indicates that there are no energy costs to the DRE beneficiaries

The cases of Río Bravo and El Bote are further explained in Chap. 6

[a] Indicates these projects are good examples of the use of DRE during climate hazards, or that performed well during a climate hazard

[b] Indicates these projects are good examples of the use of the inappropriate use of DRE during climate hazards, or that were damaged during a climate hazard

[c] Indicates these projects have not suffered through a climate hazard

[d] Indicates these projects are good examples of DRE as an energy savings measure (as compared to their previous energy source, mainly traditional biomass)

[e] Indicates these projects are good examples of DRE reducing GHG emissions

[f] Indicates these projects included domestic applications

[g] Indicates these projects included income generation applications

[h] Indicates these projects included communal applications

[i] Indicates these projects caused a savings on the monthly income spent on energy sources

[j] Indicates these projects are considered as socially acceptable by the users and/or these projects have not caused social tensions to arise

[k] Indicates these projects are good examples of how communal governance structures have been a DRE project success driver

[l] Indicates these projects are good examples of how communal governance structures have not been a DRE project success driver

[m] Indicates these projects lack a communal governance structure

been expected, what is surprising is that technical design and standards didn't play a significant role and, contrary to expectation, the centralized nature of the infrastructure did have an impact through the respective governance structures. That is, decentralized infrastructure projects tend to have weak communal governance structures that aren't conducive to good adaptation strategies, while the opposite proved to be true. This fact highlights the important correlation between infrastructure centralization and robust governance structures, which was not originally hypothesized.

11.6 Renewable Energy and Climate Mitigation

Community scale DRE projects have encountered difficulties with the CDM despite them meeting emissions reductions goals. Besides the well documented barriers of a lengthy process not understood at the community level, and the high transactions costs, it is very difficult to calculate the net amount of emission reductions because of deficiencies in baseline emissions calculations. This is particularly true in projects where a subset of the beneficiaries enjoyed some kind of modern energy source, whether it was grid electricity or a diesel or gas generator. One project in Nicaragua exemplifies this as the baseline is calculated with the emissions factor of the country's energy mix, even when 18 of the 20 beneficiary communities used traditional energy sources, in which their emissions are considerably lower. Other projects highlight the finding that DRE projects can increase emissions: as electricity demand increases through the use of new appliances, use of fossil fuels tends to increase when the DRE system can not supply electricity for those new appliances. The most common cases seen were in stores that relied on refrigerators, whether through a PV system or a hydroelectric plant. In one community, their own DRE system no longer has sufficient capacity to meet the community's demand and they are now thinking of a grid connection. These results are similar to those found in India (Reddy et al. 2006), which also highlight both the needs for involvement of local communities and of vulnerability and sustainability analysis of local resource management. The latter were missing from most of the case studies analysed in this research.

Table 11.6. describes the changes in supply, infrastructure, and demand that occurred with each project. The changes in demand reflect the changes the systems were designed for.

11.7 Renewable Energy and Sustainable Development

DRE projects were found to have a positive impact on livelihoods assets by improving its five capitals: financial, physical, human, social, and environmental. Financial capital was enhanced by energy cost savings, productive use and alternate

Table 11.6 Changes in energy supply, demand and infrastructure of case study projects

Project	Change in supply	Change in infrastructure	Change in demand
Nueva a Alianza	From traditional biomass to biodiesel and micro-hydroelectric plant	Construction of the communal electricity grid and installation of connections	Use of household and office appliances, and implementation of productive use of projects
San Isidro	Grid-connected small-scale hydroelectric plant	None	None
Chel	From traditional biomass to micro-hydroelectric plant	Construction of a communal electricity grid, public lighting, and installation of internal electric home connections	Use of household and office appliances, and implementation of productive use projects
Cahabón	From traditional biomass to solar PV home lighting	Internal electric home installations	Use of basic household appliances (2 CFL's, radio, cell phone charging, and occasionally a black and white TV)
Chap in Abajo	The workshop had no energy source prior to the system (work was mainly carried out during daylight)	Internal electric home installations	Use of two CFL's
Cancuén	Ecotourism project started with PV system	Internal electric installations in offices and tourist bungalows	Use of radio-communications, CFL's and cell phone charging
Guatemala PVEWS	PVEWS formerly used diesel generations	Internal electric home installations	Radio-communications and CFL's
Guatemala MEM	From traditional biomass to solar PV home lighting	Internal electric home installations	Use of basic household appliances (2 CFL's radio, cell phone charging, and occasionally a black and white TV)
ADM	From traditional biomass to solar PV home lighting	Internal electric home installations	Use of basic household appliances (2 CFL's, radio, cell phone charging, and occasionally a black and white TV)
PVBCS	From traditional biomass to battery systems	Internal electric home installations	Use of basic household appliances (2 CFL's, radio, cell phone charging, and occasionally a black and white TV)
EI Trapi che	Project started with the PV system	Pumping system, piping, tap	Pumping system
EI Bote	Micro-hydroelectric plant, grid connection and traditional biomass	Construction of the communal electricity grid, public lighting, and installation of internal electric home connections	For those that had traditional biomass, they currently use basic household appliances (2 CFL's, radio, cell

(continued)

Table 11.6 (continued)

Project	Change in supply	Change in infrastructure	Change in demand
			phone charging, and occasionally a black and white TV)
Rio Bravo	From traditional biomass to mini-hydroelectric plant	Construction of the communal electricity grid, public lighting, and installation of internal electric home connections	Use of household and office appliances, and implementation of productive use projects
Women's solar center	Project started with the PV system	Internal electric installations in homes and the solar center	Use of basic household appliances (2 CFL's, radio, cell phone charging, and occasionally a black and white TV). The solar center powers several computers, printers and modem

income sources and the creation of savings mechanisms, although these were for system maintenance. Financial capital, however, was also harmed by failures in the DRE systems: during blackouts some lost refrigerated products or had to spend on fossil fuels in order to avoid losing them. Physical capital was improved by the introduction of the DRE infrastructure and other infrastructure that was enabled through the DRE projects, such as roads. Human capital saw improvements at the domestic, productive and communal levels. The DRE system enabled other services, such as better education and health, and created more unity among neighbors. Lastly, social capital was impacted through the social acceptance of the projects, especially in projects that had a strong communal participation component which, in part, lead to robust governance structures that proved to be important for the DRE projects to meet climate change adaptation goals.

11.8   Cross Cutting Factors

Factors including the centralized or decentralized nature of the technology or the institutions, and governance and funding entities, can enable projects to meet their stated goals, and therefore, to meet the triple objective. For the former, all projects with a centralized infrastructure, with the exception of one project, had functional governance structures. On the contrary, with the exception of one community, all communities benefited with PV systems financially managed their systems individually and the governance structures set in place ceased functions relatively soon after the installation of the systems. For the latter, the primary goals and objectives of the donor and development entities and their interaction with the communities, was key in promoting, or not, proper understanding and upkeep of the systems. This also leads to the conclusion that when there are multiple institutions involved in the implementation of a project, coordination among them needs to be planned from the outset. Some of the case studies presented problems because they lacked such coordination. Cultural, political, economic and social differences also play a role and can be bridged through long-term social interaction and trust building. Although this is possible and increasingly recognized, not all donor and development entities understand its importance or the need to allocate appropriate resources. Social interaction should be a two-way learning process: the community learns about the project and the means to achieve it, and the donor/developer learns about the community including its needs and background, among other information (GEF 2006).

The case studies analysed did not have a 'social funding' to help the poorest people, which still can not count on renewable energy as a modern energy option. Some projects would like to have one but presently can not afford one. Implementing one would require increasing the electricity tariff which is not possible. The lack of this 'social funding' mechanism is considered by some to increase the inequality gap. During the planning and execution phases of projects, social and economic differences among the population are not always considered, leaving the poorest population vulnerable. As Krause and Nordström also found, the high costs of renewable energy systems can also increase the inequality gap (Krause and Nordström 2004) as the poorest segment of the population remains unelectrified and unable to benefit the systems.

Technical quality was important in enabling project success. Poor technical designs and lack of appropriate operation and maintenance protocols have prevented some DRE projects from meeting stated goals: if the systems do not work as expected, people will continue to use torch pine, candles, gas lamps, or diesel gensets and will not be able to carry out the productive and social activities the electricity has enabled. As was also seen in some of the case studies, poor technical quality can also make the DRE systems, together with the users, vulnerable to extreme climate events, perhaps defeating their main purpose. In a subset of the communities, systems used very low quality components, including, for example, non-listed and non-certified PV panels that were peeling within 5 years of being installed (when their expected life ranges between 20 and 30 years) and car batteries labeled as solar deep cycle batteries.

Monitoring and evaluations are essential to meeting the triple objective, although this was very rarely carried out. Some of the projects visited could be working today had proper monitoring and evaluation taken place. Unfortunately, a number of communities where projects have failed remain without electricity and there are few prospects for further investment.

The community of Nueva Alianza provides the best example of how meeting the triple objective is possible. In the short period of time the micro hydroelectric and biodiesel projects had been installed, the community has been able to reduce their fossil fuel consumption and therefore their greenhouse gas emissions, improve the quality of life of all the families and enable their survival and that of neighboring communities in the aftermath of Hurricane Stan. This is an excellent example of how a renewable energy source helped this community and neighboring ones that had no communication with the 'outside world' while members of the Nueva Alianza community indicate that the rest of the world was 'out of communication with them'. The unity and strength the families already had certainly enabled the development of the DRE and productive use projects, but it can also be said that the development of these projects strengthened their bonds even more.

11.9 Conditions, Circumstances and Considerations

To summarize, and as analysed throughout, there are several characteristics that indicate the triple objective is possible:

1. Communities in which adaptation goals are being met are communities in which, more often than not, development goals are also being met.
2. Communities in which there is a governance structure, or some form of community participation, will be better able to cope with a natural disaster than one in which there isn't. Projects that are not communal from the outset and beginning with community participation since the planning phases will most likely not be able to meet the triple objective (GEF 2006; Reddy et al. 2006)
3. Sound and site specific technical designs and appropriate operation and maintenance protocols that follow safety and quality codes and standards enable the triple objective.
4. Socialization needs to be considered a two-way learning process and community involvement and participation ought to happen from the beginning (GEF 2006)
5. Monitoring and evaluation are essential.

11.9.1 Implications for Policy, Practice and/or Research

Below I list a series of policy recommendations that can help put DRE projects on a path in which they can simultaneously achieve the triple objective, taking into consideration the cross-cutting elements necessary to success.

1. Disaster reconstruction programs are implemented in a considerably shorter period of time than development and rural electrification programs. This causes basic socialization, community participation, and training to be cut short because of timing and/or budget constraints. Recognizing that there are projects and infrastructure that need to be implemented in the short term, and that the priority is to benefit the largest number of people, the main policy recommendation is to ensure that reconstruction programs be designed to respond to future extreme climate events and other hazards to increase the community's adaptive capacity.
2. Policy makers and governments tend to relate the energy sector in general, and renewable energy projects in particular, to only climate change mitigation goals.

In reality, as some case studies showed, the energy sector and the DRE projects are vulnerable to extreme climate events and in consequence can increase or decrease the vulnerability of the populations they serve. DRE projects are vulnerable to extreme weather events, but can also be designed to enable adaptive capacity for example through coordinated and equitable use of water in a watershed.

3. Poor technical designs and lack of appropriate operation and maintenance protocols and practices have prevented DRE projects from meeting their stated goals. This issue highlights the importance of government regulation or certification that ensures quality and safety codes and standards to avoid deceitful practices such as selling bad quality and/or pirate/fake components. Even if the systems are privately owned, there should be government controls in place and an accountability system so not 'anybody' can install systems without having the appropriate knowledge, training and licenses. Most of the Central American countries have adopted the US National Electric Code (NEC) although not all have implemented it. Donors and governments implementing DRE projects should require the compliance with such codes and standards as well as product listing. Besides requiring the use of code-compliant components and equipment, donors and governments should ensure that project installers are also licensed and certified, ensuring project sustainability and a better use of limited development budgets. Moreover, code compliance will ensure that users will not be harmed in any way, nor taken advantage of monetarily.

4. One common response received from many system users and technicians was the need for more intense and periodic training sessions to ensure systems remain functioning. Two policy recommendations are suggested:

   • Set a minimum required budget for socialization and training activities as the current spending level for this topic is not sufficient to cover users' needs. Some government officials interviewed indicated the need to spend up to 10 % of the total infrastructure budget on training.

   • Aid program indicators tend prioritize first and foremost the number of beneficiaries or system users. Because of this, donors are reluctant to allocate additional budget towards training activities. During the interviews, some indicated this was unrealistic as there were specific goals for system beneficiaries and re-allocating budgets would mean a smaller number of systems installed which might be interpreted as inefficient use of the budget. This point has greater implications if program evaluation was carried out more periodically: when systems stop working and communities rely once again on traditional energy sources, statistics are not modified to reflect this and aid programs do not target these populations anymore as they are already considered 'electrified' or 'benefitted'. To be most effective, indicators must be both qualitative and quantitative (GEF 2005; Krause and Nordström 2004).

5. In rural indigenous populations in which other belief systems exist, such as with Mayan populations, donors and developers need an understanding of the cultural, political, economic and social differences to ensure that the appropriate

ideas and expectations are being transmitted. In these cases, it would be appropriate for the Ministry of Culture, or its equivalent, to be involved so no rules, customs or traditions are being violated or misinterpreted.

6. Especially for hydroelectric projects of any scale, an integrated watershed management vision ought to be implemented to ensure the well-being of the entire watershed and that users of the lower watershed do not suffer negative impacts of activities carried upstream.

7. Despite the poverty alleviation goals of many DRE projects, this objective is not always achieved. In some cases, the poverty level of the beneficiaries hampers the long-term sustainability of the projects. In the case of solar PV projects, users are not always able to maintain and/or replace their batteries or other components. In the case of hydroelectric projects, the poorest families can not afford the initial connection cost. Some of the case studies showed how this can increase the inequality gap and leaves open the question if another aid program will eventually provide the service for those unserved homes. Based on this, governments might need to consider subsidizing the electricity service for the poorest segment of the population to avoid increasing inequality in rural communities. Likewise, a subsidy for social services can also be considered. As seen in the project of El Bote, rural schools can not benefit from the electricity service because the parents can afford neither the connection nor the monthly bills and the Ministry of Education rules indicate they can only cover the costs of schools located in municipalities (Krause and Nordström 2004).

8. Some case studies pointed to one key element that is often times missing from projects and which can prevent them from attaining the triple objective: monitoring and evaluation. In one of the programs evaluated, the ADIM PV project in Guatemala, I was able to see the evolution of projects of one developer over 10 years and such lessons learned do exist.

I identified five main reasons why the projects did not meet the triple objective. The first one is *level of poverty* as people are too poor to afford the service (in the case of the hydroelectric plants) or save for operation and maintenance (in the case of solar systems) and access to available capital becomes important, if not necessary, for system upkeep. Government schemes, such as the loan of solar systems in Guatemala, seemed to work very well, except that the poorest people cannot afford necessary battery replacement. Whether government or privately owned, an important factor is the inclusion of productive use applications that can help families gain more income that could help maintain an available cash flow. The second reason is *inconsistency between users' expectations and donor's objectives.* If users are not happy; it can create conflict, leading to systems neglect. The third reason is *lack of community involvement*: users were not satisfied mainly in those projects in which community involvement was minimal or non-existent, as in the bigger projects with funding from multilateral development entities or private sector. Based on the different conceptions of community involvement, a recommendation is to gauge with the community how they envision their role to be throughout the project. *Unreliable energy* is the fourth reason: with multiple or constant blackouts, the intended goals of the projects are not entirely met and in some cases, can cause more difficulties or pose a danger to the users. Last but not least, *perceptions* had a clear impact. People form perceptions about renewable energy systems and their functioning from their own and other users' experiences. Such perceptions can make them wary of using these technologies without adequate socialization and training. For example, in Nicaragua, a system with a bad design led to two trees being hit by lighting, and as a consequence, the family is afraid of using the system and has recommended against their use to others. It is also important to note here that positive experiences also enable greater use of DRE technologies. I also saw users purchasing their own system after seeing their neighbors' system or heard from some indicating they would purchase a new system if theirs failed.

References

Adger, W. N., Huq, S., Brown, K., Conway, D., & Hulme, M. (2003). Adaptation to climate change in the developing world. *Progress in Development Studies, 3*(3), 179.

Architectural Energy Corporation. (1991). *Maintenance and operation of stand-alone photovoltaic systems*. Albuquerque: Sandia National Laboratories Photovoltaics Design Assistance Center.

Birkmann, J. (2006). *Measuring vulnerability to natural hazards: Towards disaster resilient societies*. Tokyo: United Nations University.

Blaikie, P. M., & Brookfield, H. C. (1987). *Land degradation and society*. London: Methuen.

Bradshaw, B., Dolan, H., & Smit, B. (2004). Farm-level adaptation to climatic variability and change: Crop diversification in the Canadian prairies. *Climatic Change, 67*(1), 119–141.

CEPAL. (2007a). *La Energia y las Metas del Milenio en Guatemala, Honduras y Nicaragua*. Mexico: CEPAL.

CEPAL. (2007b). *Istmo Centroamericano: Estadísticas del subsector eléctrico (Datos actualizados a 2006)*. Mexico: CEPAL.

Crichton, D. (2007). What can cities do to increase resilience? *Philosophical Transactions of the Royal Society A: Mathematical, Physical and Engineering Sciences, 365*(1860), 2731–2739.

Downing, T. E., & Patwardhan, A. (2005). *UN adaptation policy frameworks for climate change: Developing strategies, policies and measures* (Technical Paper 3. Assessing Vulnerability for Climate Adaptation, pp. 93–115). Cambridge: Cambridge University Press.

Eakin, H., & Luers, A. L. (2006). Assessing the vulnerability of social-environmental systems. *Annual Review of Environment and Resources, 36*, 365–394. doi:10.1146/annurev.energy.30.050504.144352

Eriksen, S., & O'Brien, K. (2007). Vulnerability, poverty and the need for sustainable adaptation measures. *Climate Policy, 7*(4), 337–352.

GEF. (2005). *OPS 3: Progressing toward environmental results. Third overall performance study of the GEF*. Washington, DC: Global Environment Facility and ICF Consulting.

GEF. (2006). *The role of local benefits in global environmental programs* (Evaluation report no. 30). Washington, DC: Global Environment Facility Evaluation Office.

Great Britain. Department for International Development. (2000). *Sustainable livelihoods guidance sheets*. London: Dfid.

Great Britain. Department for International Development. (2004). *Climate change and poverty*. London: Department for International Development.

GVEP (2008). Available: www.gvep.org

Inversin, A. R. (1999 [1986]). *Micro-hydropower sourcebook: A practical guide to design and implementation in developing countries*. Arlington: NRECA International Foundation.

Johnson, C., Penning-Rowsell, E., & Parker, D. (2007). Natural and imposed injustices: The challenges in implementing 'fair' flood risk management policy in England. *Geographical Journal, 173*(4), 374–390.

Krause, M., & Nordström, S. (2004). *Solar photovoltaics in Africa: Experiences with financing and delivery models* (Monitoring & evaluation report series issue 2). New York: UNDP-GEF.

Linnerooth-Bayer, J., & Mechler, R. (2006). Insurance for assisting adaptation to climate change in developing countries: A proposed strategy. *Climate Policy, 6*(6), 621–636.

Mills, E. (2007). Synergisms between climate change mitigation and adaptation: An insurance perspective. *Mitigation and Adaptation Strategies for Global Change, 12*(5), 809–842.

Moser, S.C., Kasperson, R.E., Yohe, G., & Agyeman, J. (2007). Adaptation to climate change in the Northeast United States: Opportunities, processes, constraints. *Mitigation and Adaptation Strategies for Global Change,* 1–17.

Najam, A., Rahman, A. A., Huq, S., & Sokona, Y. (2003). Integrating sustainable development into the Fourth Assessment Report of the Intergovernmental Panel on Climate Change. *Climate Policy, 3*(Suppl 1), S9–S17.

Naylor, R. L., Battisti, D. S., Vimont, D. J., Falcon, W. P., & Burke, M. B. (2007). Assessing risks of climate variability and climate change for Indonesian rice agriculture. *Proceedings of the National Academy of Sciences of the United States of America, 104*(19), 7752–7757.

Olsen, K. H. (2007). The clean development mechanism's contribution to sustainable development: A review of the literature. *Climatic Change, 84*(1), 59–73.

Ostrom, E. (2002). *The drama of the commons, National Research Council. Committee on the Human Dimensions of Global Change.* Washington, DC: National Academy Press.

Reddy, V. R., Uitto, J. I., Frans, D. R., & Matin. (2006). Achieving global environmental benefits through local development of clean energy? The case of Small Hilly Hydel in India. *Energy Policy, 34*(2006), 4069–4080.

Risser, V., & Post, H. (1991, November). *Stand-alone photovoltaic systems: A handbook of recommended design practices* (SAND87-7023). Sandia PV Design Assistance Center.

Romilly, P. (2007). Business and climate change risk: A regional time series analysis. *Journal of International Business Studies, 38*(3), 474–480.

Sperling, F. (2003). *Poverty and climate change: Reducing the vulnerability of the poor through adaptation.* Washington, DC: World Bank.

Swart, R., Robinson, J., & Cohen, S. (2003). Climate change and sustainable development: Expanding the options. *Climate Policy, 3*(Supplement 1), S19–S40.

Thomas, D. S. G., & Twyman, C. (2005). Equity and justice in climate change adaptation amongst natural-resource-dependent societies. *Global Environmental Change Part A, 15*(2), 115–124.

Tol, R. S. J., Van der Grijp, N., Olsthoorn, A. A., & Van der Werff, P. E. (2003). Adapting to climate: A case study on riverine flood risks in The Netherlands. *Risk Analysis, 23*(3), 575–583.

UNDP. (2007). *Human Development Report 2007/2008. Fighting climate change: Human solidarity in a divided world.* New York: Palgrave Macmillan.

Venema, H. D., & Rehman, I. H. (2007). Decentralized renewable energy and the climate change mitigation-adaptation nexus. *Mitigation and Adaptation Strategies for Global Change, 12*(5), 875–900.

Wisner, B., & Blaikie, P.M.. (2004). *At risk: Natural hazards, people's vulnerability and disasters* (2nd ed.). London: Routledge.

Chapter 12
Unpacking the Black Box of Technology Distribution, Development Potential and Carbon Markets Benefits

Jasmine Hyman

Abstract In 2005, the international carbon market was launched under the Kyoto Protocol, creating an innovative financing design for low-emissions development initiatives. Just over 10 years after its inception, the carbon market can now provide insight on the opportunities and limitations of "blended finance" approaches, whereby private-public partnerships are employed to pursue global development goals such as poverty alleviation and development. Utilizing process-tracing and value chain methods, this chapter adds granularity to debates on whether and how carbon markets can support local economic development, as measured through the creation of local enterprises and the support of local livelihoods. It offers a "Livelihood Index" to assess the employment impact of the carbon intervention in order to address the core question: how is the carbon credit pie divvied up? Three carbon projects in Cambodia, aimed at household level interventions (water filters, biodigesters for cooking and fertilizer production, and fuel-efficient cookstoves) are evaluated through the livelihood index and results indicate that distribution strategies matter for local economic gains. Distribution strategies to deliver low-carbon technologies within the carbon market are currently a "black box", understudied and undocumented in the project pipeline; this paper argues that opening the black box may be useful for policymakers, standard setting organizations and academics interested in promoting pro-poor impacts through carbon market interventions.

Keywords Carbon markets • Waterfilters • Cookstoves • Biodigesters • Climate finance • Climate change • Market mechanisms

J. Hyman (✉)
School of Forestry & Environmental Studies, Yale University, New Haven, CT, USA
e-mail: jasmine.hyman@yale.edu

© The Author(s) 2017   213
J.I. Uitto et al. (eds.), *Evaluating Climate Change Action for Sustainable Development*, DOI 10.1007/978-3-319-43702-6_12

12.1 Introduction

Efforts to provide clean cooking and water filtration facilities to the poor have been pursued in earnest by aid agencies, government ministries and the non-governmental sector for decades, though many initiatives have been stymied by inadequate and inconsistent funding, the introduction of inappropriate technologies, and a lack of follow-up (Clasen et al. 2004; Baumgartner et al. 2007; Lantagne et al. 2008).

In 2005, the international carbon market was launched under the Kyoto Protocol and the concept of "carbon finance" entered the world stage. Carbon finance marked an innovative approach to development finance in that it was designed to harness the motor of private finance to goals for the public good by awarding fungible "carbon offsets" for the delivery of development services that displaced activities that would otherwise generate greenhouse gas emissions (UNFCCC 1997). Two years later, the voluntary carbon market was launched and remained a viable channel for financing low-carbon projects even as support for the Kyoto Protocol's market mechanisms waned (Peters-Stanley 2013) The projects analysed in this chapter draw from both the Kyoto Protocol's market mechanism for developing countries, the "Clean Development Mechanism" (CDM) and the similarly structured voluntary carbon market. While the CDM and the voluntary market are both undergoing transformation as the Kyoto Protocol's implementation period draws to a close, consensus on the Paris Agreement at the 21st Conference of the Parties to the UNFCCC in December 2015 indicates that market mechanisms will continue to play a role in the upcoming climate regime. As such, lessons derived from the first generation of carbon market efforts under the Kyoto Protocol are relevant towards the design of the next generation of market-oriented climate finance tools.

Projects that aim for a high social and local development component are called "pro-poor carbon projects" (Verles and Santini 2012), "charismatic carbon projects" (Cohen 2011), "premium carbon" (The Gold Standard 2010) or "carbon with a human face" (World Bank 2002). These terms encompass carbon projects targeting the least well-off, either by introducing technological innovations to underserved households or by being physically located in Least Developed Countries where the emissions footprint is already low and investment risks are high (and therefore the incentive to invest in carbon reductions is minimal).

The majority of pro-poor projects are household-level interventions for responding to basic needs, such as fuel-efficient cook stoves, water filtration devices, and mini biodigesters that convert livestock and organic household waste into gas for cooking and household lighting. Significantly, pro-poor projects emphasize "co-benefits," or sustainable development deliverables, to the project recipients beyond offsetting emissions alone: they promise the creation of skilled job opportunities, increased household income, improved health outcomes, etc. Premium certification schemes, such as the Gold Standard for both the CDM and the voluntary carbon market, specialize in verifying that both emissions reductions and co-benefits have been achieved (though the Gold Standard does not hold a monopoly on pro-poor projects).

There is an underlying development narrative associated with pro-poor carbon projects, namely, that market-driven development tools can attract private resources into public services resulting in a win-win outcome for the environment and for the poor. The premise of the "win-win" outcome has been challenged (Simon et al. 2012) and the need to add granularity and precision to discussions on private-public partnerships is also well-established (Kwame Sundaram et al. 2016). This chapter builds upon these discussions to identify some of the conditions that might make "win-win" outcomes more likely: what kinds of elements determine the likelihood of local economic benefit when aid organizations, donor agencies, and private actors join together? Analysis reveals that the technology dissemination strategy is a significant, yet presently invisible, driver for pro-poor outcomes. Administratively, dissemination strategies are absent from project design documents; as a research topic, they are under-represented in the literature. This chapter argues that technology dissemination strategies merit more focus and attention given its bearing on livelihood outcomes for market-driven climate projects targeting the poor.

The chapter is structured as followed. A literature review on household interventions in the carbon market establishes that critiques of win-win market approaches and public-private partnership models are well documented and that there is an established need for further research on the conditions and variables that determine whether innovative financing partnerships will lead to their intended outcomes. The literature review also reviews current tools for evaluating low-emissions development projects and presents an adapted version of an evaluation tool forwarded by the Global Alliance for Clean Cookstoves. This adapted version of the tool, named the "Livelihood Index," provides a rough indicator on projects' local economic impact, specifically on a project's ability to catalyze skilled and long-term employment opportunities at the local level. The second section describes the methods of analysis and the parameters for case study selection. Next, the cases are described. The final section applies the Livelihood Index to the cases, alongside an analytical discussion as to the implications of each distribution strategy. Finally, the chapter concludes by arguing that the success or failure of a green technology to benefit its target population relates as much to the question of "how is the technology distributed?" as to "what is distributed in the first place?" The conclusion addresses areas for further research and suggests a new round of questions for a continued exploration of the conditions for designing climate finance projects that benefit the poor.

12.2 Literature Review

Carbon projects are, by definition, complicated subjects for impact evaluations. They represent dense policy experiments due to their pursuit of multiple goals, i.e. to support local sustainable development while mitigating global climate change. It follows that "project success" is a multifaceted term that can be measured in terms of avoided greenhouse gas emissions, expanded economic opportunities within the host country, improved local health outcomes or even in terms of social ideals such as increased gender equity or enhanced participation in decision making processes. The promise of "win-win" outcomes associated with environment and development projects is readily critiqued (Visseren-Hamakers et al. 2012; Mayrhofer and Gupta 2016). To add further to the conceptual tangle, the success of the project is contingent upon the household's willingness to utilize the technology, a behavioral feature that involves considerations such as cultural appropriateness (Troncoso et al. 2007; Shankar et al. 2014), intra-household dynamics (Shankar et al. 2014), and aftercare (Levine et al. 2013).

Globally, Wang et al. (2015) tracked 277 cookstoves, 134 biodigesters projects and 11 water filter projects that were either preparing for registration, registered, or issuing credits with both CDM and other voluntary standards as of June 2014 (Wang et al. 2015). Of this total, 112 projects had issued credits at least once and 222 projects were registered, with the remaining 88 projects in various stages of preparation (idem).

Given that these carbon projects have multiple goals, it is likely that evaluations for their "success" can differ greatly, depending on the goal of interest. The likelihood of unintended negative consequences resulting from a development intervention have been well documented in the general development literature (Ferguson 1994; Scott 1998) and in specific assessments of carbon credit projects. However, existing studies tend to focus on the theoretical merits and pitfalls of market-based approaches either by providing a global assessment of the market (Abadie et al. 2012; Kossoy and Guigon 2012; Climate Policy Initiative 2014; Climate Funds Update 2016) or by utilizing illustrative case studies to bolster a position on the carbon market's merits in general (Haya 2007; Bumpus and Cole 2010) or that achieving climate and development co-benefits is context dependent (Simon et al. 2012). Rather than condemn or condone carbon markets as a concept, there is a need to uncover causal mechanisms that can explain variations in development outcomes between carbon project types and designs.

12.2.1 Conceptualizing Local Economic Development Impacts for Carbon Finance Projects

There are numerous attempts in the academic and gray literature as to how one might approach evaluating the sustainable development impact of a household intervention. Household interventions which are subsidized by carbon finance are often called "charismatic carbon" "premium" or "pro-poor" projects (The Gold Standard 2010; Cohen 2011; Verles and Santini 2012) given that they directly address the development needs of the rural and urban poor and are therefore assumed to have higher sustainable development impact than projects which focus on reducing industrial gas or manufacturing emissions. While inconclusive on best practices, the academic literature provides the contours of how program design features may engage with intended outcomes (Bailis et al. 2009; Mobarak et al. 2012). This body of research has informed the policy-making community, most notably with the development of the Gold Standard certification scheme for best practices in carbon offset project design (The Gold Standard 2010) and the Global Alliance for Clean Cookstoves' (GACC) recent presentation of a conceptual framework on how to measure and monitor sustainable development against project indicators (GACC 2014).

12.2.2 Measuring Sustainable Development in Carbon Interventions

Most practical attempts to measure sustainable development impacts across the market landscape mirror or modify the Gold Standard's sustainable development matrix, which identifies environmental, economic and social indicators and asks the project developer to rank the project's impact using a scaled score chart from -2 to 2. Numerous academic and gray assessments of carbon projects utilize a portfolio analysis approach in which they conduct a textual analysis of the project's benefits, extracting information from the sustainable development matrix (Olsen and Fenhann 2006; Sutter and Parreño 2007). A limitation across these assessments is an absence of information on the causal pathways that link the indicator of interest to a development outcome.

The GACC is currently working with the International Center for Research on Women to create conceptual frameworks that link project indicators with three development outcomes of interest: women's empowerment; the pathway between technology adoption and social/economic wellbeing and finally, the pathway between project implementation and livelihood enhancement (Fig. 12.1). These conceptual frameworks are based upon the GEF's Theory of Change, a policy design paradigm that makes transparent the assumed relationships between policy actions (indicators), policy impacts (components) and outcomes (goals).

An earlier GACC publication by Troncoso presents an adoption index and project impact index for comparing project effectiveness within a portfolio (Troncoso 2014). Troncoso's approach simply identifies key variables for the outcome of interest and weights them according to relevance. Adapting Troncoso's general method for creating an impact index derived from the GACC's conceptual framework results in the creation of a new tool – a Livelihood Index (LI) – for valuing livelihood impacts from carbon-financed interventions.

Before delving further into the assumptions underlying and the application of the livelihood index, it is worth addressing why local economic impacts matter. The vast majority of studies on carbon markets and environment-development projects more generally focus on the user experience: how and why users adopt a new

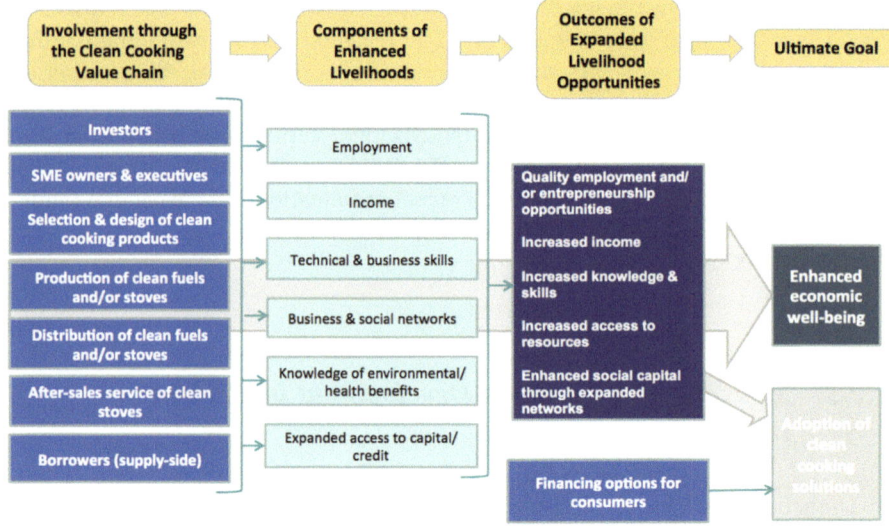

Fig. 12.1 GACC social impact, conceptual framework

technology, whether or not they replace it, whether or not it is appropriate for their local settings, and how their livelihoods are enhanced in terms of social, environmental and economic outcomes. This body of literature is crucial for the discussion of carbon finance evaluation and effectiveness, as it addresses whether and how climate-compatible technologies can enhance the lives of target communities while supporting the global goal of climate change mitigation and adaptation. However, user-focused studies cannot address a primary assumption within the environmental markets policy narrative, namely, that market approaches support the development of local economies and are therefore more empowering than the traditional aid model.

The Livelihood Index is derived from the GACC's conceptual framework on social impacts from cookstove projects, which are often carbon finance projects as well. It adapts GACC's broad notion of "involvement in the value chain" on the far left of the diagram, and converts the listed categories of project involvement into "actors" whose jobs are assessed for evidence of an enhanced livelihood and social impact: Investors; SME owners and executives; selection and design of clean cooking technologies; production of clean cooking fuels/stoves; distribution of clean fuels/stoves; after sales service of clean stoves; borrowers (supply-side). These categories are adapted to the broader range of project types and the specific range or actors when carbon credit creation is involved: (1) Carbon credit buyers/ investors; project developers (i.e. executives) and SMEs (when applicable); clean technology producers; clean technology distributors; after sales service agents; borrowers and users. In addition, we have added the third party validator and verifier to the supply chain, an actor whose role is specifically created by the carbon market to validate the quantifications associated with greenhouse gas emissions reductions. Within a carbon offset project, the project developer selects the clean cooking technology design, so this category has been eliminated. In all, the far left column of the GACC's conceptual framework translates intuitively into carbon offset project's value chain, determining the categories of actors that we assessed for evidence of access to and gains associated with utilizing the carbon market.

The second column of the GACC framework, "components of enhanced livelihoods" includes the following categories: employment; income, technical and business skills; business and social networks; knowledge of environmental health/benefits; expanded access to health and credit. The semi-structured interviews with actors in the first column touched upon all of these elements of an enhanced livelihood, and aspects of these interviews will be discussed in the case analysis. However, due to variability in the categories that were relevant for all actors in the value chain, the livelihood index we utilize here references those aspects of an enhanced livelihood that were pertinent in every single interview: steady and predictable employment; income for labor, enhanced opportunities engendered by skilled labour and enhanced opportunities engendered my managerial positions (i.e. positions with some degree of decision making power). The need for expanded access to capital and credit was not always a prerequisite for acquiring the new technology; in some cases, households were given the technology for free. The relationship between users, borrowers and the local impact of integrating them into the formal economy through enhanced credit options is significantly complex that it is the subject for another paper.

The third column articulates varying "outcomes of enhanced livelihoods:" quality employment and/or entrepreneurship opportunities; increased income; increased knowledge and skills; increased access to resources; and enhanced social capital through expanded social networks. Quality employment and entrepreneurship opportunities arguably encompasses other outcomes, such as increased skills and resources, increased networking opportunities and enhanced social capital and status. Another outcome worth further investigation would be increased employment choices. For example in addition to the outcomes identified within the GACC framework, avoided sacrifices where money was not the priority outcome were also positively mentioned; i.e. "employment with the carbon offset project enables me to work close to my village, and without this job I would be forced to live far away from my family.

Pressed further, this particular interviewee admitted that he could earn a better income in Vietnam, but the benefit of living with his family at home in Cambodia and engaging with the environment-development project far outweighed the potential increase in income. This type of benefit is not clearly captured in the conceptual framework or the livelihood index as it currently stands; further research is required to establish how and under what conditions carbon finance can engender or hinder livelihood choices where income is not the salient driving factor.

Thus, the livelihood index offers a rough proxy as to the impact of a carbon finance project on local incomes and livelihoods within the economy that surrounds the carbon finance intervention; while imperfect, the livelihood index can help to

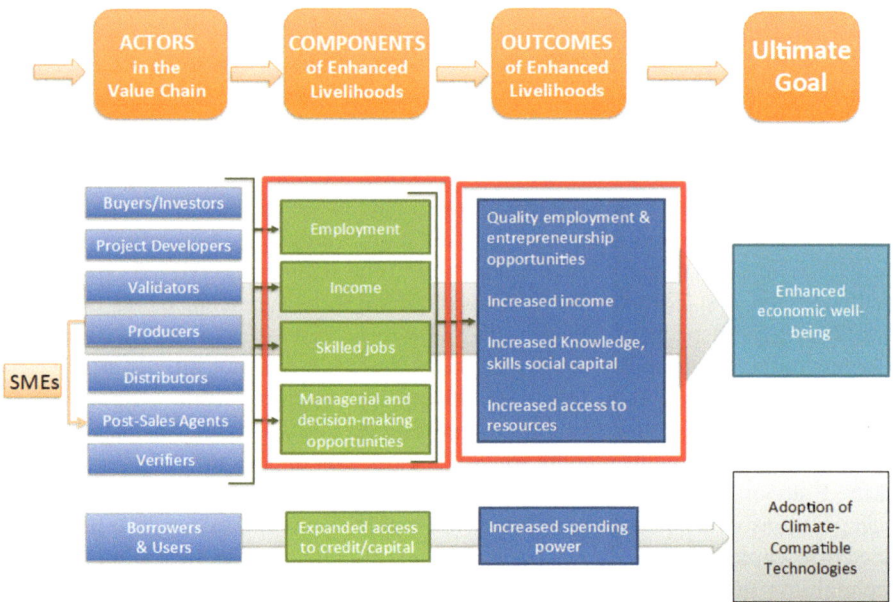

Fig. 12.2 Livelihood impact of a carbon project, conceptual framework

begin a conversation on the economic distribution of innovative environmental financing tools (Fig. 12.2).

Working from the conceptual framework, the formula for the index is as follows:

$$\text{Livelihood impact} = (2 * \text{SKL}) + (\text{PAY}) + (\text{SAT}$$

$$\text{Livelihood Index (LI)} = \frac{\text{Sum of job impact values}}{\text{Total number of jobs in value chain}}$$

The score for "Skill" includes employment in terms of jobs created and employment in terms of jobs containing skilled and managerial opportunities, thus the variable is double-weighted given that its value encompasses half of the indicators of interest in the conceptual framework. Along the employment spectrum, unskilled work means that there was no training involved for the position and managerial work implies that the employee has a degree of decision making power within the enterprise. "Pay" relates to the type of employment, given that not all of the jobs are financially compensated. Along this continuum, "Volunteer" labor includes consistent work for the carbon finance project that is paid outside of the formal economy (i.e. through company swag or promises of future employment). Commissioned labor lies at the midpoint of the "PAY" valuation scale given that there is a predictable financial gain from work effort, but the risks of project failure are born by the employee. While there is indeed the possibility of high reward through commissioned work, of the 88 people interviewed who are directly involved with the field-level implementation of the projects under review, only one person cited "commissioned work" favorably. Payment structures varied among the three projects, but there was near consensus from workers that a salary was preferable to commissioned pay. The single respondent who positively described the commissioned payment structure had been hired just 2 months prior to the interview. "Salaried" work receives the highest score in the index.

Evidence of job satisfaction (SAT) is a qualitative assessment based on the open-ended interviews wherein the self-reported ability to save and/or self-reported personal benefits from doing the job are volunteered within the interview process. All interviewees were asked to nominate their favorite and least favorite aspects of their job: mention of looking for a new job ranked at zero, while apparently genuine and detailed feelings of pride in the work and specific reasons that the job was appreciated (i.e. job location and the ability to achieve work/life balance) garnered the full rating of 1. The LI's maximum score is 4, while each variable has a scale between 0 and 1 (Table 12.1).

The LI's main utility is in comparing – rather than determining in absolute terms – the ability of a project to distribute economic benefits across the value chain. The strength of the index is that it accounts for equality – a few elite members within the value chain have little influence on the LI if the majority of workers are undercompensated. A more nuanced livelihood index would better capture how expanded access to credit, business and social networks and knowledge relate to improved livelihoods; this rough index assumes that skilled jobs will include some degree of technical and business skill, and that managerial jobs will include some component of training, networking and increased opportunity. While there are surely examples where these assumptions prove faulty, the presence of skilled

Table 12.1 Values of the livelihood index

Value Scale	0	.25	.5	.75	1
Quality employment/ Skilled Labour (SKL)	Unskilled work	Semi-skilled labor at minimum wage equivalent	Skilled, manual labor	Skilled, white-collar work	Managerial position
Employment Type/Income Type (PAY)	Unpaid, uncompensated labor		Commissioned Labor		Salaried Labor
Evidence of satisfaction through enhanced personal options (SAT)	Mentions or demonstrates desire to leave job	Explains why current job is favorable to past work	Mentions pride in work and positive aspects of the job	Mentions lifestyle benefits associated with the job and/or describes trainings and skills acquired at job	Demonstrates signs of upward mobility (refers to savings/future investments).

labor and managerial labor is not a likely hindrance to the outcomes of interest. Thus, this preliminary livelihood index offers insight into a project's ability to improve local economic well-being by focusing on the necessary (though possibly insufficient) components of an enhanced livelihood.

12.3   Field Methods

The three projects under evaluation have been registered under the Gold Standard or Voluntary Carbon Standard since 2012, enabling adequate time for the projects to perform and to begin to make an impact on the community of interest. The projects are located in Cambodia each project is national in scope. The projects all market their carbon assets as "pro-poor", "Gold Standard" or useful for sustainable development, citing community benefits as a salient marketing feature of their project in addition to the environmental benefits.

This research is based on 144 semi-structured interviews with 91 individual carbon asset managers, project managers, and financiers. Interviewees included the full range of people involved with, and impacted by, the project, including: technology producers (including designers, factory workers, supervisors, and distributors), technology promoters, micro credit agents, local banking institutions, recipient households (both husband and wife when possible), households that opted not to participate in the project, agricultural extension workers involved in project dissemination, carbon asset managers, carbon asset brokers, financiers, foreign consultants to the projects, hedge fund managers, and researchers who had previously written on or had reportedly observed my projects of interest.

In formulating the interviews and research approach, process tracing provided the analytic basis; it is a method that focuses on identifying sequential processes and mechanisms that determine outcomes of interest (Checkel 2008; Bennett 2010). Process tracing favors "thick" (in-depth) analysis of a small set of cases because of its primary interest in sequential processes within a case, as opposed to comparing correlations of data across a large N case-set. For example, the semi-structured interviews, conducted with a translator, followed a basic template designed to quantify gains (and losses) from project participation in terms of income, time, and opportunity costs, while also covering qualitative questions on the participants' assessment of their quality of life in general terms, and the impact of the project on their livelihoods and choices. Open-ended questions such as "what is your greatest concern about the project?" helped to identify the criteria for locally-relevant project success. The theory in question here relates to the belief that carbon market projects that target households and utilize point-of-use technologies for public health are going to support local, sustainable development and are therefore worthy of premium carbon credit labels such as the "Gold Standard" or the privileged position of being named "charismatic carbon" within the carbon market community. Process tracing can dig deeply into the assumption that household-scale interventions are synonymous with local development.

In addition, value chain analysis disintegrates commodity production into discrete stages – from product design to raw material acquisition to retail – to identify where high value activities are located and how they can govern the activities in the lower-value regions (Gereffi et al. 2005). In practical terms, this means interviewing every type of worker involved in the project to determine how they benefited from project participation; their salary and method of compensation and their complaints or sources of joy and pride in their work.

The organizing idea for value chain scholars is that "disintegrated production" can explain the unintended phenomena of immiserizing growth, i.e. economic growth accompanied by increased inequality (Bhagwati 1968). Importantly, high value activities are characterized as having high entry barriers – in the case of the carbon market the largest barrier to entry is technical understanding of an opaque and highly complex commodification process (Bair and Gereffi 2001). Low value activities have low-entry barriers. Consequentially, the lower rungs are subject to excess labor supply resulting in competitive pressure on wages and output. It follows that increased productivity and employment can result in diminishing economic returns for low-value activities in the chain.

Given that the carbon market was created under the Kyoto Protocol to simultaneously reduce global greenhouse gas emissions at their point of least cost while also stimulating technology transfer and development revenue by integrating developing countries into the global marketplace for green technologies, value chain analysis is a well-tailored tool to assess how geographical position and asset accumulation relate within the carbon offset context. By mapping the different pathways for economic accumulation for a patronage and a partnership style carbon project, value chain analysis can show how and how much the distribution system actually matters.

12.4   Case Study Attributes

Cambodia is a newly graduated lower-income developing country in Southeast Asia, which was a Least Developing Country prior to 2016 when the fieldwork was conducted. It has a population of 14, 864,646 and an average income of $ 2.59 USD per day. Eighty percent of the population lives in rural conditions, and 75 % of all households lack access to grid-powered electricity (GACC 2015). Cambodia suffers from one of the highest rates of deforestation in the world, in part due to the fact that over 80 % of Cambodians rely on wood and charcoal for their daily cooking and water boiling needs. While charcoal is officially banned from use, it is the de facto fuel source of choice, and its consumption alongside woodfuel accounts for more than 4.7 million tonnes of forest mass consumed annually just for domestic cooking (Nexus 2015). The economic conditions and the degree of environmental degradation within Cambodia have made it an attractive host for carbon market investments, and as such three national programs to distribute water filters, cookstoves, and household biodigester systems have been established with head-quarters in Phnom Penh.

12.4.1 Cookstove Case

The New Laos Stove (NLS) project was managed by the French NGO, "Groupe Energies Renouvelables, Environement et Solidarites" (Geres), Cambodia. The charcoal stove, designed for urban households but almost equally utilized in rural communities, has sold over a million units since 1998. Carbon finance from the voluntary market financed the rull range of the project's operational costs from 2006 to 2013, when the carbon crediting period closed (Geres 2013).

NLS utilized a low cost technology of improved biomass cookstove, valued at approximately 5 USD per unit, which is produced in local centers in region of the country known for artisanal stove production. The project developers used an "intrinsic revenue model" (Verles 2015), whereby they fund technical workshops to teach local artisans to execute their design, and then recycle funds from carbon finance into expanding the program and monitoring the implementation. Within this model, carbon credits act as a temporary subsidy for the establishment of a long-term national industry and local supply chain (idem). Given the close alignment between project participation and livelihood incentives, the value chain and the project structure are impossible to distinguish.

Geres attributes their considerable success in technology distribution to the strategic use of already existing production and dissemination networks within Kampong Ch'nang province, the traditional ceramics region of Cambodia. Utilizing historic production channels also offered monetary benefits: distributors received the technology on good faith from the producers, pedaling their wares thousands of miles away from the home factory based on generations of trust. This social aspect of the distribution system enabled the administrators to avoid financing difficulties in disseminating the locally produced stoves nationwide. However the emissions reductions per household serviced are the low, while the breadth of the dissemination and local livelihood index score for local economic gain is the strongest in the set.

The relatively high Livelihood Index score is derived from Geres' decision to train existing ceramics factories to produce their stove model, and they achieved the transition in production type through frontloading financial incentives for the producers during the training and in the first years of production. By feeding subsidies to the producers, and not to the consumers, Geres effectively transformed the cookstove producing region of Kampong Ch'nang into their improved stove model. Notably, the 35 stove factories that are registered NLS producers are all locally owned and managed, raising the LI due to the strong presence of managers, decision makers, and skilled labor positions engendered by the project. Another advantage to utilizing fully local production and distribution methods is that risk insurance, crediting, and norms for product guarantees were already in place due to the multi-generational history between the stove distributors and producers.

12.4.2 Water Filter Case

Hydrologic Ceramic Water Purifiers (CWP), has distributed over 150,000 locally produced clay waterfilters throughout rural Cambodia and is currently undergoing its first validation for the Gold Standard voluntary credit stream. The project initially received traditional donor aid from USAID in 2002 and partnered with the Red Cross to develop the CWP model; since switching to carbon finance the project is now views the Red Cross as a competitor (Hydrologic Social Enterprise 2012).

Hydrologic also utilizes an intrinsic revenue model, locating its single water filter factory outside of the national capital, also in Kampong Ch'nang. This location is not only strategic due to the localized expertise in clayware, but it is also a more residential area than the textile factories outside of Phnom Penh where the majority of the water filter factory workers previously worked.

The filters produce a greater emission reduction per unit and a significant health benefit in terms of reducing cholera and typhoid. Factory wages are similar to garment worker wages, yet laborers unanimously agreed that working at the Hydrologic factory was preferable to working at the garment district due to strategic positioning near their home (enabling mothers to remain close to their children and spend the long lunch hour with their family) and the work conditions themselves. However, unlike the NLS project where the majority of technology producers owned their own company, the workers at the CWP factory were frequently paid on commission leading to income uncertainty and distress, accounting for the lower LI score. Only one factory worker of the 15 interviewed reported using a CWP at home, which they had won at a company party. The remaining laborers interviewed said that the CWP was "too expensive" and three of the laborers interviewed mentioned that they had missed work due to "stomach and water problems."

Hydrologic has created three distribution channels: direct sales; indirect sales; and wholesale to NGOs for charitable use/emergency aid campaigns. A sales coordinator manages inventory, communicates with headquarters and trains local villagers in sales. Sales agents are paid on commission with a 5-dollar monthly stipend for gasoline; the presence of a set gasoline reimbursement incentivizes the sales agents to stick close to home and pocket the gasoline cost savings. A problematic partnership with a microfinancing NGO also hinders sales: the microcredit organization has little incentive to travel long distances to disseminate the micro-technology widely, preferring instead to offer multiple loan types within a single village for ease of administrative follow up. The absence of a reliable financing partnership is likely to undermine the program's resilience and long-term capacity.

An indirect sales channel (aka the retail channel) is somewhat simpler and the model of choice for urban areas –project managers sell the filters directly to market vendors at bulk rates. Given that urban vendors usually lack the capital to buy the filters upfront, sales supplies the vendors with filters and pays them on commission for each sale, approximately $1.50 per $23 unit. A cheaper version of the same filter (housed in a less attractive casing) is only offered to NGOs at $13 per unit. Perverse incentives exist for pharmacists who were originally targeted for retail given the health benefits associated with the technology. The pharmacists earned less money by avoiding cholera and typhoid cases than by charging the sick for treatment.

The principle difference between the NLS and the Hydrologic distribution system is that the NLS builds upon pre-existing local networks, whereas Hydrologic has built a distribution system from ground zero. An absence in social inroads, i.e. the presence of distributors who can deliver on good faith credit given their longstanding relationship with the stove producers, means that Hydrologic company must incentivize all aspects of the supply chain. In an attempt to reduce costs from salaried work, the project managers rely on commissions for the successful sale and delivery of the filter to deleterious effect for the lowest laborers on the rung: they assume risk for product failure and high turnover rates undermine the longevity of the project.

On the other hand, the project managers are Khmer nationals and receive an extraordinary amount of networking opportunities and skill enhancement by participating in the project, including international travel, exposure to the highly-specialized carbon finance project cycle, and entrée to international conferences on environment-development project design. The water filter project manager said that he is not fully satisfied with his job at Hydrologic, but that he had been able to amass adequate savings to launch his own company in the near future. Thus, while gains were less distributed in the Hydrologic model, managerial jobs offered high reward.

12.4.3   Biodigester Case

The National Biodigester Programme in Cambodia was initiated in 2002 by Dutch development agency SNV, and is now a joint collaboration with the Cambodian Ministry of Agriculture, Forestry and Fisheries. The program has installed nearly 18,000 plants to households throughout the country, 95 % of them are still in operation. Biogas plants are locally made by Khmer-run Biogas Companies; the project was certified to the Gold Standard voluntary stream in 2011. Dutch aid agency Hivos will buy all the credits (NBP 2012).

Of all the technologies, these are the most aspirational – graduating their users from biomass burning stoves to piped indoor gas burners with accompanying light fixtures for methane-fueled indoor lighting. The project manager has a policy that it will always partner with the local government, enabling it to utilize a similar technology dissemination structure as the NLS whereby inroads into the product distribution channels are already made. Government partners hail from local ministries of agriculture and livestock, and the government contribution is training through agricultural extension workers as to how the biodigesters might benefit a family that owns at least two cows. The technology is unaffordable for the very poor; the smallest biodigester costs $400 dollars and requires dung from the equivalent of two cows or four pigs in order to run. In order to ensure that poor farmers (albeit not the poorest of the population) can access the technology, the project managers have created a flat subsidy of $150 and have partnered with local banks wherein they assist in approving regular commercial loans. The default rate on the loans is an astonishing zero percent, reflecting the high savings associated with a biodigester's ability to essentially eliminate fuel and manure costs, while contributing to indoor lighting needs.

In addition to partnering with the government, the project developers in the NBP utilize a local NGO that assists in the training of masons and technicians to install the biodigesters. This workforce is trained by the project manager, and is paid on commission – though commission is substantially higher ($90 per unit) than for the Hydrologic sales agents. By training and employing masons, technicians, and involving local agriculture extension services in their marketing strategy, the NBP has managed to achieve national coverage with a seemingly unaffordable product. However, since 2012 the subsidy is being phased out and uptake has drastically declined (Tables 12.2 and 12.3).

Table 12.2 Case study attributes

Case Study Snapshot	New Laos Stove	Hydrologic	National Biodigester Program
Households serviced	2- 2.5 million	65,064	23,000
Technology deployed	Cookstoves	Water filters	Biodigesters
Total emissions reductions to date	1,200,000 tons	146,378	335,519
Certification Type	Verified Carbon Standard	Gold Standard	Gold Standard
Unit cost in dollars[a]	$5	$23	$250
Satisfaction rate[b]	Unknown	94.10%	97%
Last mile distribution mechanism	NA	Yes, for 30%	Yes, subsidy
Distribution Strategy	Local technology production and local markets for distribution	Local production and assisted distribution (markets and some subsidized market channels)	Local production and subsidized distribution
Livelihood index score	**2.53**	**2.19**	**1.80**

[a]All projects are located in a Least Developed Country except project D. Monthly income is 60–120 dollars a month in the communities of interest

[b]As evidenced by drop out rate in user surveys, reported in project documents by developers

Table 12.3 Livelihood index calculations by case

Case	Employment Functions	Jobs (#)	SKL	PAY	SAT	Job Impact	Total job impact	LI
New Laos Stove	Supplier	253	0.25	0.5	0.5	1.5	1303	2.53
	Producer	84	1	1	1	4		
	Distributors	171	1	0.5	0.75	3.25		
	Administrator	8	1	1	1	4		
Hydrologic Water Filters	Field Manager	4	1	1	1	4	378	2.19
	Carbon Sales Manager	8	1	1	1	4		
	Field Sales Agent	50	0.5	0.5	0.75	2.25		
	Distributor	30	0	0.5	0.5	1		
	Urban Sales Agent	10	0.5	0.5	0.75	2.25		
	Retailer	30	0.75	0.5	0.75	2.75		
	Factory Manager	1	1	1	1	4		
	Laborer	39	0.25	1	0.5	2		
National Biodigester Program	Administrator	6	1	1	1	4	1499	1.80
	Construction Managers	252	0.75	0.5	0.5	2.5		
	Labor Assistants	504	0.25	0.5	0.25	1.25		
	Technicians	66	1	0.5	0.75	3.25		

12.5 Discussion

There is considerable variation in the projects previously described, in terms of their approach to existing inroads in distribution networks, existing entrypoints to local markets, the quality of the technology on offer, and the unit cost. All of these projects offer a carbon-saving technology, an aspect of sustainable development benefit, and a focus on poor communities. However, by peering into the blackbox of project design and distribution strategy, it becomes apparent that dissemination method is an invisible and meaningful factor in determining a carbon project's ability to promote livelihood enhancement in the global south.

While carbon offset projects are often presented as win-win solutions, the cases presented here support an entirely different notion: development outcomes may compete rather than compliment one another. The cases with the most aspirational technologies have the lowest LI value, and the biodigester program is reliant on a donor subsidy to stimulate the local market that it creates. Further, the highest economic benefit from a carbon offset project (NLS) utilizes the least effective emissions reduction technology. While the carbon market was originally created to promote both sustainable development at the local level while reducing global greenhouse gas emissions, does this very design mask hard trade offs between the creation of locally appropriate market mechanisms and the short term delivery of modern energy technologies?

Further research is necessary to add granularity on local acceptance of the technologies, and on the long-term prospects for the technology to be adopted and disseminated. The LI must be further explored in longitudinal studies in order to determine if distribution networks with strong emphasis on local livelihood enhancement do indeed lead to longer project lifelines. Furthermore, the livelihood index may be refined to better capture prospects for upward mobility, aspirational employment, and entry into high level networks – all features of gainful employment mentioned by the GACC in their conceptual framework but poorly captured here. Still, the LI is useful as a starting point for considering how and why seemingly similar projects perform so differently in the field. These cases give weight to the view that distribution models deserve more attention in pro-poor policy design.

References

Abadie, L. M., Galarraga, I., & Rübbelke, D. (2012). An analysis of the causes of the mitigation bias in international climate finance. *Mitigation and Adaptation Strategies for Global Change, 18*(7), 943–955.

Bailis, R., Cowan, A., Berrueta, V., & Masera, O. (2009). Arresting the killer in the kitchen: The promises and pitfalls of commercializing improved cookstoves. *World Development, 37*(10), 1694–1705.

Bair, J., & Gereffi, G. (2001). Local clusters in global chains: The causes and consequences of export dynamism in Torreon's blue jeans industry. *World Development, 29*(11), 1885–1903.

Baumgartner, J., Murcott, S., & Ezzati, M. (2007). Reconsidering 'appropriate technology': The effects of operating conditions on the bacterial removal performance of two household drinking-water filter systems. *Environmental Research Letters, 2*(2), 024003.

Bennett, A. (2010). Process tracing and causal inference. In H. E. Brady & D. Collier (Eds.), *Rethinking social inquiry: Diverse tools, shared standards* (pp. 207–219). Lanham: Rowman & Littlefield. 362 p.

Bhagwati, J. N. (1968). Distortions and immiserizing growth: A generalization. *The Review of Economic Studies, 35*(4), 481–485.

Bumpus, A., & Cole, J. (2010). How can the current CDM deliver sustainable development? *WIREs Climate Change, 1*(July/August), 541–547.

Checkel, J. T. (2008). It's the process stupid! Tracing causal mechanisms in European and International Politics. In A. Klotz & D. Prakash (Eds.), *Qualitative methods in international relations a pluralist guide* (Vol. xii). Basingstoke: Palgrave Macmillan. 260 p.

Clasen, T. F., Brown, J., Collin, S., Suntura, O., & Cairncross, S. (2004). Reducing diarrhea through the use of household-based ceramic water filters: A randomized, controlled trial in rural Bolivia. *The American Journal of Tropical Medicine and Hygiene, 70*(6), 651–657.

Climate Funds Update. (2016). *Climate funds update data.* From http://www.climatefundsupdate.org/

Climate Policy Initiative. (2014). *The global landscape of climate finance 2014.*

Cohen, B. (2011). *Charismatic carbon-offset projects with co-benefits.* Triple Pundit.

Ferguson, J. (1994). *The anti-politics machine: "Development," depoliticization and bureaucratic power in Lesotho.* Minneapolis: University of Minnesota Press.

GACC. (2014). *Webinar: Defining and measuring social impact of clean cooking solutions.* Washington, DC: GACC.

GACC. (2015). Cambodia country profile, global alliance for clean cookstoves.

Gereffi, G., Humphrey, J., & Sturgeon, T. (2005). The governance of global value chains. *Review of International Political Economy, 12*(1), 78–104.

Geres. (2013). 2 million improved new Laos stoves sold in Cambodia. Retrieved December 17, 2015.

Haya, B. (2007). *Failed mechanism: How the CDM is subsidizing hydro developers and harming the Kyoto protocol* (p. 12). Berkeley: International Rivers.

Hydrologic Social Enterprise. (2012). Hydrologic social enterprise. Retrieved December 18, 2012, from http://www.hydrologichealth.com/

Kossoy, A., & Guigon, P. (2012). *State and trends of the carbon market 2012*. Washington, DC: The World Bank.

Kwame Sundaram, J., Chowdhury, A., Sharma, K., & Platz, D. (2016). *Public private partnerships and the 2030 agenda for sustainable development: Fit for purpose?* (D. W. P. N. 148). New York: UN.

Lantagne, D., Meierhofer, R., Allgood, G., McGuigan, K., & Quick, R. (2008). Comment on "Point of use household drinking water filtration: A practical, effective solution for providing sustained access to safe drinking water in the developing world". *Environmental Science & Technology, 43*(3), 968–969.

Levine, D., Beltramo, T., Harrell, S., Toombs, C., & Young, J. (2013). A guide to optimizing behavior change in fuel efficient stove programs.

Mayrhofer, J. P., & Gupta, J. (2016). The science and politics of co-benefits in climate policy. *Environmental Science & Policy, 57*, 22–30.

Mobarak, A. M., Dwivedi, P., Bailis, R., Hildemann, L., & Miller, G. (2012). Low demand for nontraditional cookstove technologies. *Proceedings of the National Academy of Sciences of the United States of America, 109*(27), 10815–10820.

NBP. (2012). *National biodigester programme*. Retrieved April 4, 2012, from http://www.nbp.org. kh/page.php?fid=3

Nexus. (2015). Cambodia improved cookstoves Phnom Penh, Nexus carbon for development

Olsen, K. H., & Fenhann, J. (2006). *Sustainable development benefits of clean development projects* (p. 28). Roskilde: UNEP-Risoe.

Peters-Stanley, M. (2013). *Maneuvering the Mosaic: State of the voluntary carbon markets 2013*. Washington, DC: E. Marketplace, Ecosystem Marketplace.

Scott, J.C. (1998). Seeing like a state how certain schemes to improve the human condition have failed. *Yale Agrarian studies* (xiv, 445 p). New Haven: Yale University Press

Shankar, A., Johnson, M., Kay, E., Pannu, R., Beltramo, T., Derby, E., Harrell, S., Davis, C., & Petach, H. (2014). Maximizing the benefits of improved cookstoves: Moving from acquisition to correct and consistent use. *Global Health: Science and Practice, 2*(3), 268–274.

Simon, G. L., Bumpus, A. G., & Mann, P. (2012). Win-win scenarios at the climate–development interface: Challenges and opportunities for stove replacement programs through carbon finance. *Global Environmental Change, 22*(1), 275–287.

Sutter, C., & Parreño, J. C. (2007). Does the current Clean Development Mechanism (CDM) deliver its sustainable development claim? An analysis of officially registered CDM projects. *Climatic Change, 84*(1), 75–90.

The Gold Standard. (2010). *The Gold Standard annual report 2009* (p. 1). Geneva: The Gold Standard Foundation.

Troncoso, K. (2014). *A recipe for adoption and impact indices*. GACC, GACC.

Troncoso, K., Castillo, A., Masera, O., & Merino, L. (2007). Social perceptions about a technological innovation for fuelwood cooking: Case study in rural Mexico. *Energy Policy, 35*(5), 2799–2810.

UNFCCC. (1997). *Kyoto protocol to the United Nations framework convention on climate change* (p. 23).

Verles, M. (2015). Correspondence with Gold Standard CEO Marion Verles. Geneva.

Verles, M., & Santini, M. (2012). Pro-poor carbon projects: Challenges and perspectives (Newsletter #20). Carbon Market Watch.

Visseren-Hamakers, I. J., McDermott, C., Vijge, M. J., & Cashore, B. (2012). Trade-offs, co-benefits and safeguards: Current debates on the breadth of REDD+. *Current Opinion in Environmental Sustainability, 4*(6), 646–653.

Wang, Y., Bailis, R., & Hyman, J. (2015). *Carbon for clean cooking: A review of household energy interventions under the carbon markets*. New Haven: Yale School of Forestry of Environmental Studies.

World Bank. (2002). *Community development carbon fund*. World Bank.

Part III
Climate Change Adaptation

Chapter 13
What Do Evaluations Tell Us About Climate Change Adaptation? Meta-analysis with a Realist Approach

Takaaki Miyaguchi and Juha I. Uitto

Abstract Evaluating climate change adaptation (CCA) interventions has yet proved to be a difficult task, as they involve a number of different stakeholders, time and geographical scale and political jurisdictions. As one effort to shed light on the subject, this paper presents the methodology and the results of a meta-analysis of ex-post evaluations of CCA programmes using a realist approach. This paper analyses CCA programmes in nine countries: Armenia, Egypt, Malawi, Mozambique, Namibia, the Philippines, Tanzania, Turkey and Zimbabwe. Together with their respective host governments, these programmes were implemented by either UNDP or various United Nations partner agencies and have already been evaluated by independent evaluators. Based on the analytical frameworks for evaluating CCA interventions, the authors hypothesized a number of key context, mechanism, and outcome configurations, which are considered vital in realist evaluation approach but have not yet been widely tested in the field of CCA. Although ex-post evaluations of multi-donor funded projects tend to be prepared out of bureaucratic requirement, the analytical method used in this paper, if used carefully, can unearth otherwise hidden important lessons and provide useful explanations. The results of the analysis can indicate that adopting a realist approach to complex development projects, such as these CCA programmes, is indeed a useful way of providing applicable explanations, rather than judgments, of what types of interventions may work for whom, how and in what circumstances for future CCA programming.

Keywords Realist approach • Climate change adaptation • Meta-analysis

T. Miyaguchi (✉)
Ritsumeikan University, Kyoto, Japan
e-mail: takaakinet@gmail.com

J.I. Uitto
Independent Evaluation Office, Global Environment Facility, Washington, DC, USA
e-mail: juitto@thegef.org

© The Author(s) 2017
J.I. Uitto et al. (eds.), *Evaluating Climate Change Action for Sustainable Development*, DOI 10.1007/978-3-319-43702-6_13

235

13.1 Introduction

Climate change is a reality. Although it is important to acknowledge that the evidence of the linkage between rising economic loss of disasters and climate change has not been statistically established (Pielke 2014), changing precipitation and temperature patterns, as well as occasional hydro-meteorological extreme events, such as floods, droughts and landslides, have been hitting people especially at the community level, who have to rely on natural resources for their daily substance (Global Humanitarian Forum 2009). Reflecting the urgency and importance of climate change, the donor community for the past decade has been funding a number of climate change programmes in developing countries in close collaboration with host governments and various UN agencies. And it is in recent years that their initial implementation cycles have been completed and subsequently their ex-post evaluations have been conducted. In the meantime, discussions regarding evaluation practice, its criteria and framework specifically tailored to climate change projects and programmes have taken place, most notably through such communities of practice as Climate-Eval, the International Development Evaluation Association, and United Nations Evaluation Group.

Discussions in such arena have highlighted a number of difficulties related to evaluating climate change projects and programmes, including shifts in the objects of evaluation, new metrics, and greater focus on risk, uncertainty and complexity (Picciotto 2009). More specifically, evaluation of climate change adaptation (CCA) projects and programmes poses a number of difficulties and complications. For example, Valencia (2009) lists five types of such features: (1) "success" of CCA is when nothing happens; (2) evaluation of CCA occurs too early to tell whether the intervention has successfully withstood the projected impacts; (3) there are uncertainties of climate scenarios; (4) short-term weather variability disguises effectiveness of adaptation measures; and (5) contribution rather than attribution should be emphasized, because of the complexity of "overall adaptation process that is largely shaped by external factors" (Bours et al. 2014).

Even though very few evaluations on CCA have been conducted so far (Feinstein 2009), Uitto (2014) emphasizes the need of the evaluation community to start building "an adequate body of evaluative evidence" from this area in order to synthesize the lessons.

13.2 Approach and Study Material

In light of such background, the purpose of this paper is to adopt and test a certain philosophical lens, called critical realism, to a meta-analysis of CCA evaluation reports and to show implications of this approach for the current as well as future CCA programming.

The study material used was the evaluation reports of those CCA programmes that: (1) have been implemented by UNDP and other United Nations agencies; (2) have finished initial implementation cycles; and (3) have been subject to terminal evaluations. One of the unique aspects of these identified CCA programmes is that they represent the first evaluation results of the completed CCA programmes within the UNDP system (as of November 2014). Out of a total of 11, nine CCA programmes were selected based on the criterion that the quality of the evaluation reports was rated to be moderately satisfactory or higher by the UNDP Independent Evaluation Office.[1] The authors conducted a meta-analysis of those ex-post evaluations by closely examining and comparing the contents of the evaluations by applying the philosophical lens of critical realism.

The nine programmes included were implemented in the following nine countries: Armenia, Egypt, Malawi, Mozambique, Namibia, the Philippines, Tanzania, Turkey and Zimbabwe (see Table 13.1 for summary). As the table shows, within the context of UN programming, these programmes vary in many aspects: the funding source (such as Global Environment Facility, Millennium Development Goals Achievement Fund, and United Nations internal resources); types of beneficiaries, target audiences and geographic regions (ranging from local vulnerable communities to inter-ministerial mainstreaming at the government level); and implementation modalities (including UNDP stand-alone, United Nations interagency joint programming and Delivering as One[2]).

This paper presents the findings of the meta-analysis conducted of the nine evaluation reports. Although the programmes evaluated vary from one another in many aspects, what is common is the structure of the evaluation reports. Each report consists of four major sections, each of which covers a specific evaluation criterion: relevance, efficiency, effectiveness and sustainability.

The evaluators who conducted the nine CCA programme evaluations all utilised the definitions of each criterion in Table 13.2, which are based on the OECD evaluation criteria adapted by UNDP and its partners (OECD 2002).[3]

---

[1] It was done through UNDP IEO's quality assurance exercise. It is concerned with the quality of how evaluation report is written by checking whether the structure of evaluation reports includes the necessary sections and a proper evaluation framework has been put in place. Thus "moderately satisfactory" or above rated evaluation reports do not necessarily mean high quality of project *activity results* themselves.

[2] Although there is no unified definition of Delivering as One modality (UN 2012), it should entail "Four Ones", i.e. one leader, one programme, one budget and one office amongst different agencies of the UN system. Joint Programm*ing*, is often contrasted with Joint Programm*es*, where the latter implies a set of discrete but related programmes by UN agencies and the former implies joint efforts even from the stage of planning and designing of a programme, which is also to be implemented together.

[3] The authors are aware of criticism pertaining to the rather narrow application of the criteria internally towards interventions (for instance, relevance could include whether the intervention is contributing to positive change and the achievement of impact; and sustainability should include not only the continued benefits from the intervention but whether the intervention contributes to broader sustainable development). However, as these criteria are widely used in the evaluations in the narrow sense, this understanding is appropriate for our analysis.

Table 13.1 List of the CCA programme/project evaluation reports reviewed

Country	Programme/project title	Duration (months)	Implementation modality (funding source)
Armenia	Adaptation to climate change impacts in mountain forest ecosystems of Armenia	May 2009 – Jun 2013 (50 m)	UNDP (GEF)
Egypt	Joint programme: climate change risk management in Egypt	Oct 2008 – Apr 2013 (55 m)	JP (MDG-F)
Malawi	The national programme for managing climate change in Malawi	Apr 2010 – Dec 2012 (33 m)	UNDP (AAP)
Mozambique	Joint programme on environmental mainstreaming and adaptation to climate change in Mozambique	Sep 2008 – Aug 2012 (48 m)	JP (MDG-F)
Namibia	Namibia country pilot partnership programme; adapting to climate change through the improvement of traditional crops and livestock farming	Jun 2007 – Dec 2011 (55 m)	UNDP (GEF)
Philippines	Joint programme: strengthening the Philippines' institutional capacity to adapt to climate change	Dec 2008 – Dec 2011 (37 m)	JP (MDG-F)
Tanzania	Joint programme on environment with a focus on climate change, land degradation/ desertification and natural resources management	Oct/Dec 2009 – Jun 2011 (21 m)	JP (MDG-F)
Turkey	Joint programme on enhancing the xapacity of Turkey to adapt to climate change	Apr 2008 – Dec 2011 (45 m)	JP (MDG-F)
Zimbabwe	Coping with drought and climate change in Zimbabwe project	May 2008 – Sep 2012 (53 m)	UNDP (GEF)

JP Joint Programme, *MDG-F* Millennium Development Goal Achievement Fund, *GEF* Global Environment Facility, *AAP* Africa Adaptation Programme

Table 13.2 Definitions of evaluation criteria

Criteria	OECD definition
Relevance	The extent to which the objectives of a development intervention are consistent with beneficiaries' requirements, country needs, global priorities and partners' and donors' policies
Efficiency	A measure of how economically resources/inputs (funds, expertise, time, etc.) are converted to results
Effectiveness	The extent to which the development intervention's objectives were achieved, or are expected to be achieved, taking into account their relative importance
Sustainability	The continuation of benefits from a development intervention after major development assistance has been completed. The probability of continued long-term benefits. The resilience to risk of the net benefit flows over time

Source: OECD (2002)

13.3   Realist Approach

This meta-analysis was conducted using a philosophical lens called critical realism. In evaluation, the realist approach emphasizes underlying assumptions about the way certain interventions are expected to yield certain outcomes in a certain context (Pawson and Tilley 2004). It thus defies the deterministic worldview which is symbolized as "if X happens, it automatically produces outcome Y." Such a linear, sequential worldview is considered deterministic or *positivistic*, in that hypothesized theories of change are thought to work *regardless of the context* within which theories of change are situated. In other words, deterministic theory of change does not give us the explanations as to "for whom such interventions may work, in what circumstances, and how" (Pawson and Tilley 1997). Moreover, although the deterministic findings can tell us what interventions may have worked in certain countries under certain conditions (*"there"*), they may not tell us for whom these successful interventions are expected to work, under what circumstances, and how (*"here"*). The realist approach thus resonates with evidence-based policy making in that it is thought to be useful in answering the important evaluation question, i.e. "it worked *there*, but will it work *here*?" (Cartwright and Hardie 2012).

The following sections, however, first present the results of the meta-analysis that are considered *deterministic* in nature, immediately followed by non-deterministic ones and how the realist approach is applied. The intention behind this structure is to emphasize the characteristics of critical realism philosophy. Deterministic findings *appear* to help evaluators to know whether certain interventions work or not for achieving key outcomes, but such a deterministic approach is what a realist approach attempts to defy.

The realist approach belongs to the school of theory-based evaluation (Stern et al. 2012). The realist approach is based on a school of thought in a philosophy of science, called critical realism. The concept of critical realism has been most significantly developed by Roy Bhaskar.[4] Critical realism can provide a useful lens especially in social sciences for the world that is "structured, differentiated, stratified and changing," and recognizes the shift of emphasis "to what produces the events – not just to the events themselves." (Danermark et al. 2002). An evaluation approach based on critical realism is thus an "intuitively appealing approach to those trying to expose and unpack the complexities of contexts and interrelated mechanisms underlying implementation activity" (Rycroft-Malone et al. 2012). The use of this evaluation approach is thus considered appropriate in the complex experience of CCA projects. Adoption of critical realism in evaluation field (principally in public health and criminology) has significantly progressed thanks to the work of Pawson (2013), Pawson et al. (2004), Pawson and Tilley (1997, 2004), and Wong et al. (2013) and other scholars.

---

[4]His most notable works include *The Possibility of Naturalism* (1979) and *A Realist Theory of Science* (2008).

However, a realist approach has not been widely conducted in international development, although some cases are found in a type of systematic reviews, e.g. Betts (2013). This meta-analysis is one such attempt. Quite unlike the conditions in making laboratory type experiments possible ("closed system"), critical realism acknowledges that the world is an "open system" consisting of things possessing causal powers (and also their potentialities) situated within many layers of structures (Bhaskar 2008). And because the world that people live in is an open system, it tells us that, unlike natural science, social science cannot predict things or present the world with successionist, cause-and-effects sequences.

The realist approach pays close attention to "contextual conditions" and how they influence mechanisms that generate (different) outcomes. It is a continuous, not a one-off, process of identifying specific contexts that may trigger some generative mechanisms to generate an outcome. Realist approach is thus about hypothesizing, selecting and refining so-called CMO (Context + Mechanism = Outcome) configurations.

13.4 Meta-analysis Conducted

The structure of the evaluations of the nine CCA programmes is based on the four evaluation criteria, i.e. relevance, efficiency, effectiveness and sustainability. Within this analysis framework, these criteria are considered as "outcomes" that lead to the ultimate CCA programme objectives. Within each outcome, there are several important intermediate outcomes (IOs) identified through the meta-analysis. Each IO is reported to have been influenced by a number of interventions on the ground.

According to Weiss (1997), a theory of change consists of two kinds of theories, i.e. implementation theory and programme theory. Implementation theory mainly pertains to programme activities or interventions themselves. It represents the assumptions that if certain interventions are implemented as planned, they are thought to generate desired results. Programme theory on the other hand represents the "ideas and assumptions [that] link the programme's inputs to attainment of the desired ends" (Weiss 1997). It is not just what the programme activities are expected to achieve, but also *how*. The essence of such interventions and programme theories can be considered as a generative mechanism according to the realist approach and within CMO configurations.

The authors first extracted every single evaluative remark of these evaluations, each of which is categorized either 'positive' or 'negative'. It altogether resulted in a total of 577 remarks gleaned out of the nine evaluations. Each of these remarks belonged to one or multiple evaluation criteria (i.e. relevance, efficiency, effectiveness, and sustainability). These remarks were then clustered according to: the evaluation criteria (i.e. outcomes[5]); intermediate outcomes (IOs) that lead to each evaluation criterion; and types of programme interventions implemented in achieving each IO. What this step enabled was a comparative analysis of the CCA programmes where similar interventions or activities across different CCA programmes were implemented. In other words, the meta-analysis conducted the following steps: identification and extraction of key IOs toward an outcome (each evaluation criterion); categorization of interventions to generate the corresponding IOs; development of hypothesis of programme theories that necessarily lead to an IO. And since this meta-analysis is based on the realist approach, it then sought contextual conditions that may or may not activate an underlying mechanism in generating IOs, and thus outcomes. It sought to identify theories of change for each outcome (evaluation criterion).

The following sections present first the M-O (mechanism = outcome) combinations for each criterion that can be estimated from analysing the CCA evaluations; and second, C (context) conditions which may or may not activate these M-O combinations, thereby showing a set of hypothesized CMO configurations. Each criterion is presented first only with M-O sequences, which represents a deterministic view. The latter half of the sections presents the contextual conditions, thereby completing the presentation of the hypothesized CMO configurations. Tables in the following sections present the summary of C-M-O configurations.

13.5 Mechanism-Outcome Sequences

13.5.1 Relevance M-O Sequences

Overall, a high degree of relevance is seen in all the studied CCA programmes. The joint programme for managing climate change risks in Egypt is found to be highly relevant in supporting Egypt to develop its climate change adaptation strategies. The programme in Mozambique is also found to be highly relevant to the national policy context, responding to the necessity to support institutional progress on CCA. Armenia's programme focusing on its mountain forest ecosystem was evaluated to be well aligned with the national needs and priorities. Nonetheless, the aspect of relevance does not end with alignment at a national level. Tanzania's programme has addressed problems of fuelwood availability and other means of

---

[5]Note that these four evaluation criteria are used as "outcomes (O)" within the CMO configurations. In each of the four criteria, the authors have hypothesised certain sets of CMO configurations. For example, efficiency criterion – which itself is the relationship between inputs and outputs – a CMO configuration will treat efficiency itself as "O" (outcome) that is achieved through several key IOs, through generative mechanism ("M"), under certain context, ("C"). Thus within each evaluation criterion, CMO configurations were constructed, even when one criterion is not related to (project's overall) outcome.

improving livelihoods amongst local communities, reflecting the issues that had been considered high priority at a local level.

Through comparing the interventions taken place in each of the nine programmes from the point of view of the relevance criterion, the following theory of change was developed: "close coordination and working relationship with the national and local government enables both partners (government and United Nations implementing agency) to develop an appropriate CCA programme." Here, the implementation theory part represents the type of similarly implemented interventions, and the programme theory part is a hypothesized mechanism of change attached to such implementation theory.

13.5.2 Efficiency M-O Sequences

Unlike relevance, for which it was relatively straightforward to construct a theory of change, all the other evaluation criteria were not necessarily straightforward, since each of the criteria can contain a number of different IOs to achieving a high level of an outcome. For the efficiency criterion, a number of IOs that helped achieve a high level of efficiency outcome were identified. The analysis was done by comparing similar interventions that were reported to have worked across the nine programmes.

As a result of a meta-analysis, stakeholder involvement at an early stage was identified as the first "recommended practice" to ensuring a high level of efficiency. In the Armenia, Mozambique and Zimbabwe programmes, there was active engagement of the stakeholders at a programme identification and planning stage. A corresponding hypothesis (i.e. programme theory) is that such an intervention activity fosters a high level of motivation and sense of ownership to the programme.

Four programmes, i.e. Egypt, Turkey, Armenia and Namibia, were reported to have achieved a high level of efficiency through strong financial controls, swift reporting, clarified roles and responsibilities and adaptive management through which the programmes were quick in responding to the changing needs and priorities of the beneficiaries on the ground. One way to achieving a high level of efficiency can thus be such interventions as adoption of adaptive management and clearly defined roles and responsibilities for involved parties. A corresponding programme theory can be that such adaptive management activities enable the programmes to attend to the needs and demands of the local beneficiaries whilst maintaining the ultimate programme goal.

13.5.3 Effectiveness M-O Sequences

The effectiveness criterion presents one of the most important aspects of programme's success. Analysing the positive remarks found in the evaluation reports of the studied programmes has revealed that a high level of effectiveness is achieved, amongst others, through an IO of development of adaptive capacity and utilization of adaptive measures introduced by the programmes.

As a means to achieve such IO, training and transfer of techniques and practices for reducing the stakeholders' vulnerability seemed to have ensured a high level of effectiveness of CCA programmes. Eight out of the nine programmes reported such activities and thus were evaluated positively for their effectiveness. For example, in Egypt, adaptive capacity was further enhanced within the Ministry of Agriculture and Land Reclamation in order for government staff to be able to forecast future scenarios in water and agriculture sectors. In Zimbabwe, a more accurate system of weather forecasts was introduced and capacity to manage the system was developed, thereby enabling high quality crop planting advice given to farmers. In Tanzania, the establishment of an environmental information system and a national environmental web portal were considered to be highly relevant adaptive measures that were introduced by the programme. The Namibia programme introduced such adaptive measures as dryland crop farming, conservation agriculture and improved seeds, and a drip irrigation system, all of which are reported to have played an important role in achieving a high level of effectiveness. A corresponding theory of change can thus be hypothesized as follows: "introduced adaptive measures and developed adaptive capacity facilitate these skills, techniques and knowledge to be kept applied and used."

Realizing a wide range and level of mainstreaming is considered to be another IO in making a programme more effective. For example, in Turkey, a national climate change adaptation strategy and action plan was drafted and henceforth expected to be approved by a high level climate change coordination board. In Armenia, the introduced adaptive measures by the programme were successfully incorporated into an existing infrastructure that manages mountain forest ecosystems, including policy, legislation, institutions, procedures and mechanisms. In order to achieve such IO, provision of relevant technical, policy and advisory support to relevant stakeholders, from government staff to rural farmers have been reported to be effective. The corresponding programme theory here can be that provision of technical, policy and advisory support facilitates integration with "business-as-usual" infrastructures.

Another important IO that can lead to high effectiveness is a high level of awareness amongst the general public. Development and dissemination through documentary films, social network groups, large scale public events, TV and newspapers were seen in Egypt, Zimbabwe, the Philippines, Tanzania and Armenia. All these activities were reported to have contributed to realizing a high level of effectiveness by increasing awareness amongst the general public. One can thus infer that, in order to ensure a high level of effectiveness of a CCA programme, it is important to utilize various media, including face-to-face events, for wider publicity. A hypothesized programme theory here is that these events can attract attention and boost interest toward CCA amongst citizens.

13.5.4   Sustainability M-O Sequences

Since the studied evaluation reports were prepared right after the completion of programme activities, which corresponds to the second challenge discussed by Valencia (2009), it poses a significant challenge to evaluating the programme's long-term sustainability. The meta-analysis nonetheless could identify some of the pertinent IOs and interventions, even if these were not explicitly identified in the evaluation reports.

The first IO for sustainability is "sustained built adaptive capacity, and a high utilization level of introduced adaptive measures." Here an emphasis should be placed for *sustaining* (and not just one-off training of) the adaptive capacity that is built through programme activities, and a high level of *utilization* (and not just mere introduction) of adaptive measures. Hypothesized programme theory to ensuring them seems that such interventions foster a sense of ownership towards built capacities and introduced adaptive measures.

Sustained and high level of stakeholder engagement was identified as the second IO toward sustainability. The CCA programme in the Philippines has made sure that national and local partners continue similar activities and outputs that have been introduced by the programme. A hitherto non-existent network of environmental specialists was formed under the programme in Tanzania which since enabled all partners to work collaboratively.

The third IO identified was that mainstreaming at central policy and planning level is successful and sustained. The CCA programme of Tanzania has implemented its activities within the national institutional framework fully aligned with their national environmental policies. The programme also adopted a crosscutting framework in order to mainstream environment and climate change issues into plans and policies of multiple sectors in the country. Similarly, in Mozambique, the programme has successfully integrated CCA activities in the country's district-level strategic development and socioeconomic plan, the land use plan as well as integrated waste management plan. A theory of change, which is the combination of implementation theory and programme theory, can thus be hypothesized that CCA programme activities that are implemented within the local/ national and existing institutional frameworks can foster a sense of ownership and trigger smooth integration in the target country's planning and policies.

Fourth, high likelihood of generating broader adoption and replication is considered to be another IO that leads to a high level of sustainability. Introduction of adaptive measures to the stakeholders and institutions with relevant mandates seems to have yielded favourable results in achieving this positive IO. The programme activities in Egypt were well embedded into the work of the Agricultural Research Centre, whose relevant mandate successfully incorporated the new climate change risk research. A partnering technical university in Turkey is reported to be continuing to conduct a CCA related certification course which had been developed as part of the programme. A theory of change corresponding to this IO generation can be that the introduction of adaptive measures to the institutions already with relevant mandates can realize 'rooting' of such measures inside the institutions.

13.6   Contextual Conditions

Presented above was a series of M-O sequences without taking the contextual conditions into consideration. Such M-O only sequence, if used as it is, presents a deterministic view. Under such view, an underlying mechanism in generating above-mentioned IOs, namely the essence of programme theory, is believed to function everywhere, anytime, regardless of varying contexts. However, realist approach pays closer attention to the contextual conditions that necessarily allow such mechanism to function. In order to identify the contextual conditions, one needs first pay attention to those incidences where the identified theory of change did *not* work, i.e. those that have generated *negative* IOs. A general tendency amongst many meta-analyses of evaluation reports is to report what has *worked* in the effort to present so-called "best practices" by paying close attention to successful interventions and their programme theories. That approach risks missing lessons from failed interventions or strategies that may have worked only under specific conditions. The section below presents the findings about contextual conditions that have enabled (and not) a certain theory of change to work.

13.6.1   Context for Relevance

Almost all the evaluative remarks pertaining to the relevance criterion reported positive outcomes. But when focusing on those few incidences that were reported to have yielded slightly negative IOs, one can unearth the contextual conditions that may have helped this theory of change to trigger more successful IOs. In the case of Mozambique, even though there had been close coordination and working relationship with the national and local governments, relevance at a sub-national level was not considered high. In this case, local CCA priorities may not have been identified by the local governments and local partners. Similarly, in Turkey, because of abrupt insertion of carbon-footprint offsetting activities as part of CCA vulnerability reduction (though it is essentially for climate change mitigation), the relevance level of this programme was not evaluated to be high.

From those incidences, one can hypothesize another contextual condition that may have allowed a theory of change (in this case in generating positive IOs for securing a high level of relevance) to work, i.e. that host government and line ministries have identified national and sectoral CCA priorities, or fully internalized the programme objectives specifically targeting adaptation. A set of identified CMO configurations for relevance criterion is shown in Table 13.3.

Table 13.3 Identified CMO configurations for relevance criterion

Context		Theory of change		Intermediate outcome	Outcome/ criterion
Host government and line ministries have already identified national and sectoral CCA priorities, and understand programme objective	+	Close coordination and working relationship with the national and local government enables both partners (government and United Nations implementing agency) to develop an appropriate CCA programme	=	High relevance of programme strategy and intervention components with national and global priorities	Relevance
Local CCA priorities are identified by the local government and local partners					

Here, a theory of change as a whole is categorically treated as CMO's "M". In developing this table, the authors have referred to the way Pawson in his work illustrated, e.g. in Chapter 5 of Pawson and Tilley (1997). However the authors are of the view that the identity of so-called "generative mechanism" is the essence of programme theory; thus a theory of change itself is not the same as "M", the mechanism. A similar argument is developed by Blamey and Mackenzie (2007)

13.6.2   Context for Efficiency

Referring to the estimated theory of change for realizing high stakeholder involvement, which is considered to be one of the key IOs in securing a high level of efficiency. Building partnerships at an early stage seems a common-sensical intervention to yield this IO. However, as reported in the case of Zimbabwe, even if partnerships are established at an early stage, when participating stakeholders are not well aware of CCA issues and risks and the CCA programme's objectives, it is not likely for this corresponding theory of change to trigger a positive IO. Another contextual condition which can be identified for this theory of change from all of the studied evaluations is that the programme design is sector specific and focused rather than broad. Though this may not be a "recommended" context for a CCA programme because it can seem to be promoting a "silo" or sector-driven programme design, the degree of programme interventions' focus seems to have enabled this theory of change to realize a high level of stakeholder involvement.

The second theory of change relates to another IO, i.e. level of programme management achievements. When a national programme management team (case of Tanzania) or national steering committee (case of Malawi) have not shown adequate leadership, the corresponding programme theory did not produce positive results. The more sector specific and focused the programme design is, the more positive patterns of results concerning this theory of change seem to be generated. Through a deterministic meta-analysis represented by mere M-O sequence, one could have ended the analysis in recommending adaptive management and clarified roles and responsibilities of the involved parties. The realistic approach can facilitate our thinking regarding the necessary contextual conditions and their

Table 13.4 Identified CMO configurations for efficiency criterion

Context		Theory of change		Intermediate outcome	Outcome/ criterion
Relevant stakeholders are supportive of United Nations and well aware of CCA issues and risks	+	Partnerships with stakeholders are built at an early stage, where they feel more motivated to participate in the programme	=	High stakeholder involvement	Efficiency
Sector specific and focused programme design					
Strong leadership from national executing agency	+	Adaptive management and clearly defined roles and responsibilities to each party enable the programme to attend to the needs and demands of the local beneficiaries whilst maintaining the ultimate programme goal	=	High level of programme management achievements	
Sector and region specific scope of programme					

hypotheses. A set of identified CMO configurations for efficiency criterion is shown in Table 13.4.

13.6.3 Context for Effectiveness

There are three theories of change identified for the criterion of effectiveness. First one refers to the positive IO of a high level of adaptive capacity built and utilisation of adaptive measures. Contrary to the deterministic approach which automatically assumes the power of a mechanism (of generating an outcome) fully exercised regardless of context, realist approach pays close attention to the very structure wherein a mechanism is situated. For example, one contextual condition is the level of awareness of a local government partner. Local government partners can play a critical role in translating introduced adaptive measures and built adaptive capacity into actual benefits of the vulnerable people on the ground, such as rural farmers. If the introduced adaptive measures or built capacity is not clear to such partners, their utilization level can be quite limited. This refers to a case of Turkey where seasonal weather forecasts information provided over internet was introduced and related know-how taught. But since the end-users, e.g. rural farmers, were not reached, even though implementation theory may have held, the corresponding programme theory was not realized. Another contextual condition is where the types of adaptive capacity and adaptive measures are clear and well understood by those involved parties. In Namibia, the meteorological climate decision support tools were introduced to a government agency, but since the types of adaptive measures were not clear, introduced adaptive measures or built capacity did not generate a positive IO.

Second, in order for the theory of change for the wide range of mainstreaming to work, one can hypothesize, as part of the necessary contextual conditions, that relevant ministries and stakeholders should be highly aware of the climate risks and the vital importance of reducing vulnerability. A relevant contextual condition that is applicable for this theory of change is where government officials understand the actual need to integrate CCA issues in their business-as-usual activities. A case of Zimbabwe described the situation that, even though relevant and technical support was introduced, senior government officials did not fully appreciate the significance of such support, which thus did not yield a positive IO.

The third theory of change is about the raised level of awareness amongst the general public and government staff, since the level of awareness amongst them is considered key to achieving a high level of effectiveness. A relevant contextual condition for this theory of change that may alter the results of IO (i.e. high/low level of awareness) is that the general public is relatively unaware or lack knowledge of climate change risks. This condition should also be recognized as an important baseline situation under which planned interventions may trigger the corresponding programme theory in generating a positive IO. A set of identified CMO configurations for effectiveness criterion is shown in Table 13.5.

13.6.4   Context for Sustainability

For this criterion, a high likelihood for sustaining built adaptive capacity and high utilisation level of adaptive measures introduced is considered to be one of the important IOs. In order for the corresponding theory of change for this IO to work, it is first necessary for the introduced adaptive capacities and measures to be those types that are needed and requested by end-users themselves (which was not the case in Mozambique). Sustained political interest towards the CCA programme's intended objectives also need be present as another contextual condition that helps this theory of change to exercise its generative power.

Another IO that can contribute to achieving a high level of effectiveness is high likelihood for sustained, high level stakeholder engagement. One hypothesis for the contextual condition is where beneficiaries on the ground and government continue to be present and see the need and benefits in engaging themselves in the CCA programme's intended objectives. This context can be hypothesized since there was one country case (Egypt) where the ultimate beneficiaries of the CCA programme, i.e. farmers, had not been in the programme activity process, which has negatively contributed to the sustainability element of this programme. Under such circumstances, though the corresponding implementation theory was held in all the programmes, the programme theory did not get to generate a positive outcome, if such contextual condition was not met.

The third IO in this criterion is about sustained level of mainstreaming at central policy and planning level. The corresponding implementation theory makes

Table 13.5 Identified CMO configurations for effectiveness criterion

Context		Theory of change		Intermediate outcome	Outcome/criterion
Specific types of skills that they need to acquire are clear to them	+	Training and transfer of techniques and practices for the relevant people facilitate these skills, techniques and knowledge to be applied and used	=	High level of adaptive capacity and utilisation of adaptive measures	Effectiveness
Specifically identified types of participants are well aware of the climate risks					
Relevant ministries and stakeholders are highly aware of the climate risks and the vital importance of reducing vulnerability	+	Provision of relevant technical, policy and advisory support to relevant people (from government staff to rural farmers) facilitates its integration with their "business-as-usual" activities	=	Wide range of mainstreaming	
General citizens are relatively unaware or lack knowledge of climate change and associated risks	+	TV, newspaper and symposium for wider publicity attract attention and boost curiosity in citizens about CCA issues	=	Raised level of awareness amongst the general public	

intuitive sense in that in order to mainstream CCA programme activities, they ought to be implemented within an existing local or national framework. However, in order for the corresponding programme theory to function and exercise its power, it seems to require a certain contextual condition where government counterparts understand the need of mainstreaming and a relatively high motivation level is found amongst government officials. One case (Zimbabwe) is reported to have designed and implemented a set of mainstreaming activities at central government level, but due to a lack of motivation of government counterparts, this theory of change did not see its generative power exercised.

The fourth IO pattern identified is about a high likelihood of generating broader adoption and replication in the long term. There are several cases identified through the meta-analysis where the corresponding theory of change did not generate such positive IO. The contextual conditions that can be extracted from these cases (Egypt and Mozambique) are that relevant stakeholders, such as government counterparts, have a strong sense of ownership, adequate resources and capabilities. Through the analysed cases, rooting of programme activities and intended directions within host government and agency seems well achieved under such contextual conditions. A set of identified CMO configurations for sustainability criterion is shown in Table 13.6.

Table 13.6 Identified CMO configurations for sustainability criterion

Context		Theory of change		Intermediate outcome	Outcome/criterion
Key government counterparts, end-users and beneficiaries have relatively high levels of understanding of CCA programme's intended objectives, and have clear ideas as to what types of adaptive capacity or measures they need	+	Development of adaptive capacities and introduction of new adaptive measures that are requested by the end-users and can yield tangible results foster a sense of ownership towards built capacities and introduced measures	=	High likelihood for sustaining built adaptive capacity and high utilisation level of adaptive measures introduced	Sustainability
Sustained political interest towards the CCA programme's intended objectives					
Beneficiaries on the ground and government continue to be present and see the need and benefits in engaging themselves to the CCA programme's intended objectives	+	Formulation of communities of practice for developing and implementing new initiatives provides a useful platform for the committed partners/stakeholders to continue to be active for the CCA matters	=	High likelihood for sustained, high level stakeholder engagement	
Government counterparts understand the need of mainstreaming	+	Programme activities implemented within the local/national and institutional existing framework foster a sense of ownership and trigger smooth integration of planning and policies	=	High likelihood for sustained level of mainstreaming at central policy and planning level	
Institution's sufficient resources and motivation level of government officials					
Relevant stakeholders have strong sense of ownership and have adequate resources and capabilities	+	Introduction of adaptive measures to the stakeholders and institutions with relevant mandate enables 'rooting' of these measures inside the respective stakeholders and institutions	=	High likelihood of generating broader adoption and replications in the long term	

13.7 Methodological Implications

The purpose of this meta-analysis was to apply a critical realism philosophical lens and realist approach proposed by Pawson and Tilley (1997, 2004). Concretely, the purpose thus was to introduce and apply the method to extracting and hypothesising theories of change and contextual conditions under which programmes are expected to generate results through an underlying mechanism. In addition to a focus on programme theories, the realist approach pays close attention to the kinds of contextual conditions which enable (but not necessarily determine) a programme's IOs to be realised. Therefore, the first implication of adopting a realist approach in a meta-analysis of CCA programmes is its focus on enabling contextual conditions. It can be a significant element since non-realistic evaluations often focus on the aspects that are only related to programme interventions and their programme theories and not such contexts.

Second, the contextual conditions that are identified and hypothesized in this meta-analysis can be useful for future CCA programming, particularly since similar types of interventions are often designed without necessarily thinking of the contexts. A realistic approach can provide explanations (rather than deterministic "answers") as to what type of programme interventions may work under what type of conditions, and for whom. CCA programmes are embedded in quite a complex environment, e.g. involving a number of stakeholders and beneficiaries, implementing partners, funding sources and their requirements, and differing programme goals and local priorities, on top of the five types of challenges identified by Valencia (2009). All of these aspects can further be influenced by the country's culture, history and socio-economic conditions. These are also important context aspects to explore in further deepening the CMO configurations for CCA programming. By paying close attention to such contextual conditions, the realist approach can thus be considered useful for knowing how, when and where to place the relevant interventions in a relevant context.

Third, this type of meta-analysis based on a realist approach may be able to shed new light onto a number of ex-post evaluations that have been already prepared. Though it will be difficult to prove quantitatively, there seems to be a tendency in the development practitioner's community to pay inadequate attention to such ex-post evaluations, since they may be simply perceived as a mere requirement routinely asked by sponsoring agencies and donors. Since only in recent years have we started to complete ex-post evaluations of multilateral CCA programmes, a realist approach can provide a good analytical lens in fully utilizing those evaluations to better inform future CCA programming.

13.8 Conclusion

This paper presents a case of meta-analysis using a realist approach, the evaluation approach based on a philosophy of science called critical realism. The authors have adopted this approach in the meta-analysis of the nine CCA programme terminal evaluations, paying special attention to the context under which a mechanism is triggered to generate an IO. As a result, it could identify a number of pertinent programme theories and specific contextual conditions for each type of implemented interventions. This approach encourages the evaluator to go beyond deterministic cause-and-effect world and can provide *explanations* (rather than *judgments*) about what may work for whom, under what circumstances. CCA programmes by nature are quite complex, and are characterised by "multi-sectoral nature, cross-thematic focus, and long timeframes" (Bours et al. 2014), whilst impact of climate change felt differently in a different location and context. Thus simply collecting "best practices" of CCA interventions will not help policy makers and stakeholders to know what may work under their own circumstances, and how they are supposed to work for whom. What this analysis has revealed is that it is not just about "doing right things" or about "doing things right"; but it is also about "doing right things right, in right context".

Some of the findings of this meta-analysis can indeed help provide useful explanations. For example, a rather usual intervention of closely coordinating with national and local government may not automatically produce the anticipated result of a higher level of relevance should the priorities of CCA not be identified by host government or line ministries prior to the programme. A result of an increased level of stakeholder involvement may not be guaranteed by simply building partnerships at an early stage; as it may depend on how specific and focused programme design is. Ensuring an increased level of adaptive capacity and a high level of utilisation of introduced adaptive measures is what virtually all CCA programmes wish to achieve through, e.g., facilitating training and transferring techniques and know-how. But even this may not work if specifically identified targeted groups of people are not well aware of climate risks, or cognisant of specific skills that they themselves want to acquire. Moreover, fostering a sense of ownership towards built capacities and introduced adaptive measures is key in generating the linkage between the programme's inputs and attainment of the desired ends, in this case high likelihood of sustainability. But such generative mechanism may not be triggered under the context where key partners do not have a high level of understanding of programme's intended overall objectives (as opposed to, e.g., their understanding toward introduced adaptive measures).

The CMO configurations presented in this paper should not, however, be considered a mere check-list for future CCA programming. Rather, they provide a good platform through which policy makers, programme designers and implementers can be *guided*, in order for them to make better decisions and develop CCA programmes that are suited for the respective circumstances.

Finally the authors would like to emphasize the point that adoption of realist approach in international development is still at its nascent stage. Exactly how critical realism should be adopted in international development evaluation still remains to be discussed and a challenge. Closer comparative examination of the framework put forward by Pawson (2013) and Wong et al. (2013), and its research implications in social sciences explained by Danermark et al. (2002) should be done to identify the methodological gaps (and potentially misapplied parts in our analysis), so that a realist approach can be more readily applied in evaluation of CCA and, more broadly, in international development evaluation.

References

Betts, J. (2013). Aid effectiveness and governance reforms: Applying realist principles to a complex synthesis across varied cases. *Evaluation, 19*(3), 249–268. doi:10.1177/1356389013493840.

Bhaskar, R. (2008). *A realist theory of science*. Oxon: Routledge.

Blamey, A., & Mackenzie, M. (2007). Theories of change and realistic evaluation. Peas in a pod or apples and oranges? *Evaluation, 13*(4), 439–455. doi:10.1177/1356389007082129.

Bours, D., McGinn, C., & Pringle, P. (2014). *Monitoring & evaluation for climate change adaptation and resilience: A synthesis of tools, frameworks and approaches* (2nd ed.). SEA Change Community of Practice and UKCIP.

Cartwright, N., & Hardie, J. (2012). *Evidence-based policy: A practical guide to doing it better*. Oxford: Oxford University Press.

Danermark, B., Ekström, M., Jakobsen, L., & Karlsson, J. C. (2002). *Explaining society. Critical realism in the social sciences*. New York: Routledge.

Feinstein, O. (2009). Monitoring and evaluation of adaptation to climate change interventions. In R. D. van den Berg & O. Feinstein (Eds.), *Evaluating climate change and development* (pp. 237–239). New Brunswick: Transaction Publishers.

Global Humanitarian Forum. (2009). *Human impact report climate change*. The Anatomy of A Silent Crisis.

OECD. (2002). *Glossary of key terms in evaluation and results based management* (The Development Assistance Committee (DAC) Working Party on Aid Evaluation). Paris: Organisation for Economic Co-operation and Development.

Pawson, R. (2013). *Evidence-based policy: A realist perspective*. London: SAGE.

Pawson, R., & Tilley, N. (1997). *Realistic evaluation*. London: SAGE.

Pawson, R., & Tilley, N. (2004). *Realist evaluation*. (http://www.communitymatters.com.au/RE_chapter.pdf)

Pawson, R., Greenhalgh, T., Harvey, G., & Walshe, K. (2004). *Realist synthesis-an introduction*.

Picciotto, R. (2009). Evaluating climate change and development. In R. D. V. D. Berg & O. Feinstein (Eds.), *Evaluating climate change and development*. New Brunswick: Transaction Publishers.

Pielke, R. (2014). *The rightful place of science: Disasters & climate change*. Clarleston.

Rycroft-Malone, J., McCormack, B., Hutchinson, A.M., DeCorby, K., Bucknall, T.K., Kent, B., ... Wilson, V. (2012). Realist synthesis: Illustrating the method for implementation research. *Implement Science, 7*, 33. doi:10.1186/1748-5908-7-33.

Stern, E., N., Stame, J., Mayne, K., Forss, R., & Davies, B. B. (2012). *Broadening the range of designs and methods for impact evaluations* (Report of a study commissioned by the Department for International Development DFID Working Paper). London: Department for International Development.

Uitto, J. I. (2014). Evaluating environment and development: Lessons from international cooperation. *Evaluation, 20*(1), 44–57.

UN. (2012). *Independent evaluation of delivering as one*. New York: United Nations.

Valencia, I. D. (2009). Lessons on M&E from GEF climate change adaptation projects. In R. D. V. D. Berg & O. Feinstein (Eds.), *Evaluating climate change and development* (pp. 265–283). New Brunswick: Transaction Publishers.

Weiss, C. (1997). *Evaluation: Methods for studying programs and policies* (2nd ed.). Upper Saddle River: Prentice Hall.

Wong, G., Westhorp, G., Pawson, R., & Greenhalgh, T. (2013). *Realist synthesis*. RAMSES Training Materials.

Chapter 14
Adaptation Processes in Agriculture and Food Security: Insights from Evaluating Behavioral Changes in West Africa

Jacques Somda, Robert Zougmoré, Issa Sawadogo, Babou André Bationo, Saaka Buah, and Tougiani Abasse

Abstract This chapter focuses on the evaluation of adaptive capacities of community-level human systems related to agriculture and food security. It highlights findings regarding approaches and domains to monitor and evaluate behavioral changes from CGIAR's research program on climate change, agriculture and food security (CCAFS). This program, implemented in five West African countries, is intended to enhance adaptive capacities in agriculture management of natural resources and food systems. In support of participatory action research on climate-smart agriculture, a monitoring and evaluation plan was designed with the participation of all stakeholders to track changes in behavior of the participating community members. Individuals' and groups' stories of changes were collected using most significant change tools. The collected stories of changes were substantiated through field visits and triangulation techniques. Frequencies of the occurrence of characteristics of behavioral changes in the stories were estimated. The results show that smallholder farmers in the intervention areas adopted various characteristics of

J. Somda (✉)
Planning, Monitoring, Evaluation and Learning, International Union for Conservation of Nature, Ouagadougou, Burkina Faso
e-mail: jacques.somda@iucn.org

R. Zougmoré
Africa Program, CGIAR Research Program on Climate Change, Agriculture and Food Security (CCAFS), Bamako, Mali
e-mail: r.zougmore@cgiar.org

I. Sawadogo • B.A. Bationo
Environment and Agricultural Research Institute, Ouagadougou, Burkina Faso
e-mail: sawissa2001@yahoo.fr; babou.bationo@gmail.com

S. Buah
Savanna Agricultural Research Institute, 494, Wa, Ghana
e-mail: ssbuah@yahoo.com

T. Abasse
National Agricultural Research Institute of Niger, Niamey, Niger
e-mail: abasse.tougiani@gmail.com

J.I. Uitto et al. (eds.), *Evaluating Climate Change Action for Sustainable Development*, DOI 10.1007/978-3-319-43702-6_14

255

behavior change grouped into five domains: knowledge, practices, access to assets, partnership and organization. These characteristics can help efforts to construct quantitative indicators of climate change adaptation at local level. Further, the results suggest that application of behavioral change theories can facilitate the development of climate change adaptation indicators that are complementary to indicators of development outcomes. We conclude that collecting stories on behavioral changes can contribute to biophysical adaptation monitoring and evaluation.

Keywords Behavioral changes • Climate change • Monitoring • Evaluation

14.1 Introduction

Adaptation to climate change refers to adjustments of physical, ecological and human systems that increase societies' abilities to cope with the change (see Box 14.1). This may involve any adjustment to the physical systems, social or environmental processes, or perceptions of climate risk, practices and functions that reduce risks and increase exploitation of new (or previously overlooked) opportunities. Agriculture is particularly sensitive, because it will be significantly affected by climate change through effects on water availability, temperatures, soil processes, pests, pathogens and competitors, which in turn will influence crop productivity at farm level (Turral et al. 2011).

> **Box 14.1: Adaptation, Adaptive Capacity and Food Security**
>
> **Adaptation** is an adjustment in natural or human systems in response to actual or expected climatic stimuli or their effects, which moderates harm or exploits beneficial opportunities (IPCC 2014b).
> **Adaptive capacity** is the ability of systems, institutions, humans, and other organisms to adjust to potential damage, to take advantage of opportunities, or to respond to consequences of climate change (IPCC 2014b).
> **Food security** exists when all people at all times have physical or economic access to sufficient, safe, and nutritious food to meet their dietary needs and food preferences for an active and healthy life (FAO 2008).

At the center of climate change adaptation efforts are interventions intended to boost adaptive capacity and/or stimulate adaptive action (Pringle 2011). Fortunately, there are several categories of adaptive options in agriculture, including: technological developments, government programs, insurance, and modifications of farm production and/or financial management practices (Smit and Skinner 2002). Nevertheless, to date agricultural adaptation initiatives have mainly focused on mitigating risks to crop productivity associated with changing climatic conditions. Furthermore, links between climate change and food productivity have been largely explored by analyzing the relationships between climatic and agricultural variables (Di Falco et al. 2011).

In practice, continued refinement of soil, water, tree and crop management practices will contribute much of the required adaptation, except in systems that are already water stressed (Turral et al. 2011). However, while it is globally acknowledged that food productivity contributes to food security, post-harvest processes are also important. Furthermore, since their own agricultural activities are the primary sources of food for many people in developing countries, effects of climatic changes on crop productivity (and the people's responses to them) will strongly influence their overall food security (Ingram et al. 2008). Hence, efforts to ensure food security must include strengthening of the adaptive capacity (Plummer and Armitage 2010) of individuals, households and communities by improving their access to, knowledge of, and control over natural, human, social, physical and financial resources (Pramova and Locatelli 2013). For these reasons, several authors (Pittock and Jones 2000; Stafford Smith et al. 2011) have argued that adaptation to climate change needs to be seen as an iterative process. If so, monitoring and evaluation (M&E) of adaptation and/or progress towards it are clearly important to assess the effectiveness of adaptation interventions, options and technologies (UNFCCC 2010).

However, there are uncertainties regarding appropriate adaptation indicators. Ideally, they should be different but complementary to development variables, but current approaches to adaptation M&E do not take this distinction into account. This chapter describes efforts to improve the design and implementation of adaptation M&E, at program and project levels, undertaken in a CGIAR Research Program (CRP7). Specific objectives were: (i) to demonstrate the applicability and utility of the theory of planned behavioral changes for adaptation M&E, focusing on adaptive capacity, and (ii) contribute to the development of an integrated biophysical-behavioral changes approach to adaptation M&E.

14.2 Approach

14.2.1 The Intervention

The efforts to improve adaptation M&E reported here were part of the Consultative Group on International Agricultural Research (CGIAR) Research Program CRP7, on climate change, agriculture and food security (CCAFS), a strategic collaboration between CGIAR and the Earth System Science Partnership (ESSP). The overarching objectives of CRP7 are: (1) to identify and test pro-poor adaptation and mitigation practices, technologies and policies for enhancing food systems, adaptive capacity and rural livelihoods; and (2) to provide diagnosis and analysis that will ensure cost-effective investments, the inclusion of agriculture in climate change policies, and the inclusion of climate issues in agricultural policies, from the sub-national to the global level in ways that benefit the rural poor (CGIAR 2011).

The program encompasses four research themes, being addressed from 2011 to 2015, designed to enhance adaptive capacity in agricultural, natural resources management and food systems, thereby leading to improvements in environmental health, rural livelihoods and food security through diverse trade-offs and synergies. The four themes are: (i) adaptation to progressive climate change, (ii) adaptation through managing climate risk, (iii) pro-poor climate change mitigation, and (iv) integration of decision-making processes.

Research and development activities under this CCAFS program were place-based and undertaken at several spatial levels within so-called "target regions". West Africa region was one of the places where the research and development activities were undertaken in five countries: Burkina Faso, Ghana, Mali, Niger and Senegal. A participatory action research (PAR) approach (led by the International Center for Research in Agroforestry, ICRAF, in collaboration with the five countries' national agricultural research systems) was used to promote agricultural technologies (assisted natural regeneration, composting, tree planting, etc.), practices, policies and capacity enhancement (on-farm application trainings) for adaptation to progressive climate change. The participatory action research has contributed to the CCAFS's planned 5-year output, as stated in the Research Proposal (CGIAR Research Program 7 2011; output 1.1.1): *"Development of farming systems and production technologies adapted to climate change conditions in time and space through design of tools for improving crops, livestock, and agronomic and natural resource management practices."*

Parallel to this participatory action research on adaptation, a capacity enhancement action on planning, monitoring and evaluation of climate change adaptation (led by the International Union for the Conservation of Nature, IUCN, in collaboration with the five national agricultural research systems) was conducted. Thus, prior to the development of the M&E plan, vulnerability assessments were conducted and adaptation actions planned in a participatory action research framework (Somda et al. 2014). Four of the five West African countries (Burkina Faso, Ghana, Niger and Senegal) were involved in the participatory action research of the CGIAR's CCAFS program.

14.2.2   The Monitoring and Evaluation Approach and Technique

The framework for monitoring and evaluating adaptive capacity was developed based on the theory of planned behavior (TPB) by Ajzen (1991), which proposes a model that can help efforts to measure the effectiveness of interventions designed to guide human actions. It has been applied to adaptation M&E because adaptation requires technological and/or behavioral changes that are consistent with the sustainable livelihood framework (IPCC 2014a). Hence, climate change adaptation interventions are designed not only to implement adaptation actions, but also to change behavior at individual, household, community, country and international levels. The TPB holds behavior to be an outcome of competing influences balanced and decided upon by the individual. Direct influences are the behavioral intentions, which are also influenced by attitudes towards the interventions, subjective norms and perceived behavioral control. It should be noted that the TPB helps efforts to identify cognitive targets for change, rather than offering suggestions on how these cognitions might be changed (Hardeman et al. 2002; Morris et al. 2012).

In this project, researchers, governments and NGOs' extension officers and stakeholder communities' members were convened in workshops to plan the adaptation M&E, with the intention to use the most significant change technique. These workshops allowed stakeholders in each country to discuss various domains where intentional changes of behavior of participants in the planned field adaptation activities were expected, and plan M&E activities accordingly. Stakeholders in each country were asked to identify domains of their lifestyles that would change if the CCAFS program was successful. The identified domains of change were deliberately left fuzzy to allow people to have different interpretations of what constitutes a change in that area (Davies and Dart 2005). Table 14.1 summarizes the M&E plans that emerged from the countries' workshops.

The predefined domains of changes are inevitably context-specific, reflecting expectations regarding focal communities' likely changes and evolution during

Table 14.1 Summary of the adaptation monitoring and evaluation plans that emerged for each country

Key elements of M&E plans	Burkina Faso	Ghana	Niger
Intentional domains of changes	D1: Partnership	D1: Partnership	D1: Partnership
	D2: Knowledge	D2: Knowledge	D2: Knowledge
	D3: Practices	D3: Practices	D3: Food security
	D4: Organization	D4: Food security	
Behavioural changes collection methods	Focus group and Individual discussion	Focus group and Individual discussion	Focus group and Individual discussion
Types of behavioural change to collect	Individual and collective behaviours	Individual and collectives behaviours	Individual and collective behaviours
Technique for selecting most significant changes	Iterative voting	Iterative voting	Iterative voting
Number of stories of changes collected (experimental)	2 collective changes (men and women)	2 collective changes (men and women)	2 collective changes (men and women)
	34 individual changes (men and women farmers)	12 individual changes (men and women farmers)	16 individual changes (men and women farmers)

Sources: Reports from workshops on adaptation M&E in each country

adaptation-intervention cycles. However, communities in different contexts or locations may often share similar domains of change. Hence, using predetermined domains of change should be considered advisable rather than compulsory. Furthermore, changes that have occurred outside predefined domains should also be collected (i.e. identified and characterized) for learning purposes in order to improve future adaptation action M&E.

Purposive sampling was then used to collect individual level stories of changes through interviews. The sample size for individual interviews was kept small for experimental reasons. Purposive sampling was preferred to random sampling because the ultimate objective of our adaptation M&E was to learn from stories of changes, and ultimately move agricultural extension practices more towards success and away from failure. However, to improve the validity and reliability of the purposive sampling, discussions were conducted to collect stories of changes of male and female groups of farmers.

The most significant change technique (Davies and Dart 2005) was used to collect stories of changes of both individual farmers and gender-based groups. The technique is not based on predefined performance indicators, but on "field-based stories" that give meaning to people's reality and effects of projects on that reality. It allows the story tellers (individuals or groups) to describe what has happened in their lives and practices (particularly, in this project, the way they farm) in conjunction with the participatory action research adaptation action. Scientists from the respective countries' national agricultural research systems collected the stories of change.

The collected significant stories were subjected to participatory processing, in which characteristics of behavior changes in the stories were counted, and then the most significant changes were selected, substantiated and validated. To select the most significant changes participants read the stories one by one and discussed the characteristics of changes described by the individuals or gender-based groups. The substantiation involved field visits and triangulation processes including discussion with resource persons and groups in the communities to ascertain whether behavioral changes noted in the stories had effectively occurred. Such substantiation has two objectives: (i) to verify the effectiveness of the occurrence of the change characteristics with the story tellers, other community members and fieldworkers who have worked with the selected communities, (ii) to gather additional data to complement information obtained during the story collection step.

The characteristics of behavior changes were counted by extracting all identified characteristics in the collected stories, then calculating their frequencies of occurrence, in terms of the percentages of people whose stories included them. This also allowed the identification of domains of life where changes had been induced in the selected communities by the participatory action research of the CCAFS program. In this chapter we have chosen to present frequencies of occurrence of behavioral change characteristics, but not the selection and substantiation results (which can be obtained from the authors on request).

14.3 Analysis

14.3.1 Consistency Between Planned Behavioral Theory and the CCAFS Program's Objectives

The plans developed for adaptation M&E suggested that involving farmers at the onset would help to clarify the domains of life that adaptation activities can influence. It allowed researchers to become aware of aspects of the beneficiaries' lifestyle that the technology and training activities they offered were likely to change. This is often lacking in traditional adaptation M&E, which is usually based on biophysical performance indicators. Thus, pre-identifying domains of behavioral change has added value to the quantitative biophysical performance indicators. The results clearly showed that if the CCAFS program resulted in successful adaptation of farming systems and production technologies to changing climatic conditions, farmers would put in place changes in domains including partnership, knowledge, practices, organization, and food security. This was consistent with expectations as adaptation is a process, and the development of adapted farming systems and production technologies requires communities' members to continuously improve knowledge, work in partnership and an organized manner, adopt new practices and (thus) enhance their food security.

14.3.2 Identified Behavioral Changes Induced by the CCAFS Program in West Africa

In line with the theory of planned behavior, outcomes were defined following Earl et al. (2001), as changes in the behavior, relationships, activities, or actions of the people, groups, and organizations with whom the CCAFS program directly engages. In West Africa, the CGIAR's program for climate change, agriculture and food security works through national agricultural research systems to help farmers develop climate-smart farming systems, through participatory vulnerability assessment and adaptation planning, on-farm trials, training, monitoring and evaluation. The results of behavioral changes M&E presented here can be seen as early or short-term outcomes of the program (or outcomes to which it has contributed). Table 14.2 summarizes the characteristics of behavioral changes extracted from the stories of changes gathered in 2013.

These findings show that both men and women farmers have put in place initial changes in knowledge, agricultural practices, organization, partnership, access to productive assets and food security. Analysis of the collected stories of changes identified a domain of change that was not included in the set identified in the planning stage. This was access to productive resources, in Burkina Faso and Niger, where the CCAFS's adaptation activities have contributed to improve access to on-farm and medicinal trees for both men and women. Further the results show that

Table 14.2 Characteristics of behavioral changes identified in individual farmers' stories (% of respondents)

Domains of changes/characteristics	Burkina Faso		Ghana		Niger	
	Men	Women	Men	Women	Men	Women
1. Changes in knowledge						
Knowledge about agricultural techniques (relationships between climate change and improved varieties, plowing flat and row planting, compost preparation, etc.)	84.21	60.00	100	100	100	100
Knowledge about implementing on-farm assisted natural regeneration techniques	57.89	46.67	a	a	100	100
Knowledge of trees (planting and utilization)	36.84	62.50	33.33	33.33	10	16.67
2. Changes in agricultural practices						
Agricultural practices (use of improved seeds, row planting, compost application, fertilizer use, etc.)	57.89	73.33	100	100	100	83.33
Practicing on-farm assisted natural regeneration of trees (associated with anti-erosion sites)	5.26	13.33	33.33	33.33	100	83.33
Planting trees	26.32	40.00	a	a	a	a
3. Organizational changes						
Relationships among farmers	36.84	6.67	16.67	16.67	a	a
4. Changes in partnering	57.89	66.67	66.67	66.67	60.00	33.33
In-community collaboration (exchange of information, services and goods)	57.89	66.67	66.67	66.67	60.00	33.33
5. Access to productive resources (on-farm trees, etc.)						
Access to on-farm and medicinal trees	31.58	80.00	a	a	a	16.67
6. Changes in food security						
Diversity of diets and early harvests from early maturing crops	a	13.33	50.00	a	a	83.33
Total surveyed sample	**19**	**15**	**6**	**6**	**10**	**6**

Source: Authors' counts from the stories of changes (2013)
[a]Indicates that the characteristic was not found in the significant change stories told by farmers

involving men and women in the process of developing climate-smart agriculture has changed attitudes of both men and women to on-farm tree planting and management. Similar changes were mentioned in the Ghanaian women's group discussions (not reported in detail here). For example, a group of women of the Doggoh community in Ghana said they did not know before that women can plant trees, as they had not seen any women in the community doing it before the CCAFS program's intervention. This had restricted the access of women in most rural communities to on-farm trees for their own purposes until their attitudinal change towards such trees.

The results also suggest that in the adaptation process farmers exhibit different stages of behavioral chances in various livelihood domains. For example, in rural communities in Burkina Faso, 84 % and 58 % of the story-tellers respectively expressed changes in knowledge of agricultural techniques and practicing improved agricultural techniques. In the Doggoh community in Ghana, none of the interviewed farmers expressed changes in knowledge about implementing assisted natural regeneration techniques, but 33 % of interviewed women and men farmers revealed changes in applying on-farm assisted natural regeneration. These differences reflect the likelihood that farmers in a community will be in different stages of behavioral changes in early parts of adaptation initiatives such as the CCAFS program.

Finally, some characteristics of changes were not identified in the individual stories of changes. This should not necessarily be interpreted as an absence of such changes, because the M&E questions only asked the farmers to report the significant changes they had experienced through participation in the CCAFS program's adaptation activities. Thus, they may have considered some changes too insignificant to describe in their stories of change.

Overall, the results indicate that participating farmers have initiated behavioral changes in various domains. Furthermore, the application of planned behavior theory allowed identification of the initiation of behavioral change at both individual and group levels in communities participating in the intervention in all three countries. Thus, the applied technique has clear potential utility for monitoring the implementation of farming systems and production technologies adapted to climate change, the spatial and temporal dissemination of adaptations, and the sustained changes in people's livelihoods and lifestyles that may be required to reduce vulnerability to its impacts.

These results are consistent with findings of innovative adoption studies, unsurprisingly as changes in behavior represent adoption of new behaviors and/or innovative practices, which is one of the most frequently advocated strategies for adapting agriculture to climate change. It should be noted that numerous variables will influence results of initiatives to foster changes. Notably, Rogers (1983) reported that factors such as attitudes, values, motivations, and perceptions of risk differ between decision-makers (producers) who are 'innovators' and those who are 'laggards' with respect to the adoption of particular innovations. In addition, according to Rothman (2000), individual or group decisions regarding behavioral initiation depend on people holding favorable expectations of the future outcome of the new pattern of behavior. However, maintenance of these new behavior patterns will mostly depend on farmers' satisfaction with the outcome they obtain (Rothman 2000).

14.3.3 Learning Opportunities from Applying Behavioral Changes Theory in Adaption Processes

Application of the theory of planned behavior has valuable potential to complement and extend the monitoring and evaluation of biophysical changes (the foci of previous agriculture and food security adaptation efforts). Three major learning opportunities can be identified from its use to monitor and evaluate adaptation processes reported here. As outlined below, the interviewees' stories of changes provided evidence of: (i) behavioral changes induced by adaptation activities; (ii) a need to maintain new patterns of behaviors and (iii) possibilities to identify adaptation-based metrics from behavioral change stories.

- *Evidence of various new behavior patterns*: Stakeholders including researchers and extension officers from both governmental and nongovernmental organizations have learned the existence of a wide range of changes in farmers' behavior. It was particularly easy for them to identify adaptation-relevant behavior. Furthermore, the most significant change technique allowed farmers to learn how to own the adaptation process and express views about potential barriers to adaptation outcomes or maintaining initiated behavioral changes. It provided opportunities for other farmers to learn about types of changes that are occurring in their community. In this manner it can help remove barriers related to attitude, subjective norms and perceived behavioral control within farmers' communities and enhance community and other stakeholders' engagement in the CCAFS program.
- *New behavior patterns need maintenance*: The results also suggest that initiating new behavior patterns may expose farmers to new challenges. Their stories of change provided researchers with insights into barriers related to assets and/or additional adaptive capacities after the farmers' initiation of adaptation-relevant behavioral changes. Such insight will facilitate discussion by researchers, farmers and extension officers regarding additional support farmers may require to maintain effective new behavior patterns, and avoid potential reversion to old practices that are considered inappropriate for adaptation to climate change. Furthermore, addressing the additional burdens faced by farmers after they have initiated relevant changes is important to minimize the risk of maladaptation to climate change.
- *Developing adaptation-related metrics from behavioral change stories*: Characteristics of behavioral changes portrayed in the stories of change could be readily identified, classified, counted, and used in designing metrics that effectively reflect progress towards adaptation. For instance, evidence that farmers have changed their agricultural practices to include assisted natural regeneration of trees on their farmland indicates that the adaptation initiative has contributed to increases in: (i) the area of land under this practice, (ii) the agricultural productivity and production of that land, and (iii) the food security of farm households involved. This is highly significant, because assuring traceability of

biophysical outcomes from adaptation activities has been the most controversial aspect of monitoring and evaluating adaptation. Because adaptation takes place in an economic development context, adaptation metrics should not be defined in isolation from changes in farmers' behavior. Otherwise, there is a high risk of measuring development indicators rather than adaptation indicators. Knowing domains where adaptation-relevant behavioral changes have been initiated and maintained would be helpful for evaluators to trace adaptation components in development outcomes, and reduce risks of confounding adaptation and development effects.

14.4 Needs for Incorporating Behavioral Theory into Adaptation M&E Approaches

Several authors Olivier et al. (2012) and Bours et al. (2013) have recognized the need for modifying conventional M&E approaches to meet the needs of climate change adaptation programs. They advocate a greater results-orientation in climate change adaptation interventions. However, there have been minor differences between most attempts to do so and conventional interventions. This may be because designing adaptation projects and appropriate M&E systems requires robust understanding of both adaptation to climate change (Olivier et al. 2012) and behavioral theory. In fact, the differences between adaptation-related and development outcomes will depend on whether new patterns of behaviors, actions, activities and relationships have been initiated and maintained by stakeholders, including smallholder farmers, policy-makers, researchers and agricultural extension officers.

It appears important to mainstream behavioral theory into results-based monitoring and evaluation of adaptation, because adaptation comes through various domains of behavioral changes. Behavioral theory is compatible with any existing tools, frameworks and approaches used in adaptation intervention programs and the associated M&E. In addition to assisting project managers to refine existing M&E frameworks, the application of behavioral theory will contribute to strengthening communities' ownership of the biophysical changes induced by adaptation actions. Results of this research are consistent with conclusions by Gifford et al. (2011) that behavior science is crucial for confronting the complex challenges posed by climate change. Knowledge of human behavior, cognitions, and psychological adaptation can also help the integration of derived adaptation-relevant indicators with those produced by researchers in related social and natural science disciplines.

Three major conclusions can be drawn from this research. First, an adaptation process leads to behavioral changes of the beneficiaries. These changes span various domains of community life, which may go beyond adoption of technologies in the targeted sector. They may or may not be adaptation-relevant, but all must be addressed to strengthen adaptation capacities or avoid mal-adaptation. Secondly, domains of behavioral changes can be identified before or after collecting stories of changes. These domains are useful for refining metrics of adaptation indicators.. In fact, although attributes of individual behavioral changes may vary widely, both within and among communities, they can always be located in relatively stable domains of changes. Thirdly, although claims about the generalization of changes' characteristics must be tempered by consideration of the contextual socioeconomic factors, behavioral theory can clearly add value to the existing adaptation M&E framework.

14.5 Implications for Policy, Practice and Research

14.5.1 Improving Adaptation Policy with Behavioral Theory and Models

Adaption and economic policies are subject to a number of biophysical, social and psychological influences, which future policies must consider. Thus, there are urgent needs for governments to improve the application of social research to enhance and evaluate policy, and measure longer-term trends, if adaptation policies, plans and programs are to achieve positive outcomes (i.e. enhance adaptation capacities and economic development). Behavioral change theory is one of the most promising elements of social sciences in terms of potential for improving policy outcomes. Indeed, changing individual and group behavior appears to be crucial for the effective delivery of policy outcomes, particularly in the context of climate change adaptation and mitigation. Therefore, designing adaptation policies that incorporate relevant aspects of behavioral change theory into biophysical frameworks will improve their outcomes by helping to ensure that adaptive behavior is initiated and maintained, while reversion to unhelpful behavior patterns is avoided.

14.5.2 Fitting the Human Behavior Framework into Adaptation Works

In light of the above results, current procedures for formulating and implementing adaptation options and strategies need to be revisited to tackle food insecurity more effectively in the face of climate change. To date, most adaptation programs in developing countries, from national to local, have neglected the behavior component of vulnerability analysis and adaptation action. Of course, the socioeconomic context of vulnerability is addressed together with the environmental context, but questions remain about whether current behaviors of community members are supportive of desired biophysical adaptation outcomes. There is therefore an urgent need to consider behavioral changes when planning adaptation activities, which implies a participatory approach involving appropriate stakeholders, particularly the local communities. It also requires analysis of the current context of community members' behavior, for which knowledge of behavior theory and models is essential.

14.5.3 Strengthening Human Behavior Elements of Participatory Action Research

Participatory action research (PAR) is an approach to research in communities that strongly recognizes the importance of participation and action. It seeks to understand the world by trying to change it, collaboratively and following reflection. This approach appears consistent with research focusing on adaptation of agriculture to meet challenges posed by climate change and enhance food security. However, to increase the relevance of this approach specifically in the context of climate change, adjustment of action research aspects is required, including research designs, implementation of actions, data collection and analysis methods, reporting and learning. In the research designs it is essential to include both biophysical and behavioral components, and equal attention should be paid to activities that will influence biophysical and behavior components during implementation of the actions. The data used to evaluate success of adaptation research actions should also include biophysical and behavioral indicators, or parameters. Thus, robust conceptualization of the data collection and analysis procedures is required at the start of the participatory action research to ensure that the collected data are properly analyzed and reported, and that lessons are drawn for learning by the PAR stakeholders and other scientific communities.

References

Ajzen, I. (1991). The theory of planned behavior. *Organizational Behavior and Human Decision Processes, 50,* 179–211.

Bours, D., McGinn, C., & Pringle, P. (2013). *Monitoring & evaluation for climate change adaptation: A synthesis of tools, frameworks and approaches.* Oxford: SEA Change CoP, Phnom Penh and UKCIP.

CGIAR. (2011). *CRP7 Proposal: Climate change, agriculture and food security.* http://www.cgiar.org/wp-content/uploads/2011/08/CRP7-Proposal-Final.pdf

Davies, R., & Dart, J. (2005). *The 'Most Significant Change' (MSC) technique: A guide to its use.* http://www.mande.co.uk/docs/MSCGuide.pdf

Di Falco, S., Veronesi, M., & Yesuf, M. (2011). Does adaptation to climate change provide food security? A micro-perspective from Ethiopia. *American Journal of Agricultural Economics*, 1–18; doi:10.1093/ajae/aar006.

Earl, S., Carden, F., & Smutylo, T. (2001). *Outcome mapping: Building learning and reflection into development programs*. International Development Research Centre (IDRC). http://www.idrc.ca

FAO. (2008). *Climate change and food security: A framework document*. Summary. Interdepartemental Working group on climate change: 1–107.

Gifford, R., Kormos, C., & McIntyre, A. (2011). Behavioral dimensions of climate change: drivers, responses, barriers, and interventions. *WIREs Clim Change*, 2011. doi:10.1002/wcc.143.

Hardeman, W., Johnston, M., Johnston, D., Bonetti, D., Wareham, N., & Kinmonth, A. L. (2002). Application of the theory of planned behaviour in behaviour change interventions: A systematic review. *Psychology & Health, 17*, 123–158.

Ingram, J. S., Gregory, P. J., & Izac, A. (2008). The role of agronomic research in climate change and food security policy. *Agriculture Ecosystems and Environment, 126*, 4–12.

IPCC. (2014a). Summary for policymakers. In O. Edenhofer, R. Pichs-Madruga, Y. Sokona, E. Farahani, S. Kadner, K. Seyboth, A. Adler, I. Baum, S. Brunner, P. Eickemeier, B. Kriemann, J. Savolainen, S. Schlomer, C. von Stechow, T. Zwickel, & J. C. Minx (Eds.), *Climate change 2014, mitigation of climate change. Contribution of Working Group III to the Fifth Assessment Report of the Intergovernmental Panel on Climate Change*. Cambridge, UK: Cambridge University Press.

IPCC. (2014b). Annex II: Glossary [K. J. Mach, S. Planton, & C. von Stechow (Eds.)]. In: *Climatte change 2014: Synthesis report* (Contribution of working groups I, II and III to the fifth assessment report of the intergovernmental panel on climate change [core Writing team R. K. Pachauri, & L. A. Meyer (Eds.)]). Geneva: IPCC, pp 117–130

Morris, J., Marzano, M., Dandy, N., & O'Brien, L. (2012). *Theories and models of behavior and behaviour change*. Forestry, sustainable 14 behaviours and behaviour change: Theories. http://www.forestry.gov.uk/pdf/behaviour_review_theory.pdf/$FILE/behaviour_review_theory.pdf

Olivier, J., Leiter, T., & Linke, J. (2012). *Adaptation made to measure: A guidebook to the design and results-based monitoring of climate change adaptation projects*, Manual. Deutsche Gesellschaft für Internationale Zusammenarbeit (GIZ). Available from: www.seachangecop.org/node/1661

Pittock, B., & Jones, R. N. (2000). Adaptation to what and why? *Environmental Monitoring and Assessment, 61*, 9–35.

Plummer, R., & Armitage, D. (2010). Integrating perspectives on adaptive capacity and environmental governance. In R. Plummer & D. Armitage (Eds.), *Adaptive capacity and environmental governance* (pp. 1–22). Berlin: Springer.

Pramova, E., & Locatelli, B. (2013). *Guidebook on integrating community-based adaptation into REDD+ projects: Lessons from Indonesia and the Philippines*. CIFOR, Indonesia, p. 72, 2013,

Pringle, P. (2011). *AdaptME: Adaptation monitoring and evaluation*. Oxford: UKCIP.

Rogers, E. M. (1983). *Diffusion of innovations* (3rd ed.). New York: The Free Press of Glencoe, 453p.

Rothman, A. J. (2000). Towards a theory-based analysis of behavioral maintenance. *Health Psychology, 19*(1), 64–69.

Smit, B., & Skinner, M. W. (2002). Adaptation options in agriculture to climate change: A typology. *Mitigation and Adaptation Strategies for Global Change, 7*, 85–114.

Somda, J., Sawadogo, I., Savadogo, M., Zougmoré, R., Bationo, B. A., Moussa, A. S., Nakoulma, G., Sanou, J., Barry, S., Sanou, A. O., & Some, L. (2014). *Analyse participative de la vulné rabilité et planification de l'adaptation au changement climatique dans le Yatenga, Burkina Faso*. Programme de recherche du CGIAR sur le Changement Climatique, l'Agriculture et la Sécurité Alimentaire. Available from www.ccafs.cgiar.org

Stafford, S. M., Horrocks, I., Harvey, A., & Hamilton, C. (2011). Rethinking adaptation for a 4 °C world. *Philosophical Transactions of the Royal Society, 369*, 196–216.

Turral, H., Burke, J., & Faurès, J-M. (2011). *Climate change, water and food security* (FAO Water Reports 36). FAO Land and Water Division. http://www.fao.org/docrep/014/i2096e/i2096e00.pdf

UNFCCC Adaptation, Technology and Science Programme. (2010). *Adaptation assessment, planning and practice: An overview from the Nairobi work programme on impacts, vulnerability and adaptation to climate change.*

Chapter 15
Using Participatory Approaches in Measuring Resilience and Development in Isiolo County, Kenya

Irene Karani and Nyachomba Kariuki

Abstract This article highlights the process of using participatory approaches in measuring resilience using the Tracking Adaptation and Measuring Development (TAMD) Framework. The utilization of participatory approaches in Isiolo County using the TAMD framework is aligned to the recent thinking of measuring 'subjective resilience' using people's perceptions to quantify household resilience. This article outlines the process of developing subjective indicators with communities, collection of baseline, monitoring and early outcome data by communities who were assisted in the development of their own adaptation theories of change. It also highlights the lessons and implications for policy if the approach is to be replicated at sub-national and community levels.

Keywords Resilience • Participatory • Evaluation • Theory of change • Development

15.1 Introduction

There is no commonly accepted definition of resilience across all disciplines.[1] However the International Panel for Climate Change (IPCC) in its Annual Report 5, builds on the definition used by the Arctic Council in 2013 and defines resilience as the capacity of social, economic, and environmental systems to cope with a hazardous event or trend or disturbance, responding or reorganizing in ways that maintain their essential function, identity, and structure, while also maintaining the capacity for adaptation, learning, and transformation.[2] With this in mind, there is a need to measure the impact effectiveness of adaptation actions and how they

---

[1]Community and Regional Resilience Institute (2013). Definitions for community resilience: An Analysis. A CARRI report. Page 10.

[2]https://ipcc-wg2.gov/AR5/images/uploads/WGIIAR5-AnnexII_FINAL.pdf

I. Karani (✉) • N. Kariuki
LTS Africa, Nairobi, Kenya
e-mail: irene-karani@ltsi.co.uk; nyachomba-kariuki@ltsi.co.uk

© The Author(s) 2017
J.I. Uitto et al. (eds.), *Evaluating Climate Change Action for Sustainable Development*, DOI 10.1007/978-3-319-43702-6_15

contribute to a population's resilience. Measuring resilience also contributes to measuring people's ability to respond to and accommodate adverse events.[3]

Isiolo County is located in upper eastern Kenya covering an area of 25,336.1 km². Most of the county is a flat low lying plain. Isiolo is regarded as one of the arid counties and is hot and dry for most of the year with two rainy seasons; short rains (October and November) and long rains (March–May) with average rainfall of 580 mm. The main ethnic groups found in the county are Borana, Turkana, Samburu, Somali and Meru. The main economic activities practiced in the county include pastoralism, subsistence agriculture, small-scale trade, and limited harvesting of Gum Arabica resin. Over the years, its communities have continued to feel the increasing impacts of climate variability due to the increasing frequency of drought episodes and their negative impacts.[4] These impacts include: longer trekking distances for women and girls, over dependence on humanitarian aid, infrastructure destruction due to flash flooding, changing livelihoods as communities are unable to recover from the increasing frequency of drought episodes amongst others.

The county was chosen for the TAMD feasibility testing, as it was the first county to receive climate financing from the Department for International Development (DFID) for the establishment of a County Adaptation Fund (CAF).[5] The objective of the CAF is to finance public good investments for improved resilience to climate change through the County government and six ward adaptation planning committees (CAPCs and WAPCs respectively) through the Adaptation Consortium.[6] The six wards are Kinna, Garbatulla, Sericho, Oldonyiro, Merti and Chari.

Resilience in Isiolo, according to the resident communities is equated to long term development outcomes such as sustainable livelihoods due to better livestock production which leads to increased incomes, improved human health, access to natural resources/pasture, food security and access to education.

Thus the main question was whether investing climate finance in public investment goods was going to elicit resilience measures as described by the communities. For this the TAMD framework developed by Brooks and others[7] was chosen as the tool that would be used to test whether resilience measures defined by the

---

[3]Bene, C. 2013. Towards a Quantifiable Measure of Resilience. Brighton, UK: Institute of Development Studies.

[4]Republic of Kenya (2013). Isiolo County: First County integrated development plan (2013–2017). Kenya, Nairobi: Government of the Republic of Kenya.

[5]This fund is managed under the DFID's Strengthening Adaptation and Resilience to Climate Change in Kenya (STARCK+) with funds from the International Climate Fund.

[6]Adaptation Consortium (2014). Adaptation consortium bulletin (online newsletter). Retrieved from http://adaconsortium.org/images/publications/Briefing-Paper.pdf

[7]Brooks, N., Anderson, S., Burton, I., Fisher, S., Rai, N., & Tellam, I. (2013). *An operational framework for tracking adaptation and measuring development*. Climate change working paper no. 5. London, UK: International Institute for Environmental Development (IIED). Retrieved from http://pubs.iied.org/pdfs/10038IIED.pdf

communities themselves were possible at the sub-national (county) and ward (community) levels.

15.2 Approach

The TAMD framework is for use in many contexts and at many scales to assess and compare the effectiveness of interventions that directly or indirectly assist populations in adapting to climate change. It also provides an explicit framework for two tracks; Track 1 entails assessing the capacity of institutions to undertake effective climate risk management (CRM) actions (also called top-down), while Track 2 entails assessing impacts of interventions aimed at reducing vulnerability and the extent to which such interventions keep development on track (development performance or bottom-up) – Fig. 15.1.

The TAMD operational framework[8] has a set of eight commonly used indicators that can be used to measure top down/climate risk management processes being implemented by government institutions using a score card (these indicators are further described under the top-down process – Track 1). The operational framework then suggests that theories of change (ToCs) be used to measure the change pathways from adaptation interventions to development performance in bottom up processes (Track 2). It further suggests that linkages between climate risk management processes and development performance/adaptive capacity can be shown in a ToC. Thus the development of ToCs can be within one track or between tracks. The researchers therefore chose to measure top bottom processes with a score card and ToCs to show changes in adaptive capacity using bottom up approaches. In addition the researchers used a ToC to make the linkage between Tracks 1 and 2.

15.2.1 Top-Down (Track 1) Process

For Track 1, county technical officers from the departments of water, livestock, natural resource management, meteorology, planning and the National Drought Management Authority (NDMA) were brought together to identify and prioritize CRM activities required to build adaptive capacity at community level. These activities were screened from the NDMA strategic plan, the Isiolo County Integrated Development Plan (ICIDP), and sectoral plans of the county.

---

[8]Brooks, N., Anderson, S., Burton, I., Fisher, S., Rai, N., & Tellam, I. (2013). *An operational framework for tracking adaptation and measuring development* (Climate Change Working Paper No. 5). London, United Kingdom: International Institute for Environmental Development (IIED). Retrieved from http://pubs.iied.org/pdfs/10038IIED.pdf

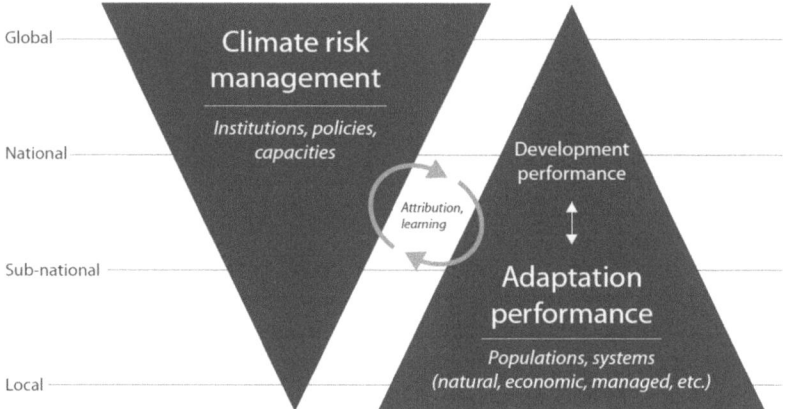

Fig. 15.1 TAMD framework (Adapted from Brooks & Fisher (2014) (Brooks, N., & Fisher, S. (2014). *Tracking Adaptation and Measuring Development (TAMD): A step-by-step guide* [Toolkit]. London, UK: International Institute for Environmental Development (IIED). Retrieved from http://pubs.iied.org/10100IIED)

The technical team assessed CRM processes through the use of Brooks score card.[9] The score card measures CRM indicators in Track 1 through 8 parameters, namely, climate change mainstreaming/integration into planning, institutional coordination, budgeting and finance, institutional knowledge/capacity, use of climate information, planning under uncertainty, participation, and awareness among stakeholders. Under each parameter, there are five questions that need to be answered before scores are assigned. The type of scoring is chosen by stakeholders in terms of weighting (0–4) or percentages. In Isiolo County, percentages were used to depict the extent to which progress against the indicator was being made. The score card and its results are shown in Table 15.1.

15.2.2 Bottom-Up (Track 2) Process

Before communities were facilitated to develop ToCs per ward, it was important that communities defined the term resilience in their own context so as to understand how their planned adaptation actions contributed to resilience. The researchers worked with six WAPCs to identify 20 ward adaptation/development interventions covering the water, livestock, and natural resource governance sectors that were in planning phases. Each of these wards was then assisted in developing their own specific ToC, identifying outputs, outcomes, long term impact, indicators and assumptions.

---

[9]See Brooks et al., 2013, p. 30–34.

Table 15.1 Indicators and assumptions for the integrated ToC

Results	Indicators	Assumptions
County level outputs	Types and number of information and communication products	There is buy-in and ownership from the different county departments
	Percentage of population reached	
	Disaster Risk Reduction (DRR) department established and operationalized	
	Policy document produced	
	Number of duplicated activities	
	Number of development agencies undertaking the same activities	
	Number of community project proposals developed and budgets justified	
	Number of dedhas (traditional natural resource governance structures) established	
	Number of natural resource management (NRM) meetings held	
Local/ward level outcomes	Types, numbers and frequency of adjustments to climate change adaptation activities	Uptake of information at community level
	Operational county contingency and DRR fund	DRR policy will be relevant and responsive to community needs
	Number of projects targeting infrastructure and services on transport, health, water and sanitation, security, education, food security and income generation	Community involvement in county coordination and planning will be done
	Number of climate change projects financed through budget allocation	Indiscriminate, fair, equitable and appropriate spread of development projects/ activities across the county
	Number of livestock with access to water and pasture during dry season	
	Number of households with access to water during dry season	
Community and county level medium and long-term impacts	Long-term Track 1 impacts can be measured through changes in resilience that are measured at the ward level. This data can be captured through aggregated data from adaptation interventions, measured through development performance indicators captured by Track 2 of TAMD	Political buy in from the county government
		Community buy in
		Financial plans of CCA activities are strictly followed/ implemented

15.2.3   Linking Track 1 and Track 2

After the top-down and bottom-up processes were completed, a composite theory of change was then developed by the county technical team and the WAPCs. This ToC

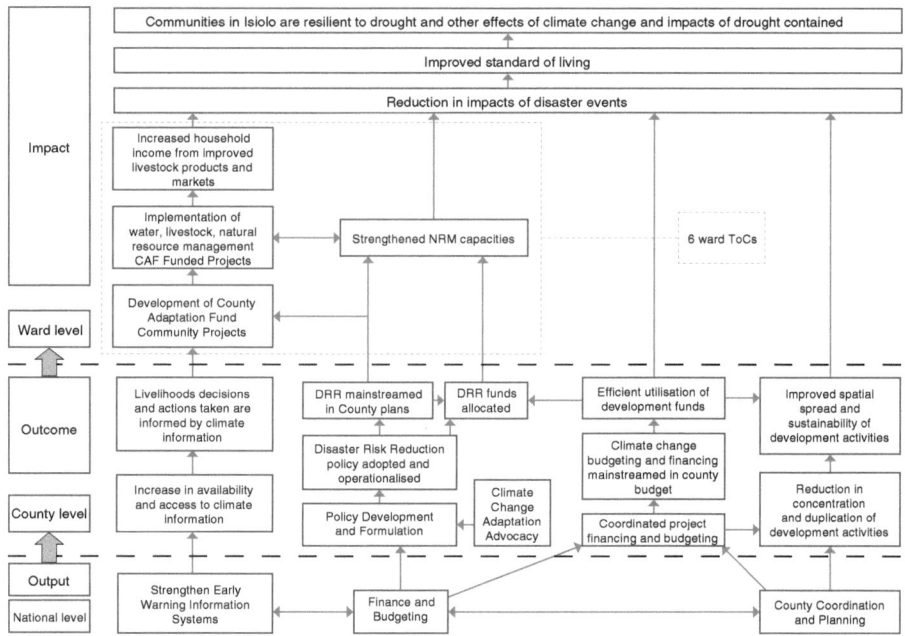

Fig. 15.2 Integrated Isiolo ToC. (Ibid)

linked the prioritized county CRM interventions identified through the score card process with the six ward ToCs as shown in Fig. 15.2.[10]

The methodology used above sought to learn lessons from two questions namely:

- To what extent can participatory processes be used in designing a ToC that links CRM activities (Track 1) with development outcomes (Track 2)?
- How can the framework be used to inform planning at sub-national and community levels?

When the composite ToC was developed and expected changes and indicators were identified in the top-down, bottom-up, Track 1 and 2 linkage processes, the County Planning Unit proceeded to integrate relevant CRM and adaptation actions into the Isiolo County Integrated Development Plan (ICIDP) in order to mainstream adaptation planning and M&E.

The use of a participatory approach in testing the feasibility of TAMD was chosen, as it sought to enhance ownership of the data collected, the analysis, and the dissemination of lessons learned. It also sought to build the evaluative capacity

---

[10]Karani, I., Mayhew, J., & Anderson, S. (2015). Tracking adaptation and measuring development in Isiolo County, Kenya. In D. Bours, C. McGinn, & P. Pringle (Eds.), *Monitoring and evaluation of climate change adaptation: A review of the landscape. New Directions for Evaluation, 147,* 75–87.

among stakeholders, an approach also supported by Preskill (2009)[11] and Preskill and Boyle (2008).[12] This was done by simplifying various climate change and M&E definitions and processes with the county officials and WAPCs e.g. climate variability, maladaptation, outputs, outcomes, impacts, indicators, evaluation and assumptions, before ToCs and M&E plans were developed with facilitation from the researchers.

15.2.4 Baseline Data

Two types of baseline data were collected from Isiolo. Track 1 (top-down) and Track 2 (bottom-up).

The sources included:

- Key informant interviews using semi-structured questionnaires
- Semi-structured group interviews
- Secondary data sources from county development plans were used for triangulating primary data collected from communities e.g. livestock numbers per ward, number of households accessing potable water.

15.2.4.1 Track 1 (Top-Down)

For the CRM processes under Track 1, the scores that were agreed upon through the use of the score card were the baseline values. The outputs of this exercise are shown in the results section. This exercise also highlighted the weak areas in CRM in the County, and as such, interventions that could address the weaknesses were prioritized. These were strengthening early warning systems, county budgeting and planning, and county coordination and planning.

15.2.4.2 Track 2 (Bottom-Up)

After the development of the ward ToCs, communities were given basic training in collecting baseline data against the indicators they had developed to measure their perceptions of resilience/adaptive capacity, for their respective ToCs, with a data collection tool that had been designed by the research team. This data was collected over a period of 3 months by the six wards.

---

[11]Preskill, H., & Boyle, S. (2008). A conceptual model of evaluation capacity building: A multidisciplinary perspective. *American Journal of Evaluation*, 29 (4), pp. 443–459.

[12]Preskill, H. (2009). Reflections on the dilemmas of conducting evaluations. In Birnbaum, N., & Mickwitz, P. (Eds), *Environmental program and policy evaluation: Addressing methodological challenges*. New Directions for Evaluation, 122, 97–103.

With respect to baseline data verification, the county officials had been tasked to verify the baseline data before the monitoring visit. However but this was not possible as Isiolo is an expansive county and the verification exercise through community visits had not been budgeted for by the county. As a result this exercise had to be done retrospectively and was conducted together with the first monitoring visit which occurred just after the commencement of interventions.

15.2.5 Output and Outcome Data

Output data, was collected after a period of 9 months, against the indicators in the ward ToCs and the county government score card (Table 15.3). Early outcome data was collected with an outcome assessment tool, after one and a half years to determine whether there were any changes being experienced from adaptation actions being implemented. This tool allowed the ward adaptation planning committees to assess the extent to which outcomes as depicted in their respective ToCs had been achieved through a scoring system. The results of this scoring are depicted in Table 15.2.

15.3 Challenges with Implementing the Methodology[13]

A few challenges were experienced when implementing the described methodology as detailed below:

- *Developing adaptation Indicators*: As stakeholders were used to developing output indicators as opposed to outcome indicators in development projects, the process of developing adaptation indicators to adequately measure resilience in the longer term proved to be a challenge.
- *Use of climate variability information in the development and adjustment of adaptation actions*: An adaptation M&E framework assumes that the design of adaptation actions has incorporated climate risk information. It also assumes that climate trends will be continuously monitored throughout project implementation in order to attribute any outcomes to enhanced adaptive capacity as a result of the interventions. However it was found that climate variability data had not been used when designing the adaptation interventions due to its unavailability during the design phase of the actions. In addition technical capacity to downscale climate trends in order to determine baseline scenarios in the county were also limited.

---

[13] Adapted from Karani, I., Kariuki, N., & Osman, F. (2014). *Tracking adaptation and measuring development. Kenya research report*. London, UK: International Institute for Environmental Development (IIED). Retrieved from http://pubs.iied.org/10101IIED.html

Table 15.2 Indicators and assumptions for the Oldonyiro ward ToC

Indicators	Assumptions
Output	Water user management committee members are able to enforce water resource management
Number of trainings held for water management committees	There are suitable areas to construct sand dams that reduce distance between water points
Number of constructed water storage tanks	
Number of sand dams constructed	The sand dam contractor has previous experience constructing sand dams and understands the intricacies of building sand dams
Number of sand dams rehabilitated	
Outcome	Sand dams constructed have the ability to hold adequate water
Number of livestock with access to water during dry season	Sand dams being rehabilitated can actually be structurally rehabilitated
Number of households with access to water during dry season	The water management committee is able to develop proper water distribution mechanisms
Number of months of 2012 that water is available in the 10 sand dams	Water management committee is recognized by community members. Community members have a proper understanding of water and sanitation
Number of hours spent walking to water point	
Number of hours spent fetching water at water point for domestic and livestock use	
Impacts	
Number of conflict incidences	
Number of families migrating	
Number of households not dependent on relief	
Access to social services schools	
Number of new permanent settlements	

- *Counterfactuals*: According to the TAMD operational framework, researchers are expected to collect data on attribution and this requires counterfactual data. This became a challenge in Isiolo because the research team had to find a community in Isiolo where climate change adaptation (CCA) interventions were not being undertaken. This proved to be difficult as there are many civil society actors undertaking CCA activities similar to the CAF in other parts of Isiolo. In addition the CAF interventions were public investment goods that were to benefit over 70 % of Isiolo's population: the remaining population comprises the urban population whose livelihoods are different from the targeted communities. They therefore, did not qualify as good counterfactuals. As such the research team made a decision to develop the before and after approach using the theory of change to measure contribution/attribution to resilience.

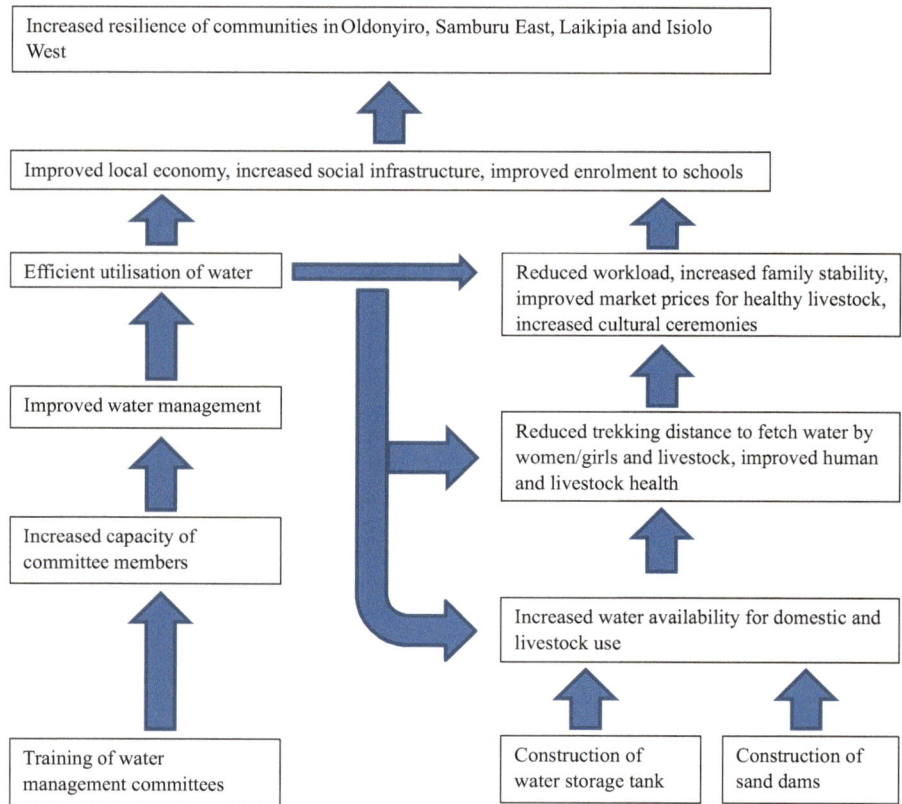

Fig. 15.3 Example of a ward/community level ToC. (LTS (2013). Ward Committees M&E report)

15.4   Results

The TAMD framework uses theories of change to measure progress towards achieving resilience. Theories of change help articulate assumptions behind smaller steps that lead to a long-term goal and the connections between these activities, outcomes and impact. They help present a visual representation on the contribution of a project or combination of projects to an intended outcome.

In Isiolo an integrated ToC at the sub-national (county level) and five theories of change at the community/ward levels was developed by both the county officials and ward committees. Figure 15.2 shows the integrated ToC at county level whilst Fig. 15.3 is an example of one ward ToC.

The indicators and assumptions for the integrated ToC are in shown in Table 15.1.

The indicators and assumptions for the Oldonyiro ward ToC in Fig. 15.3 are shown in Table 15.2.[14]

The communities were asked to discuss the assumptions described in Table 15.2 in order to develop options for risk management. These included; the legalization of traditional natural resource by-laws by the County Assembly which would assist in the enforcement of sound water resource management and would also raise the profile of the water management committees; and the strict vetting and supervision of potential sand dam contractors to enhance minimize the risk of poor dam construction.

For Track 1, Brooks et al. (2012) proposed a scoring system where each CRM indicator is scored against five questions to which the answer is yes, partially, or no, and scored 0, 1 or 2 respectively. The answers to these questions can be aggregated to yield an overall score out of 10 for each indicator, so that changes in the extent and quality of CRM over the various dimensions the indicators represent can be tracked over short time scales e.g. annually by policy and decision makers. However the scoring parameters can be changed by the users of the score card. Thus in Isiolo, the County officials changed the proposed Brooks scoring to percentages which they are more conversant with. The percentages presented in Table 15.3[15] against each CRM parameter were agreed upon by the county officials.

The county scored an average of 59.3 % for climate risk management measures with highest scores around public participation in planning and decision making in climate change adaptation as well as coordination of climate change interventions in the county. These scores provided a baseline for climate risk management activities. Subsequently they were used to develop activities needed to strengthen climate risk management and adaptation activities at county level and also formed the basis of the county's theory of change.

15.5   Track 1 Score Card Outputs

From the score card process, the county government had prioritised strengthening early warning systems, improving climate finance and budgeting and improving county coordination and planning. The CAPC was able to implement activities within two of the activity areas. The first activity involved purchasing a transmitter for the Isiolo radio station to enhance dissemination of weather and climate information. The expected output indicators for this intervention were on the types and number of information communication products and the percentage of the population reached with climate information within the whole county. Against a baseline figure of 10 % of the population coverage by the transmitter, after the intervention it was reported that the transmitter managed to enhance the coverage to 50 % of the population. However during this feasibility testing, it was not possible

---

[14]Ibid.

[15]See note 11.

Table 15.3 CRM (Track 1) scoring by county officials

CRM parameter	% Score	Reasons
1. Extent to which climate change planning is integrated in county policies or processes	20	Isiolo County does not have a climate change strategy and there is limited expertise in climate change screening of development interventions
2. Extent to which there is institutional coordination of climate change interventions	85	Climate change adaptation interventions are coordinated across sectors by the county drought coordinator from NDMA
3. Extent to which climate change financing is integrated into the county budget	55	The county had not yet budgeted or allocated finances for climate change. However CAF funding was available for adaptation activities at ward level
4. Level of institutional climate change knowledge	65	The members of the CAPC had undergone climate change training but the knowledge of technical officers within the county government was still low
5. Use of climate information	55	Some sectors of the county government (agriculture and water) took into account observational data and climate projections when planning. However there was limited capacity to interpret and use climate information for scenario planning
6. Planning under uncertainty	40	NDMA at county level updated its plans with climate information annually. However they did not use climate projections, nor did they consider maladaptation when planning
7. Extent of participation during planning and decision making processes in climate change adaptation	90	The design of ward adaptation actions took place after a highly participatory process, where women and other vulnerable groups participated
8. Level of climate change awareness amongst stakeholders	65	Only communities from 6 out of 10 wards in Isiolo County had been sensitized to climate change

to ascertain as to whether households actually received this information and how they used it.

The CAPC also collected livestock data and information to support the development of the Isiolo livestock strategy. They also conducted a workshop to integrate climate change into the Isiolo CIDP. These activities aimed at improving county coordination and planning activities climate change adaptation.

Some outcomes depicted in the integrated ToC (Fig. 15.1) have already been realised with an increase in number of projects targeting infrastructure, agriculture, health, water and sanitation, food security and income generation as well as number of climate change projects financed through county budget allocation.

Progress towards achieving outputs around the improved financing and budgeting is making slow progress however it should be noted that the county water department has provided financing for ward level adaptation activities such as rehabilitation and construction of sand dams. This indicates that implementation of climate change adaptation activities at local level have been able to influence targeted county adaptation financing.

15.6 Track 2 Outputs and Outcomes

The Track 2 adaptation interventions implemented at ward level by (WAPCs) were similar to development actions with the only difference being that they were formulated through resilience assessments conducted before the TAMD initiative begun. The ward level interventions were categorised as follows:

- Natural resource management
- Construction/rehabilitation of water structures and water management
- Strengthening of traditional resource governance structures
- Construction of other infrastructure (veterinary lab, animal holding yards).

Over 90 % of the activities were completed by the end of the study period. Early outcomes from the interventions was achieved around reduction of distances to water points, increased access to good quality water for the resident and neighbouring communities, increased capacity of traditional natural resource governance committees (dedhas), proper diagnosis of livestock diseases and strengthened local capacity for natural resource management.

With respect to measuring resilience literature has shown that there are no universal or generally applicable indicators of resilience (or of vulnerability or adaptive capacity), as these phenomena are highly context-specific. However, a number of studies have sought to define dimensions of resilience, with each dimension gathering together a suite of related factors that might be represented by context-specific indicators (Alexander 2013[16]; Nguyen and James 2013[17]). However for the purposes of this study, social or livelihood resilience as defined by Eakin (2012)[18] and Tanner et al. (2015)[19] was used as it was fit for purpose.

---

[16]Alexander, D. E. (2013) 'Resilience and disaster risk reduction: an etymological journey.' *Natural Hazards and Earth System Science* 13(11): 2707–2716.

[17]Nguyen, K. V., & James, H. J. (2013) 'Measuring household resilience to floods: A case study in the Vietnamese Mekong river delta', *Ecology and Society* 18(3): 13.

[18]Eakin, H., Benessaiah, K., Barrera, J. F., Cruz-Bello, G. M., & Morales, H. (2012) 'Livelihoods and landscapes at the threshold of change: disaster and resilience in a Chiapas coffee community, *Regional Environmental Change* 12(3): 475–488.

[19]Tanner, T.M. et al. (2015) 'Livelihood resilience in the face of climate change', *Nature Climate Change* 5: 23–26.

Additionally Brooks and Fisher (2014)[20] conducted a review of methodologies for measuring resilience and identified the following potential dimensions of resilience[21] that can be used to measure livelihood resilience:

- *Assets*: physical, financial assets; food and seed reserves, etc. (contingency).
- *Access to services*: water, electricity, early warning systems transport, knowledge and information – to plan for, cope with and recover from stresses and shocks.
- *Adaptive capacity*: to anticipate, plan for and respond to longer-term changes – for example, by modifying current practice, creating new strategies.
- *Income and food access*: the extent to which people may be poor or food insecure before the occurrence of a stress or shock.
- *Safety nets*: includes access to formal and informal support networks, emergency relief and financial mechanisms such as insurance.
- *Livelihood viability*: the extent to which livelihoods can be sustained in the face of shock/stress, or the magnitude of shock/stress that can be accommodated.
- *Institutional and governance contexts*: the extent to which governance, institutions, policy, conflict and insecurity constrain or enable coping and adaptation.
- *Natural and built infrastructural contexts*: the extent to which coping and adaptation are facilitated or constrained by the quality and functioning of built infrastructure, environmental systems, natural resources and geography.
- *Personal circumstances*: other factors that make individuals more or less able to anticipate, plan for, cope with, recover from and adapt to changes

From the descriptions of resilience above, the research team used a participatory outcome assessment tool to measure the changes anticipated from the adaptation interventions that could contribute to livelihood resilience in a pastoralist context. The results from the assessment indicated that early outcomes were already being realised and included: reduction in livestock disease cases, availability and access to water in water sources for over longer periods i.e. 3–6 months as opposed to 1–2 months, improved household hygiene and reduction in human waterborne disease incidences.

To measure outcome/adaptation benefit achievements, the wards used outcome assessment forms to provide scores on the achievement of any initial outcomes against outcome indicators. Although attaining resilience is a long term objective, ward adaptation interventions have been able to provide benefits around increasing accessibility of water, reduction of violent armed conflicts, and reduction of livelihood diseases which all play important roles in improving resilience of communities in the County. See the example of an outcome assessment form in Table 15.4 from Sericho ward.

As can be seen from the dimensions of resilience described earlier, communities in Isiolo are already beginning to experience some aspects of enhanced resilience as

---

[20]See note 7.
[21]See note 1.

Table 15.4 Outcome assessment from Sericho ward

Indicator	1 Not achieved	2	3	4	5 Fully achieved	Explanation on rating
Access to water for livestock				X		Water is available in short distances unlike in the past due to water being available in the pans. Livestock have separate water points with clean water
Access to water for domestic use			X			Only 2 out of 5 locations don't have water, water is available in the pans and wells. Humans have separate water points with clean water
NRM (dedha) meetings held				X		Dedhas have been successful in resolving conflicts. They have one official meeting day per month but can be called upon anytime there is conflict
No. of resource based conflict cases		X				The only conflicts that still exist in the area are quarrels which don't result in armed conflict. The dedha have been able to manage the conflicts arising

Source: LTS Africa (2015). Indicator Review

they are better able to cope with droughts at the household and community levels through enhanced access to clean water over longer drought periods, leading to improved household hygiene. They are also experiencing less armed conflict between communities which increases the success of any adaptation strategies they are involved in.

15.7 Lessons Learnt

- Adaptation indicators not necessarily different from development indicators depending on the context.[22] As can be seen from the indicators developed by communities and the dimensions of resilience, most of the indicators are measuring people's/communities well-being. What distinguishes the two is the contextualization of the results using climate data. However this was not possible at the time of the study.
- Communities that have been trying to adapt to a changing climate are able to understand M&E concepts when simplified, of short and long term changes to their livelihoods due to adaptation interventions and are able to assess progress in change pathways depicted in a theory of change.

---

[22]See note 11.

- A comprehensive monitoring and evaluation (M&E) system at sub-national level that is designed to collect adaptation outcomes and climate trend information ex-ante is crucial if enhanced resilience is to be proved through an adaptation evaluative framework process.
- It is possible to find and document adaptation outcomes of a community in 3 years if the interventions are designed from resilience assessments, an M&E system is established at the beginning of the intervention, baseline data is collected and verified, monitoring visits are conducted regularly to ensure that the adaptation interventions are implemented effectively and time is taken to document the changes happening in the communities through narratives. This can be regarded as M&E best practice.

15.8   Implications for Planning Policy and Practice

The results of this feasibility study have elicited a few implications for planning and investment in the use of participatory resilience M&E methodologies as detailed below:

- *Using M&E to influence planning*: TAMD in Isiolo was applied before the adaptation interventions begun. This had two advantages (a) the county adaptation committee were able to prioritise the activities that needed to be done under climate risk management and (b) the county adaptation committees was able to collect baseline data against indicators they had designed for CRM. In this way information that had never been collected before was now available for decision making and future planning on CRM. This ex ante M&E fits within the development evaluation approach described by Patton (2010)[23]. It is also proposed by the World Bank[24] for new or redesigned poverty and inequality reduction programs. This is because despite the upfront investment costs, this method can be cost effective in the long term as it allows for the adjustment and refinement of programs before implementation, and programs are likely to be better targeted as a result. This method can also provide useful information on the political consequences of new programs and therefore provide for the design of appropriate risk mitigation measures before implementation by decision makers.
- *Resilience measurement by communities and planning*: Project or program M&E is usually undertaken by independent individuals or institutions. During an evaluation exercise, it can become frustrating if the relevant data or information was not collected during the course of the project or packaged appropriately.

---

[23]Patton, M. (2010). *Developmental evaluation applying complexity concepts to enhance innovation and use*. New York, NY: Guilford Press.

[24]Busjeet, G. (undated). Planning, Monitoring, and Evaluation: Methods and Tools for Poverty and Inequality Reduction Programs. Poverty Reduction and Economic Management Unit Poverty Reduction and Equity Unit. The World Bank. Retrieved from http://siteresources.worldbank.org/EXTPOVERTY/Resources/ME_ToolsMethodsNov2.pdf

The advantage of empowering the county and ward committees with tools for collecting baseline and monitoring information increases the chances of a better quality evaluation of community resilience. This is different from normal evaluations in which the target communities are not involved in defining their indicators according to their own perceptions. This acknowledgement of the usefulness of subjective measurements of resilience is relatively recent and has been proposed as complementary to the traditional evaluation methods by Jones and Tanner (2015)[25] for planning and decision making. Through subjective resilience measurement, there is a greater understanding of household factors that contribute to resilience and policy makers/decision makers can design and plan for programs that enhance these factors in the long term and avoid introducing or planning for programs that have the potential to be maladaptive to communities.

- *Replication and scale-up of subjective resilience measurement methods*: Replication of participatory methodologies of measuring resilience such as TAMD can be beneficial for climate risk management planning by sub-national governments and adaptation planning for targeted communities. However up scaling to national level may prove challenging (Jones and Tanner 2012) especially because of initial investment. A cost and values study conducted in Kenya on TAMD concluded that the 'returns of using TAMD as a resilience M&E system are likely to be considerable, despite uncertainty. This is based only on individual indicators of avoided losses, expenditures and investment requirements. In reality, TAMD will have a system-wide impact, causing many costs to fall simultaneously and generating greater investment returns (Barrett 2014).[26] In addition Barrett states that his analysis did not factor in future escalation of climate change effects. This suggests that the likelihood of even higher Net Present Values of TAMD in the future.

---

[25]See note 1.

[26]Barrett, S. (2014). Cost and Values Analysis of TAMD in Kenya. IIED Working Paper. IIED, London. Retrieved from http://pubs.iied.org/10106IIED

Chapter 16
Evaluating Climate Change Adaptation in Practice: A Child-Centred, Community-Based Project in the Philippines

Joanne Chong, Pia Treichel, and Anna Gero

Abstract Whilst the principles of evaluating climate change adaptation are widely documented, there are many challenges in applying these principles in practice to evaluate, improve and learn from multi-sector, multi-scale and multi-stakeholder CCA initiatives with uncertain and future-oriented outcomes.

This chapter documents a research-evaluation approach applied during a 3-year, child-centred, community-based CCA project implemented in 40 barangays across four vulnerable provinces in the Philippines. The research aimed to help project implementers to learn from real-time feedback and perspectives from children and their communities and other participants. Researchers from the Institute for Sustainable Futures, University of Technology Sydney and practitioners from implementing NGOs Plan International and Save the Children collaborated on translating theory-based and development evaluation techniques into the field. We developed local-level indicators of adaptation, participatory focus group discussion and interview methods, and a guidance document for gathering and analysing evidence against these indicators.

Key to the success of this method was its participatory foundations – operationalising the principle that since ultimately adaptation is local, local voices and perspectives matter in understanding the impact of a project. Whilst there are limits to the "ideal" evaluation process, it is possible to achieve evaluative rigour in a process that is sensitive to the practical realities and pressures of project implementation. Embedding research and learning within practice – in the inherently uncertain context of supporting a community to adapt to climate change – provided new pathways for realising and sharing learnings to achieve better adaptation outcomes.

J. Chong (✉) • A. Gero
Institute for Sustainable Futures, University of Technology Sydney, Sydney, Australia
e-mail: joanne.chong@uts.edu.au; Anna.Gero@uts.edu.au

P. Treichel
Plan International, Melbourne, Australia
e-mail: Pia.Treichel@plan.org.au

© The Author(s) 2017
J.I. Uitto et al. (eds.), *Evaluating Climate Change Action for Sustainable Development*, DOI 10.1007/978-3-319-43702-6_16

Keywords Climate change adaptation • Community-based adaptation • Children and youth • Participatory approaches • Evaluation • Developmental evaluation • Indicators

16.1   Introduction

Supporting communities to adapt to climate change is a complex, uncertain exercise, and there is significant potential to learn directly from the practical applications of adaptation projects and the people whom they aim to assist. Interventions that focus on communities recognise that impacts and vulnerabilities are specific to local contexts, livelihoods are often directly dependent on local environments, and communities are at the front line of responding to the impact of climate-related disasters. Yet adaptation invariably requires concerted, coordinated action across multiple scales, sectors and actors, and community-based projects increasingly focus not only on the relationships and actions within local communities, but also whether and how government policy, planning and programs are informed by community priorities and needs.

Consequently, there is no one size fits all for community-based climate change adaptation (CCA) – and there is no singular way to evaluate or draw learnings from such projects. Nevertheless, key principles and general characteristics for monitoring and evaluating (M&E) of CCA interventions are widely articulated. In particular, it is well understood that conventional M&E approaches are ill-suited to CCA[1]; and that because adaptation constitutes pathways rather than end-points, evaluating CCA requires investigation of qualitative processes, rather than just solely relying on measurement of quantitative inputs or outputs. However, questions remain: how can project practitioners, researchers, evaluators and donors (interested in learning as well as accountability) operationalise these principles in practice? And, given the need for CCA interventions to generate and communicate transferable learnings about CCA activity design, implementation and impacts, how can we draw on good evaluation practice and theory to apply to the complex context of understanding a CCA project on the ground?

This chapter responds to a need for documented case studies of CCA evaluation in practice that generate and share methodological learnings about how to *do* rigorous, participatory and useful evaluations of CCA interventions. We share one example of an evaluation method applied over a 3-year community-based CCA project, implemented by Plan International and Save the Children in cooperation with communities and key government stakeholders, with research partner the Institute for Sustainable Futures at the University of Technology Sydney. This project acknowledged that children are amongst the most vulnerable to climate change,[2] but that they have the potential to advocate for adaptation practice and

---

[1] Intergovernmental Panel on Climate Change (IPCC). 2014. *Climate Change 2014: Impacts, Adaptation and Vulnerability. Wgii Ar5 Technical Summary.* Geneva, Switzerland.

[2] Risdell, J and C McCormick. 2013. *Protect My Future: The Links between Child Protection and Disasters, Conflcit and Fragility:* Plan International and Save the Children.

policy. The evaluation sought to help implementers understand how the project was supporting communities adapt to climate change, and also posed the question: what does successful adaptation look like *from the perspective* of children, youth and their communities?

16.2 The Project

Across the Philippines, many communities are extremely vulnerable to climate change due to high levels of poverty combined with high exposure to a wide range of climate change impacts. The Philippines was ranked 2nd on the 2014 World Risk Index[3] and 122nd out of 177 countries on the United Nations Human Development Index.[4] All areas of the Philippines are expected to see increased average daily temperatures and a spike in the number of very hot days. Rising sea levels and increased storm surges will impact coastal zones, whilst changes to seasonal rainfall patterns are likely to affect food security. The wet season is likely to become wetter, while the dry season becomes drier. However, given the diversity across the Philippines, the effects of these changes will vary across the country and will ultimately be localised and highly context-specific.[5]

The Child-Centred Community-Based Adaptation (CC-CBA) project, implemented from 2012 to 2015 and funded by the Australian Government, aimed to respond to these challenges by enhancing the resilience of children, youth, and their communities to the unavoidable impacts of climate change in 40 barangays across four vulnerable provinces (see Fig. 16.1): Aurora (led by Save the Children), Eastern Samar, Northern Samar and Southern Leyte (led by Plan). The four provinces were targeted due to their high poverty levels and vulnerability to climate change impacts. The design assessments found that the majority of the population had a low level of understanding of climate risk and vulnerability and low capacity to adapt. Likely impacts of climate change upon children include reduced ability to attend school, malnutrition, food insecurity, increased workloads, increased child abuse and increased morbidity and mortality from water and vector-borne diseases. All four provinces are located on the Eastern seaboard and are regularly subjected to extreme weather events such as typhoons, storm surges and flooding. The project areas were severely affected by Typhoon Haiyan in November 2013, as well as earlier Typhoons Utor and Nari in Aurora province.

---

[3]United Nations University – Institute for Environment and Human Security (UNU-EHS). 2014. *World Risk Report 2014*. Bonn, Germany.

[4]United Nations Development Programme. 2014. *Human Development Report 2014, Sustaining Human Progress: Reducing Vulnerabilities and Building Resilience*. New York.

[5]PAGASA. 2011. *Climate Change in the Philippines*: PAGASA, ADAPTAYO & MDG.F.

Fig. 16.1 Project sites

The project has two interconnected, long-term objectives; (1) increase the resilience of children, youth and their communities to climate change impacts across 40 Barangays (villages); and (2) develop a strengthened evidence base from the CC-CBA that informs policy and practice in the Philippines. These are in turn underpinned by three interconnected outcome areas around Knowledge, Advocacy, and Policy and Practice. The Theory of Change was built from a foundation that increasing communities' knowledge[6] about climate change, and options for adaptation and associated links with disaster risk reduction are essential to increasing resilience to climate change.

Increased knowledge also enables children and youth to take a leading role in CC-CBA activities and become climate change educators within their communities.[7] The project facilitated the active engagement of children and youth in climate change adaptation within their communities. Through supporting children, youth and the wider community to identify, develop and implement small-scale adaptation action, the project also sought to support the community to actively improve their resilience and at the same time speak with relevant decision-makers at the local, regional and national level to influence change. By working with duty bearers, the project helped to ensure these advocacy efforts do not 'fall on deaf ears'.

Project activities included school curricula development, community education, and supporting peer education and outreach.[8] Children and youth participated in training on using multimedia for communication and advocacy, including a radio media program, music, theatre and jingle-making. Local governments (LGs) and communities were supported to undertake participatory, climate change vulnerability and capacity assessments (PCVAs), which involved the participation of children and youth. From these assessments, locally developed adaptation initiatives were developed by school groups, by children and youth, as well as by adult community groups through a small grants programs. The project also supported LGs to use PCVA results to help plan, budget, design and implement local CCA activities, such as disaster preparedness and risk reduction activities. As well as working with children, the project directly supported LGs to design CCA-related local policies and regulations.

With the focus of children and communities, and through directly working with duty-bearers at various levels including within government, CC-CBA was fundamentally a human rights-based project. As Windfuhr (2000:25) notes, "a rights-based approach means foremost to talk about the relationship between a state and its citizens."[9] CCA requires actions and coordination by communities and

---

[6]Williams, Casey, Adrian Fenton and Huq Sallemul. 2015. "Knowledge and Adaptive Capacity." *Nature Climate Change* 5(February):82–83. notes the growing agreement that knowledge is an important determinant of adaptive capacity, in research frameworks, and in international policy and agreements.

[7]Children in a Changing Climate Research. 2010. *Children, Climate Change and Disasters: An Annotated Bibliography.* Brighton: Institute of Development Studies University of Sussex, Tanner, Thomas. 2010. "Shifting the Narrative: Child-Led Responses to Climate Change and Disasters in El Salvador and the Philippines." *Children & Society* 24(4):339–51. doi: 10.1111/j.1099-0860. 2010.00316.x

[8]Schoch, Corinne and Pia Treichel. 2015. *Child-Centred Climate Resilience: Case Studies from the Philippines and Vietnam*: Save the Children and Plan International.

[9]Windfuhr, Michael. 2000. "Economic, Social and Cultural Rights and Development Cooperation." in *Working Together: The Human Rights Based Approach to Development Cooperation-Report of the Ngo Workshop.*, edited by A. Frankovits and P. Earle.

governments, and in parallel rights-based approaches focus on the claims and voices of citizens, and duties of the state, and mechanisms to enable accountability and action on both sides.[10] Rights-based projects address inclusion and power imbalances and ensure poor, marginalised and vulnerable have opportunities to participate.[11] The human rights principles and standards applied in the project informed the evaluative method and sharpened the focus of participatory techniques.

The evaluative research component was explicitly built into the project design, and related to the project objective of strengthening the evidence base for child-centred, community-based adaptation. The research brief was: help project implementers understand how children and their communities are adapting to climate change by developing a set of indicators, and a process – a method – for gathering and analysing evidence. An indicator approach was selected because the implementing organisations considered that it would be a straightforward basis to systematically understand, measure and communicate outcomes.[12] At the same time, a set of indicators addresses the complexity of CCA that means there is no single appropriate metric for adaptation.[13]

As detailed in section 4, the main evaluative tool was focus group discussions (FGDs) with children. A total of 18 FGDs were conducted to pilot, develop and apply the indicators and method in Las Navas (including in Barangays of San Isidro and Hangi) in Northern Samar; Salcedo (including Barangays Matarinao, Garawon and Alog) and Hernani in Eastern Samar; and Maria Aurora (including Barangay San Joaquin), Dinalungan and Baler (including Barangay Zabali) in Aurora.

16.3 What 'Type' of Evaluation?

The research aimed to help project implementers to learn from real-time feedback about how the project was supporting children and their communities to adapt to climate change. From the outset it was clear that the approach needed to be both

---

[10]Cornwall, Andrea and Celestine Nyamu-Musembi. 2004. "Putting the 'Rights-Based Approach' to Development into Perspective." *Third World Quarterly* 25(8):1415–37.

[11]Uvin, Peter. 2007. "From the Right to Development to the Rights-Based Approach: How 'Human Rights' Entered Development." *Development in Practice* 17(4–5):597–606. doi:10.1080/09614520701469617

[12]Chong, Joanne, Anna Gero and Pia Treichel. 2015. "What Indicates Improved Resilience to Climate Change? A Learning and Evaluative Process Developed from a Child-Centred, Community-Based Project in the Philippines." *New Directions for Evaluation*.

[13]Bours, D, C McGinn and P Pringle. 2014, "Guidance Note 1: Twelve Reasons Why Climate Change Adaptation M&E Is Challenging": SEA Change CoP. (http://www.seachangecop.org/node/2728), Brooks, N, S Anderson, Jessica Ayers, Ian Burton and I Tellam. 2011. *Tracking Adaptation and Measuring Development. Iied Climate Change Working Paper No. 1*. London, United Kingdom: International Institute for Environment and Development.

rigorous *and* pragmatic, and be attuned to the realities and pressures of project implementation.

Beyond these broad aims, and unlike in many accountability-styled evaluations, the evaluative approach was not limited by pre-specified requirements in terms of questions, stakeholders or methods of inquiry. As the project itself evolved over time, we tailored the evaluative method to project needs.

The method also reflects elements of several 'types' of evaluation practices that are variously described by theorists, researchers and evaluation practitioners. This section unbundles what is meant by 'theory-based' and 'developmental' approaches to evaluation, and maps key characteristics of these approaches that were relevant to the CC-CBA project context.

16.3.1 Theory of Change Based Evaluation

'Program theory', also referred to 'theory-based' or 'theory-of-change (TOC) based evaluation' refers to developing a causal model from project activities (inputs) to a series of outcomes, then using this model as the basis for evaluation.[14] It is widely used for evaluations across sectors including to evaluate aid and development interventions.[15] Theory of change-based evaluation generally uses the theory established at program design, not just to trace if different steps actually occurred, but also to test the assumptions between the causal links in the model.[16] Findings from these types of evaluations can also be used to improve the 'quality' of theories, including by investigating alternative causal explanations to that incorporated into the initial theory of change.[17]

When framing an evaluation around a theory of change for a child-centred, community-based CCA project, the context for evaluation is a complex system that does not allow for a "neat" or "predictable" TOC to be articulated at the outset. CCA projects are usually dynamic and emergent interventions – whilst parameters and activities are set at the design stage, the exact details of implementation need to emerge and be developed over the course of implementation. For example, community participation in adaptation planning will always result in actions and

---

[14]Rogers (2000) in Rogers, P. J. 2008. "Using Programme Theory to Evaluate Complicated and Complex Aspects of Interventions." *Evaluation* 14(1):29–48. doi: 10.1177/1356389007084674

[15]Rogers, Patricia J. and Carol H. Weiss. 2007. "Theory-Based Evaluation: Reflections Ten Years On: Theory-Based Evaluation: Past, Present, and Future." *New Directions for Evaluation* 2007 (114):63–81. doi: 10.1002/ev.225

[16]White, Howard. 2009. "Theory-Based Impact Evaluation: Principles and Practice." *Journal of Development Effectiveness* 1(3):271–84.

[17]Rogers, Patricia J. and Carol H. Weiss. 2007. "Theory-Based Evaluation: Reflections Ten Years On: Theory-Based Evaluation: Past, Present, and Future." *New Directions for Evaluation* 2007 (114):63–81. doi: 10.1002/ev.225

pathways that cannot be exactly predicted at the outset of a project.[18] One of the potential applications of theory-based evaluation is to identify measures that can be used for monitoring over time.[19] But in complex systems relevant to CCA interventions, "SMART" measures may not be able to be developed in advance, making pre-and post-comparisons difficult.

Whilst some versions of program theory evaluation rely on close adherence at a detailed level to the initial theory to guide the evaluation, we do not take this strict definition. The complex character of CCA interventions far from makes theory-based evaluation redundant. Rather, flexible application is needed, and a balance struck between evaluation questions that are closely guided by the (initial) theory of change, and an approach that is open to outcomes, and the means to achieving them, emerging during implementation itself.[20]

16.3.2 Developmental Evaluation, or, Learning in Complex Systems

'Developmental evaluation' was coined by Paton to describe the types of evaluations applicable in complex situations where outcomes are emergent, where activities are not set in stone, and where it is not exactly known how, why or where activities will lead.[21] Developmental evaluation aims to "support real-time learning in complex and emergent situations" where the focus is on "adaptive learning rather than accountability."[22]

Development practitioners and researchers have widely recognised that for projects in complex environments to be successful, self-evaluation and ongoing learning is key. Developmental evaluation can be understood by considering what its purpose is not – it is not summative, in that it doesn't aim to evaluate at the end of a program and make a judgement about whether and how the program will continue;

---

[18]Rogers, Patricia J. 2011. "Implications of Complicated and Complex Characteristics for Key Tasks in Evaluation." in *Evaluating the Complex: Attribution, Contribution, and Beyond*, edited by K. Forss, M. Marra and R. Schwartz. New Brunswick, New Jersey: Transaction.

[19]Funnell, SC and PJ Rogers. 2011. *Purposeful Program Theory: Effective Use of Theories of Change and Logic Models*. San Francisco: John Wiley and Sons.

[20]Rogers, P. J. 2008. "Using Programme Theory to Evaluate Complicated and Complex Aspects of Interventions." *Evaluation* 14(1):29–48. doi: 10.1177/1356389007084674

[21]Gamble, Jamie A.A. 2008. *A Developmental Evaluation Primer*: The J.W. McConnell Family Foundation.

[22]Dozois, Elizabeth, Marc Langlois and Blacnhet-Cohen. 2010. *A Practitioner's Guide to Developmental Evaluation*: The J.W. McConnell Famliy Foundation and the International Institute for Child Rights and Development.

nor is it formative, in that it is nor primarily about setting baseline data for a future summative evaluation.[23]

These high complexity situations have characteristics such as dynamic, emergent, non-linear and uncertain[24] – a list which also fundamentally characterises climate change, its impacts on communities, and what is needed to support adaptation. Developmental evaluation has applicability where there is uncertainty, and where the program might need to change and adapt according to emerging and changing contexts. This is particularly applicable in the case of climate change adaptation, and in the case of the CC-CBA project, significant path changes were required in the aftermath of Typhoon Haiyan in December 2013.

A key characteristic of developmental evaluation is that it supports continuous learning and innovation through embedding evaluators as part of the team engaged in project delivery, in a long-term partnering relationship.[25] The CC-CBA design integrated the research component within the project and indeed it was the role of researchers to facilitate evidence-based, systematic reflection on project progress. Strong individual and organisational partnerships were successfully built.[26] The practical realities of program budgets meant that evaluative researchers could not be embedded full-time within the project implemented, but were directly involved in research design and inception and through the course of the evaluative process interacted with project implementers periodically throughout the 3-year project.

16.4 The Method: Details and Reflections

The evaluative method described in this section was developed through a collaborative effort between researchers and project implementers,[27] and field-tested in a participatory fashion through several iterations with child, youth and adult participants. The method focuses on collecting and analysing evidence against a set of indicators. These indicators were initially drafted by the team based on the theory of change and the experience of project implementers. The indicator set was revised

---

[23]Patton, Michael Quinn. 2011. *Developmental Evaluation: Applying Complexity Concepts to Enhance Innovation and Use*. New York: The Guilford Press.

[24]Ibid.

[25]Dozois, Elizabeth, Marc Langlois and Blacnhet-Cohen. 2010. *A Practitioner's Guide to Developmental Evaluation*: The J.W. McConnell Famliy Foundation and the International Institute for Child Rights and Development, Patton, Michael Quinn. 2011. *Developmental Evaluation: Applying Complexity Concepts to Enhance Innovation and Use*. New York: The Guilford Press.

[26]Treichel, Pia, Joanne Chong and Anna Gero. 2014. *A Partnership for Learning, Reflection and Evaluation in Action: Exploring Opportunities for Understanding Program Impact*: ACFID University Network.

[27]Chong, Joanne, Anna Gero and Pia Treichel. 2015. "What Indicates Improved Reslience to Climate Change? A Learning and Evaluative Process Developed from a Child-Centred, Community-Based Project in the Philippines." *New Directions for Evaluation.*

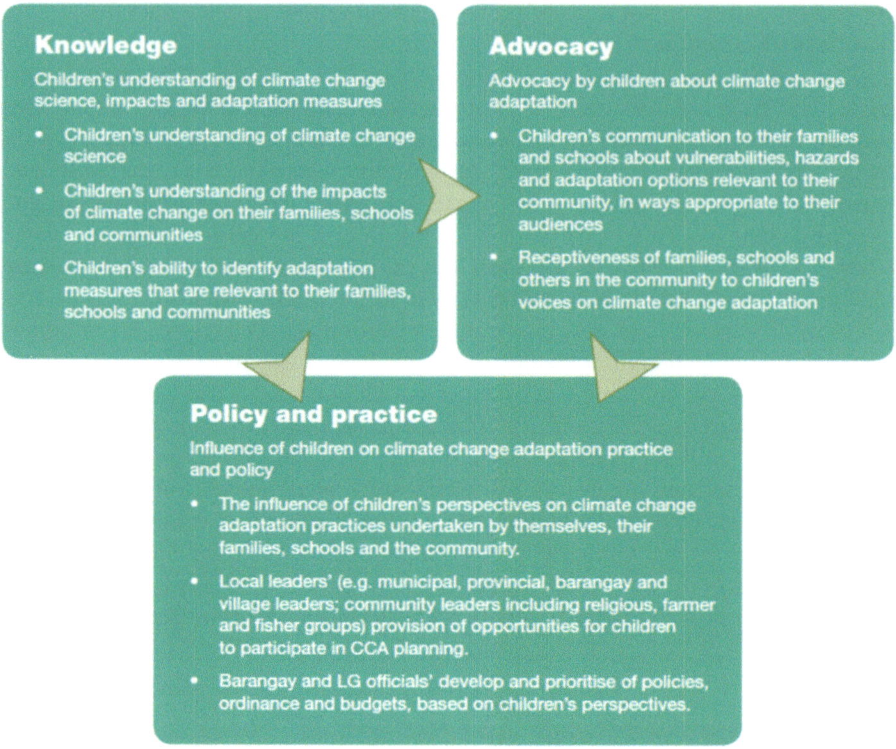

Knowledge

Children's understanding of climate change science, impacts and adaptation measures

- Children's understanding of climate change science
- Children's understanding of the impacts of climate change on their families, schools and communities
- Children's ability to identify adaptation measures that are relevant to their families, schools and communities

Advocacy

Advocacy by children about climate change adaptation

- Children's communication to their families and schools about vulnerabilities, hazards and adaptation options relevant to their community, in ways appropriate to their audiences
- Receptiveness of families, schools and others in the community to children's voices on climate change adaptation

Policy and practice

Influence of children on climate change adaptation practice and policy

- The influence of children's perspectives on climate change adaptation practices undertaken by themselves, their families, schools and the community.
- Local leaders' (e.g. municipal, provincial, barangay and village leaders; community leaders including religious, farmer and fisher groups) provision of opportunities for children to participate in CCA planning.
- Barangay and LG officials' develop and prioritise of policies, ordinance and budgets, based on children's perspectives.

Fig. 16.2 Child-centred, community based climate change adaptation indicators

following pilot, "ground-truthing" focus group discussions (FGDs) with children and youth to ensure they reflected children's experiences. From a rights-based perspective, participants should ideally be involved in setting the evaluation agenda.[28] In this case, the team sought to reflect participants' views in the indicator set through the process of iterative FGDs, and incorporating children's voices to refine the indicator set.

The indicator set is illustrated in Fig. 16.2, the method outlined below is also available in the format of a guidance document for project implementers.[29]

---

[28] See Johnson, Vicky. 2009. "Rights through Evaluation and Understanding Children's Realities." in *A Handbook of Children and Yount People's Participation*.

[29] Chong, Joanne, Pia Treichel, Gero, Anna, Rachelle Nuestro, Joseph McDonough, William Azucena, Joan Abes and Nina Abogado. 2015. *Child-Centred Commuity-Based Climate Change Adaptation in the Philippines*: *Guidance for Local Adaptation Indicators*. Institute for Sustainable Futures, University of Technology Sydney and Plan International.

16.4.1   Step A: Focus Group Discussions with Children and Youth

Participatory FGDs were the core of the method applied to evaluate, test assumptions, and hear the perspectives from children and youth about their experiences of the CC-CBA project. The focus groups were designed for small groups of 8–12 children, but in practice larger groups who attended were managed flexibly with a similar process.

The team conducting the focus group discussions comprised a facilitator, documenter and a few observers, all from the project or research team. The children were familiar with the facilitators, who were specifically selected as members of the implementation team who had worked closely with the children through various activities. Although familiarity between evaluators and participants is sometimes thought to adversely affect the "impartiality" of the process,[30] in this situation encouraging participation and ensuring children were comfortable with the adults present was considered paramount to inclusive participation, consistent with the rights-based approach to the project, and particularly important given the sensitive and potentially troubling issues discussed related to the lived experiences of children through typhoons, landslides, floods and other climate change impacts. "Bias" resulting from the familiarity was effectively managed through careful FGD design and implementation.[31]

The facilitators were well placed to encourage children to participate. However, some children were at times hesitant to speak, at least initially, when there were several adults present as observers in the background (including one to three not from the Philippines). Over repeated visits throughout the research process children became familiar and comfortable with the Manila- and Australia-based members of the team – by the end of the project activities, familiar enough to notice and ask about where we were when some or a few of us were not present. In other cases some younger children were reluctant to offer views if the groups were dominated by older children, although overall working with teachers beforehand generally ensured children within a group were fairly consistent in age. Separate FGDs were conducted with out of school youth groups. However, there were challenges in organising to hear from children with disabilities and from ethnic minorities.

Adults from the community – parents, teachers, and local government members – were generally not present at the focus groups with children, as we sought to avoid power imbalances that would discourage children from sharing their perspectives. However, in some focus groups, local government or some teachers attended, discretely in the background. In these cases, team members familiar with these adults (who were also project participants), gauged that they would not inhibit

---

[30]House, E R. 2005. "Deliberative Democratic Evaluation." in *Sage Encyclopedia of Evaluaiton.*

[31]Chong, Joanne, Anna Gero and Pia Treichel. 2015. "What Indicates Improved Reslience to Climate Change? A Learning and Evaluative Process Developed from a Child-Centred, Community-Based Project in the Philippines." *New Directions for Evaluation.*

children's participation, and considered there would be value in them hearing children's perspectives directly.

A nested approach was taken to translate the TOC indicator set into a series of questions linking knowledge, advocacy and practice and policy. The questions in the focus group were developed to investigate qualitative processes – for example, whether, how and why were knowledge and advocacy activities have influenced practices and policy? The FGDs focussed on those project activities which directly involved children, particularly on knowledge and advocacy activities. The FGDs were also used as a tool to explore whether children were aware of or involved in other participatory, planning- and policy-oriented activities such as the PCVAs conducted by local government. Beyond FGDs, policy impact was explored further through supplementary interviews with local leaders (see section).

In FGDs, children were specifically invited to share problems and barriers around communicating to their families, schools, community and government members about climate change adaptation, and project implementers found their responses crucial to fine-tune advocacy program activities with both children and duty-bearers. Children also shared with team members new stories of how they had influenced their family members (including for example, family members who were also Barangay leaders) to recognise the importance of climate change adaptation.

Attribution was a key consideration in designing questions – it was considered in such a situation that establishing precise counterfactuals was not a realistic exercise, but the questions explicitly probed fact (e.g. what children learned from a specific project activity) as well as alternative explanations (e.g. sources of information about climate change beyond the project).

The final topic of the focus groups was key to applying the rights-based approach to the evaluation. We explored with children their vision for what climate change adaptation would look like, including by asking what else they would like to do to prepare for the impacts of climate change, and what else they would like to see others do – family members, school, community and local governments. By giving children a voice on this open question, useful information was provided to the implementation team about ideas for future activities. By posing this discussion topic, it also prompted children themselves to think creatively and independently about how to adapt.

16.4.2   Step B: Supplementary Interviews with Adults

Supplementary interviews were conducted after the focus groups to gather additional perspectives on pathways of impact and changes that had occurred through the project. Attention to the responsibilities of duty-bearers is fundamental to rights-based programming and adults' attitudes and actions were explored during the supplementary interviews. Parents, teachers and local governments were asked questions that were parallel to those posed in focus groups, around knowledge, advocacy and practice and policy. For example they were asked for their perspectives as 'audience' members of children's advocacy, both formally through project activities (e.g. radio programs) and through other informal communication channels, such as at home. Local government members were asked specific questions to inform the "policy" sub-set of indicators, including about how children and their communities were involved in Barangay-level planning for disaster risk reduction and CCA, The policy impact of the project relied not solely on advocacy by children and their communities, but also critically by directly supporting local governments to: provide opportunities for children and community's to share perspectives in forums (such as PCVAs); and then use these perspectives to inform their planning and budgeting for CCA activities and development of CCA-related policies and regulations. These issues were explored during supplementary interviews.

Project implementers, reflecting on these supplementary interviews in comparison with the focus groups, noted in some cases how children and youth had developed a much more sophisticated understanding of climate change science, impacts and adaptation solutions than some of the corresponding adult participants in the project. These supplementary interviews thus provided project implementers with useful information about priorities for continuing their work with duty-bearers, including particularly on advocacy activities.

16.4.3 Step C: Reflection and Analysis via Team Debrief

The analysis of FGD and interview results was mostly conducted through structured 'debrief' sessions involving facilitators, documenters, observers and interviews closely after each community was visited. This approach to analysis was driven by the practical realities of project implementation – the busy schedules and limited time for team members to conduct further desk-based analysis – as well as recognising the value of involving the team in joint reflection exercise.

The main purpose of the debrief session was to foster learning through structured reflection on the FGDs and interviews. Through the debrief sessions, the team also captured additional observations from the FGDs that were not possible to capture in detail at the time of the FGD; to reflected on what went well and less so about the FGD and facilitation itself to inform future FGDs and briefings required; and to identify learnings from the FGD and interviews, and how these might help inform future program activities. Debriefs also involved capturing representative example quotes from children in a structured away against indicators areas that showed how well children's knowledge improved, their communication and advocacy, and the impact on practice and policy – with a reminder to link to participation in the program.

Although the emphasis was on qualitative investigation, the project team also considered it could be useful to formulate scalar measures of the indicators, potentially to enable comparison and, beyond the original thinking for applying the indicators, to assist reporting for accountability purposes. This required the team working together in the debrief sessions to articulate 'levels' of knowledge improvement, capacity to advocate, and practice and policy impact – and then assessing how many girls and boys in each focus group were at each level. Defining levels for scalar translation was a challenging process, particularly within the timeframe of a debrief session. The notion of scalar measures also prompted discussions about the appropriate baseline, adjustments for age level, and adjustments for variations in 'external' factors such as the overall level of education, access to media or other information, and whether or not children had themselves experienced climate-related disasters. Nevertheless, these discussions about how to quantify changes were in themselves valuable for the team to reflect on not only what level, but qualitatively what kind of changes were expected and could be expected as a result of the project.

16.4.4   Step D: Further Analysis

The learning and reflection aims were achieved through steps A to C, but an optional extra step was developed and trialled, and could be implemented if further resources and time are available. The aim of further analysis – detailed consideration of notes and transcripts from FGDs, interviews and the debrief sessions – is to produce written narrative that can be used to record, share, report and compare learnings, and be used as examples to inform CCA practice on the ground. Ideally, the draft narrative could be shared with those children and youth who participated in the FGDs to gather their further reflections and feedback. In practice however, time availability was a major constraint limiting this aspect of the method.

16.5   Conclusions

These FGDs have been really useful for me as a member of the project implementation team. We have had the chance to stop, reflect, and listen to the children about what they have learned about climate change adaptation and what difference the project is making. – Theresa Abogado, member of the project implementing team in the Philippines.

Key to the success of this method was its participatory foundations – operationalising the principle that since ultimately adaptation is local, local voices and perspectives matter in understanding the impact of a project. The method focused on hearing the perspectives of participants and facilitating structured, but open discussion and sharing between participants, and with project implementers. There are three main avenues by which this participatory, rights-based approach underpinned an effective evaluation that generated learnings and in itself contributed to project outcomes. Firstly, the indicators and process itself were developed through piloting in a participatory fashion with children and their communities, which in and of itself contributed to overcoming the challenges of balancing the rigour and participatory goals of evaluating a community-based CCA project. Secondly, asking questions that prompted communities to think about what it means to be more resilient, is not only a way to ascertain how the project has helped improve understanding, but is also key to enabling this resilience. And thirdly, by asking communities "what else is needed?" beyond project activities to date helps to inform the details of subsequent activities, and also helps to inform how and what changes to look out for as measures of community-defined success.

This example also illustrates that there are practical limits to the "ideal" evaluation process but that it is possible to usefully draw on key principles to inform the approach to evaluating a CCA practice. There were practical limits to full application of a right-based approach, and the extent to which children themselves are included in the development of the approach and the analysis and articulation of learnings. There were also limits to what 'can be known or found out' through an evaluation about causal relationships between activities and outcomes, when there are a myriad of interacting factors at play. It was nevertheless particularly useful to use the project's general theory of change to guide the evaluation, but allowing flexibility for the specific links and relationships – such as how knowledge, combined with formal and informal communication activities would assist communities to advocate for change and influence practice and policy – would emerge.

In practice, we developed strong team and organisational partnerships between the NGOs and research organisations involved in the evaluation and the project, which proved particularly valuable given the type of project and the project context. Whilst not 'developmental evaluation' to its full extent – researchers were not embedded in the team on a continuous basis – the approach was far from the 'conventional' end of the research spectrum where external groups of academic researchers seek out an existing applied project in order to test or calibrate a model or theory. The process involved joint learning and reflection from both implementing and research organisations throughout the project. The project was adjusted in real time to integrate lessons learned from the evaluative research; concurrently, the evaluative approach itself evolved to reflect lessons from the project's activities on the ground. Embedding research within practice – in the inherently uncertain context of supporting a community to adapt to climate change – provided new pathways for realising and sharing learnings from the ground, to achieve better adaptation outcomes.

Acknowledgements We thank the many children, youth and their communities for sharing their views and perspectives with us in: Las Navas (including in Barangays of San Isidro and Hangi) in Northern Samar; Salcedo (including Barangays Matarinao, Garawon and Alog) and Hernani in Eastern Samar; and Maria Aurora (including Barangay San Joaquin), Dinalungan and Baler (including Barangay Zabali) in Aurora.

We are grateful to Rachelle Nuestro, Joseph McDonough, William Azucena, Joan Abes, Abigail Cainglet, Reynato Cano, Arnold Peca, Jose Bacayo, Von Ami Martinez, Rochelle Agualin, May Rodriguez, Marichelle Vargas, Myrna Confesor, Edwin Elegado, Nina Abogado, Cecile Cornejo, Keren Winterford and Rohan Kent, for their dedication, enthusiasm, insights and contributions, and without whom this 'method' would never have existed in practice.

Chapter 17
Drought Preparedness Policies and Climate Change Adaptation and Resilience Measures in Brazil: An Institutional Change Assessment

Emilia Bretan and Nathan L. Engle

Abstract Brazil has historically coped with drought, a phenomenon that especially impacts the semi-arid lands of the Northeast. To deal with the various impacts of a current multi-year drought (2010-ongoing), the Government of Brazil, led by the Ministry of National Integration, partnered with the World Bank (WB) on a technical assistance program to foster proactive drought policy and management. The program works across sectors (climate/meteorology, water and sanitation, agriculture, environment, and disaster risk management) and levels (local, river-basin, urban, state, regional and federal) in relation to the outcomes and stake-holders it aims to engage and influence, and trough the integration of WB Global Practices and programs.

Inspired by successful models and lessons from other countries, the program aims to contribute to greater climate change resilience and reach a broad commu-nity of beneficiaries. To achieve these objectives, partners convened to (1) build a Northeast Drought Monitor; and (2) pilot drought preparedness plans across Northeast.

This chapter showcases the program and highlights key-milestones and direct and indirect outcomes identified by 2015. The institutional change process was assessed using qualitative analytical tools that integrate Outcome Mapping, the Capacity Development Results Framework, and Outcome Harvesting. Strengths, challenges, and outcomes (institutional changes) were identified, by tracking the program's contribution throughout its duration and at its completion.

The evidence shows that the initiative was able to convene key-regional and federal level multi-sector stakeholders at a decisive moment, resulting in an unprecedented bottom-up and regionally-led collaboration. Through the

E. Bretan (✉)
World Bank, São Paulo, Brazil
e-mail: emiliabretan@gmail.com

N.L. Engle
Climate Policy Team, World Bank, Washington, DC, USA
e-mail: nengle@worldbank.org

© World Bank 2017
J.I. Uitto et al. (eds.), *Evaluating Climate Change Action for Sustainable Development*, DOI 10.1007/978-3-319-43702-6_17

engagement and commitment of the partners, the program fostered and coordinated continuous sharing of knowledge, data, and work between service providers, secretariats, municipalities and other stakeholders from distinct sectors and scales of decision making. Thus, it influenced progress towards overcoming some of the historical challenges related to drought management in Brazil.

Keywords Drought • Climate change • Outcome harvesting • Resilience

17.1 Introduction

Extreme droughts and climate change are increasingly seen as important challenges to achieving green growth, improving agricultural livestock production, meeting water supply needs, and for residential users and commercial/industrial producers in Brazil (World Bank 2012). According to the World Bank's recent Turn Down the Heat reports, scientists expect drought phenomena to increase in frequency, duration, and intensity, ultimately translating to higher levels of evapotranspiration, reductions in arable land, and greater food insecurity in many countries and regions (World Bank 2012).

Traditional forms of dealing with drought, based on crisis management as opposed to proactive risk management (or drought preparedness) will likely not be able to tackle the devastating and long-lasting consequences expected from future climate change scenarios. In drought prone areas, such as the Brazilian semi-arid, drought preparedness appears as key to face these anticipated challenges.

Aligned with international discussions and successful initiatives from other countries, the Brazil Drought Preparedness and Climate Resilience non-lending technical assistance program (Drought NLTA), requested by the Government of Brazil (GoB), was initiated by The World Bank (WB) in July 2013 to support a process to shift the paradigm from reactive to proactive drought management.

The program aims to tackle historical challenges to the improvement of drought management in the country, through the promotion of knowledge exchange and through support for the development of drought preparedness measures and tools. Fostering drought resilience, and, as a consequence, climate change resilience, in the Brazilian case, also means promoting a strong effort of integration of institutions across sectors and levels, clarification and definition of roles, and the promotion of bottom-up and regionally-led initiatives, working towards a paradigm shift. The Drought NLTA was designed to support Brazilian partners towards these associated institutional and technical upgrades.

This chapter showcases the Drought NLTA program and highlights key-milestones and direct and indirect outcomes identified to date (the program is still being implemented at the time of drafting this chapter). It also presents the elected Planning, Monitoring and Evaluation (PM&E) approach, one that parallels the complexity of the program, integrating the variety of the multi-sector partners' perspectives. Focusing on outcomes – understood as institutional changes – the PM&E approach provides a framework to collect and analyze outcomes that looks beyond the control of the Drought NLTA and into the how the program influences its partners and stakeholders.

The evidence organized through the PM&E approach shows that the Drought NLTA initiative was able to convene key-regional and federal level multi-sector stakeholders at a decisive moment, resulting in an unprecedented bottom-up and regionally-led collaboration. Through the engagement and commitment of the partners, the program linked and promoted coordinated and continuous sharing of knowledge, data, and work between service providers, secretariats, municipalities and other stakeholders from distinct sectors, states, and governmental levels. Thus, it influenced progress towards overcoming some of the historical challenges related to drought management in Brazil.

17.2 Background

Drought, or in Portuguese, "seca", is a not a new phenomenon to the Brazilian society, especially for those living in the Northeast semi-arid region of the country. The average annual rainfall in the area is roughly 800 mm per year and is characterized not only by the minimal rainfall, but also by the timing of the rainfall (i.e., the rain typically falls only during a concentrated portion of the year). Historically, severe droughts have occurred in the Brazilian semi-arid. The semi-arid region, or the sertão is an area that reaches across nine Northeast states, covering an area of approximately 982,560 km^2, and includes more than 1,000 municipalities and 22 million inhabitants.

To combat drought, Brazil, like many nations, has invested in solutions such as increased emergency lines of credit, renegotiation of agricultural debts, expansion of social support programs, (e.g., cash transfer programs to poor families and farmers in the case of crop losses or lack of water to support plantings), and water truck deliveries of emergency drinking water to rural communities. These measures have helped to mitigate the more dramatic effects of drought, that in the past included not only economic losses but also starvation, diseases, death, losses of crops and animals, migration, pillage, and migrations. To date, however, few initiatives have been focused on adopting a long-term approach to avoid drought related losses and to promote a more resilient society.

This traditional approach to managing droughts around the world is often referred to as the "hydro-illogical" cycle (Wilhite 2011), characterized by the adoption of emergency measures when the drought hits that are quickly abandoned as the drought fades (along with the fading of decision makers' memories of the need to be better prepared for the next one).

More severe droughts are expected to happen in the Brazilian semi-arid region with climate change and increasing demand for water resources (World Bank 2014a). The most recent one, began in 2010, and has been progressing and persisting through 2015. Considered the worst drought in decades, it is costing billions of Brazilian reais for emergency and structural actions and has led to considerable crop losses, thousands of cattle deaths. The drought has been threatening the considerable gains in terms of economic, social, and human development that the region has experienced in the past several decades and placing many communities at risk of slipping back into extreme poverty[1] Reservoirs are at historically low levels, and in September, 2015, Ceará state had 80% of its municipalities depending on water trucks.[2]

Aligned with international discussions for improving drought resilience, most notably the High Level Meeting on National Drought Policy (HMNDP), in Geneva, Switzerland in March 2013, Brazil's Ministry of National Integration (MI) created an intra-ministerial work group to look critically at Brazil's drought management approaches, as well as to study the possibility of designing a national drought policy.[3] At the HMNDP, Brazil declared its commitment to improve drought planning and management in order to reduce impacts and increase resilience to future droughts and climate change.

Within this context, the MI requested the World Bank to support a process to shift the paradigm from reactive to proactive drought management. Specifically, MI requested: (i) to help with an 'institutional upgrade' through structuring and facilitating a more permanent institutional approach and response to drought, and improving integration within and between federal and state institutions; and (ii) to help with a 'technical upgrade' through developing concrete drought monitoring tools and preparedness plans/protocols. The Brazil Drought Preparedness and Climate Resilience non-lending technical assistance program (Drought NLTA) was thus designed and initiated in July 2013 to address this request.

17.3 The Drought NLTA Program Concept

The nature of the main challenge that the Drought NLTA program aims to tackle (i.e., fostering proactive drought policy and management), necessitates in its design a cross-sector program both internally to the WB and externally with the various partners involved. Water, climate, agriculture, and disaster risk management are the four key-areas involved, and the activities also involve partners from related areas, such as environment.

Adding another layer to the complexity of the program, more than 120 professionals from 50 multi-sector partners are involved with the effort: representing the federal government, federal institutions that act both nationally and regionally,

---

[1]More information can be found here: http://www.brasil.gov.br/observatoriodaseca/index.html

[2]Source: "Seca: Ceará tem 146 municípios abastecidos por carros-pipa". Available at http://www.cearaagora.com.br/site/2015/09/seca-ceara-tem-146-municipios-abastecidos-por-carros-pipa/

[3]Source: Ministério da Integração Nacional (MI) e Instituto Interamericano de Cooperação para a Agricultura (IICA). 2013. Estudos Referentes ao Diagnóstico da Política nacional de Secas no Brasil: Relatório Contendo Diagnóstico e Embasamento para a Formulação de uma Política Nacional de Secas no Brasil. Consultor, Otamar de Carvalho.

state and municipal level secretariats, technical agencies, universities and research centers, non-governmental organizations (including river basin committees), and the private sector.

Inspired by successful models and lessons from other countries[4] the key components of the Drought NLTA include: (i) developing a Northeast Drought Monitor (DM); (ii) piloting drought preparedness plans (DPPs) for different sectors across the Northeast (urban water supply, rural rain-fed agriculture, and river basin management, each at different scales of planning); and (iii) the discussion and systematization of guidelines and principles towards a national drought policy (NDP). In the Results Framework, the first two components (i, DM and ii, DPP) compose the so-called "Northeast Regional Pilot Track" and the third piece (iii, NDP) is called the "National Track". A visual summary of this structure can be seen in the Fig. 17.1. The roles of key partners in the Drought NLTA are detailed in Table 17.1, referring to the different components of the Drought NLTA (i.e., i, ii, and/or iii).

The program design is based on the "three pillars of drought preparedness" framework: (a) monitoring and early warning; (b) vulnerability/resilience and

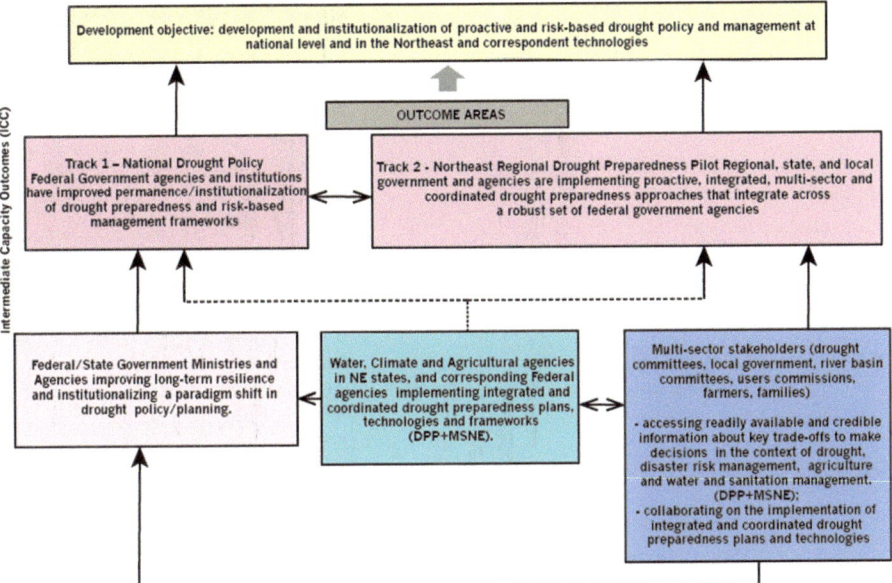

Fig. 17.1 Visual and summarized representation of the Drought NLTA results framework

---

[4]International institutions and professionals that have been developing drought preparedness plans, drought monitor and related initiatives and studies were key-partners for the Drought NLTA: US Drought Monitor and the National Drought Mitigation Center – NDMC, CONAGUA, the Mexican national water agency, academics from Spain, among others.

Table 17.1 Summary of key-stakeholders involved in the development of the program and their responsibilities

Area	Partner	Roles/responsibilities in the drought NLTA
Development	Ministry of National Integration	Supports the development of the DM (i); organized a series of consultations in the Northeast to discuss a NDP and tools (iii)
Water	National Water Agency	Central Institution/Executive Secretary of the DM (i); (i) Involved in the design and implementation of DPPs for River-Basins (ii)
Water and climate	Funceme (Ceará State Meteorological and Water Resources Foundation)	Regional leader of DM design and implementation. (i); support to the drought preparedness plans (ii)
Water and environment	INEMA – Bahia State Water Resources and Environment Institute	DM key-partner (part of the leading group) (i)
Water	COGERH – Ceará Water Resources Management Company	Member of the DM design and implementation team; (i)
Water	CAGECE – Ceará Water and Sanitation Company	Involved in the design and implementation of DPPs for Urban Water and Sanitation (ii)
Climate	APAC – Pernambuco State Water and Climate Agency	DM key-partner (member of the leading group)(i); Supported the design and implementation of DPP with COMPESA (ii)
Water and sanitation	COMPESA – Pernambuco Sanitation Company	Involved in the design and implementation of DPPs for Urban Water and Sanitation (ii)
Agriculture	EMPARN Rio Grande do Norte Agricultural Research Company	Members of the DM design and implementation team (i); Involved in the design and implementation of DPPs for River-Basin (ii)
Water	Piranhas-Açu River Basin Committee-Paraíba and Rio Grande do Norte states	Members of the DM design and implementation team (i); Involved in the design and implementation of DPPs for River-Basin (ii)
Various	Various multi-sector universities water, climate and agriculture institutions of the 9 Northeast states	Support the design and implementation of the DM. (ii); Involved in discussions of the NDP process (iii)
Various	State and municipal level water and agriculture secretariats (Piquet Carneiro Municipality -CE)	Involved in the design and implementation of DPPs for rural rain-fed agriculture (ii)
Agriculture	EMBRAPA – Brazilian	Data provider for the DM (i)
Climate	INMET	Data provider for the DM (i)

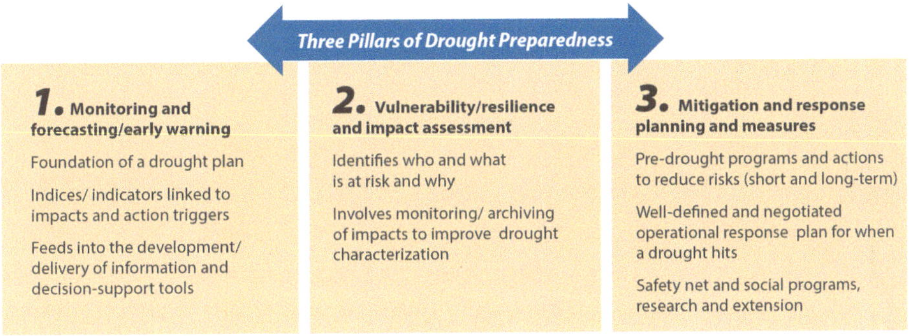

Three Pillars of Drought Preparedness

1. Monitoring and forecasting/early warning	**2. Vulnerability/resilience and impact assessment**	**3. Mitigation and response planning and measures**
Foundation of a drought plan	Identifies who and what is at risk and why	Pre-drought programs and actions to reduce risks (short and long-term)
Indices/ indicators linked to impacts and action triggers	Involves monitoring/ archiving of impacts to improve drought characterization	Well-defined and negotiated operational response plan for when a drought hits
Feeds into the development/ delivery of information and decision-support tools		Safety net and social programs, research and extension

Fig. 17.2 The 'three pillars of drought preparedness' that serve as the guiding framework for the Drought NLTA

impact assessment; and (c) mitigation and response planning and management. Fully and properly implemented, the pillars intend to contribute to better drought preparedness and build greater climate change resilience, with potential impacts in a diversity of sectors (e.g., water and sanitation, agriculture, environment, and disaster risk management), and reach a broad community of beneficiaries. Figure 17.2 provides an overview of the three pillars framework.

Elements of all the three pillars of drought preparedness are present in both the Drought NLTA tracks (the Northeast Regional Pilot Track and the National Track). The pillar that advanced more with the implementation of this Drought NLTA program was the first one, the monitoring and early warning pillar (essentially represented by the DM), followed by the third one, the mitigation and response planning and measures pillar (represented mainly by the DPPs).

The Drought NLTA implementation activities include trainings, workshops, field visits, study tours, and various meetings in and outside Brazil, with participation and guidance from numerous Brazilian national experts, and international partners, such as the National Drought Mitigation Center/US Drought Monitor (that collaborates closely with the initiative) -, the Government of Mexico (particularly Conágua – the National Water Commission) and several academic and local water utility partners from Spain, amongst others.

In these exchanges, stakeholders gather to learn, share their knowledge, communicate developments, set up priorities and agree upon responsibilities and institutional arrangements the program's phases and initiatives.

The WB team provides guidance, technical assistance, mobilization, communication, and convening services to help frame the conversation and keep the momentum of the paradigm shift, especially during potentially distracting moments, such as the October 2014 presidential and state government elections and the 2014 World Cup.

17.4   The PM&E Approach

17.4.1   Description of the Approach

As per the above description, the Drought NLTA exhibits characteristics of complex development intervention initiatives with a capacity development focus, such as:

- Multi-stakeholder context;
- Different perspectives from different actors on complex reform problems and solutions (lack of consensus about priorities);
- Distribution of the capacities to tackle the problems across actors, while no one actor is in full control (fragmented development context that makes it difficult to plan development efforts effectively with the broad ownership of stakeholders);
- Uncertainty about how to address the problems (a need for learning to adapt solutions);
- Deep-rooted institutional problems (that can impede results).

Considering the characteristics above, the WB has very limited or no control beyond the program's activities and outputs, whose outcomes are highly dependent upon the 'buy-in', initiative, and engagement of the partners involved. The design and implementation of an NDP could be supported by the WB through technical processes and capacity building, assessments from international experts, and with policy conceptualizing, and yet there is still no guarantee that by the end of the program such a policy will be in place.

The Drought NLTA then, calls for a non-traditional/non-linear (non-cause-effect) approach to PM&E. To plan, monitor, and evaluate other initiatives that have faced similar challenges within the WB Group, the World Bank Institute (WBI),[5] at the time of initiating the Drought NLTA, had been piloting tools that integrated the WB's Capacity Development Results Framework (CDRF) with Outcome Mapping (OM) and Outcome Harvesting (OH).

These three approaches were developed separately and are applied in a range of initiatives around the world, usually independently from one another. OM (Earl et al. 2001) was developed by the International Development Research Centre (IDRC), a Canadian development international non-governmental organization, to plan, monitor and evaluate some of its programs in developing countries that needed a strong participatory framework that could also engage partners in active change. OH (Wilson-Grau and Britt 2012) was developed by evaluators, strongly inspired by OM and Michael Patton's Utilization Focused Evaluation, to evaluate complex initiatives. The CDRF (Otoo et al. 2009) was developed by the WB to

---

[5]The World Bank Institute (WBI) is a global connector of knowledge, learning and innovation for poverty reduction. The WBI supports the World Bank's operational work and its country clients in this rapidly changing landscape by forging new dynamic approaches to capacity development through three areas of support: Open Knowledge, Collaborative Governance and Innovative Solutions. More information can be found at http://wbi.worldbank.org/wbi/

plan, monitor, and evaluate its capacity development initiatives. The WBI pilot brought together key-concepts from these three approaches to develop specific tools to plan, monitor, and evaluate WB's initiatives that are multi-sector, with a capacity development focus, and that operate in complex environments.[6]

This piloted approach, which operationalizes a process and framework for systematically understanding outcomes that are structured around policy, institutional, and/or behavior change, has been synthesized into a guide (World Bank 2014b) and a book (World Bank 2014c), the latter sharing experiences of implementation of the tools in a range of initiatives supported by the WBI. Another document, "Designing a Multi--Stakeholder Results Framework: A toolkit to guide participatory diagnostics and planning for stronger results and effectiveness" (WBI 2013b), and other draft documents provided by the WBI (Gold 2013, 2014; WBI 2013a, 2014), guided the design of a Results Framework (RF) for the Drought NLTA.

The first step is the design of a multi-stakeholder RF. The step-by-step process involves the identification and analysis of challenges and constraints to institutional change, followed by the development of a change process that includes a development goal, institutional change outcomes, and intermediate capacity outcomes.

Some of the questions that guide the design of the change process (and that are seen again when harvesting outcomes to monitor and evaluate the initiative) are "Who needs to drive the needed changes; what local leaders, groups and citizens?; and How and When is change expected to happen?" (WBI 2013b)

The analytical framework – that can be adapted – incorporates the lenses of institutional and policy changes. Challenges and constraints are categorized, e.g., as "weak organizational capacity", or "inefficient policy instruments", while intermediate capacity outcomes (progress markers) that are part of the change process would fall into categories such as "raised awareness, enhanced knowledge or skills, improved consensus and teamwork, strengthened coalitions" and so on (WBI 2013b). These categories and tools guide the design of the multi-stakeholder RF.

The change process is focused on behavior/policy/institutional changes driven by the partners. To capture this, partners are aggregated into groups involved in similar activities and promoting similar changes. The change process envisioned for each group of partners is grouped under a so-called "Outcome Area".

The change process, therefore, strongly based on the OM concepts, "unpacks" full theory of change to learn how milestones link to more transformative changes, creating a scale of change to measure progress along the process. Outcomes are understood as what each social actor (or change agent) did, or is doing, that reflects a demonstrated change in awareness, knowledge or skills, collaborative action, or

---

[6]More information about these approaches can be found at http://www.outcomemapping.ca and http://betterevaluation.org/plan/approach/outcome_mapping (Outcome Mapping) and http://betterevaluation.org/plan/approach/outcome_harvesting (Outcome Harvesting); Capacity Development Results Framework can be found at http://documents.worldbank.org/curated/en/2015/10/25228268/capacity-development-results-framework-strategic-results-oriented-approach-learning-capacity-development and at http://betterevaluation.org/resources/capacity_dev/results_framework

Table 17.2 Example of the progress markers for "National Track"

Change agent: water, climate and agricultural agencies in Northeast states, and corresponding Federal agencies	
Love to see (intermediate capacity outcomes)	Implementing integrated and coordinated drought preparedness plans, technologies and frameworks
Like to see (milestones)	Collaborating through established networks with defined governance rules
	Engaging in and promoting capacity development activities with multi-sector stakeholders
	Agreeing on a common agenda towards pilot drought preparedness plans and technologies/frameworks;
Expect to see (early outcomes)	Increasing knowledge, data sharing and cooperation
	Having increased know-how to plan drought mitigation and response actions

the use of knowledge or innovative solutions. Outcomes might also describe deeper institutional changes relating to policy, citizen engagement or government accountability and organizational arrangements.

Initial involvement, awareness raising and other immediate outcomes are described as something we would expect to see; deeper engagement as what we would like to see, and institutional and sustainable change as what we would love to see as the program progresses to the end and beyond its limits. Examples of the progress markers for "National Track", extracted from the RF, are demonstrated in Table 17.2.

An OH approach mainly informs the monitoring and evaluation (implementation phase), helping the gathering and analysis of information on changes influenced by the project to inform decisions and next steps. It also captures intended and unintended outcomes during implementation to inform corrections and next steps and helps to evaluate and articulate how complex projects advance toward impact. The analytical framework provided by the WBI approach helps to make sense of the outcomes, demanding each described milestone or progress-marker to be sustained by more than one source of information to be considered valid (see Fig. 17.3).

The OH process includes a rigorous check of the significance of the outcome for the development goals the initiative and the partners want to achieve ("why does the change matter?") and the identification of the contribution of the development organization and of the partners.

One of the key-elements of OM and OH methodologies that are the basis of the WBI approach, is the fact that both acknowledge *contribution* and *influence*, but not necessarily *attribution*. Policy and institutional change processes, the focus of initiatives such as the Drought NLTA, are very susceptible to the influence from many factors and actors, as well as the- political environment, to name a few. As mentioned before, the design and implementation of an NDP could be strongly

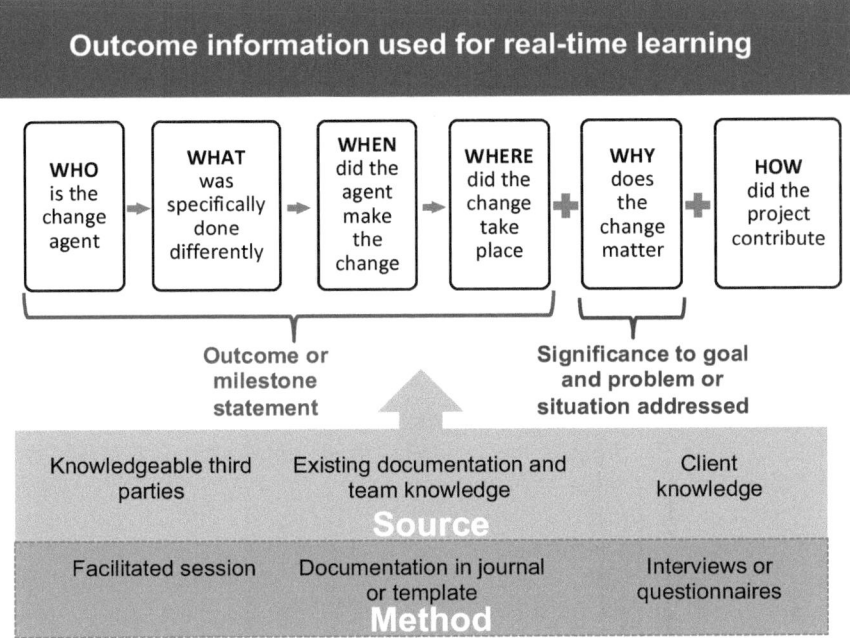

Fig. 17.3 Process for learning from outcome information, which is critical to the OH approach (Source World Bank 2014b, c)

supported by the WB's Drought NLTA, and yet there is still no guarantee that by the end of the program a national drought policy will be in place.

Sustainable changes (outcomes), therefore, are *influenced* by the Drought NLTA, but *promoted and implemented* by the stakeholders. A sustainable change (outcome) is understood as a result of complex collaboration processes that are naturally influenced (both positively and negatively) by many factors (commonly considered as "externalities" in other approaches).

The original design of the WBI pilot suggested the active engagement of the WB team and partners of the stakeholders in the design and implementation of the M&E strategy, through meetings and workshops to promote joint reflection along lines of implementation, as well as the harvesting and analysis of outcomes. The specific conditions of the Drought NLTA, once implemented however, did not allow these opportunities. Nevertheless, with the intent to preserve the participatory nature that is one of the key-features of the approach, individual and group interviews, questionnaires, formal and informal interviews were carried out in person (taking advantage of regional workshops), through e-mail (written), and Skype or telephone, and were used to collect input at key points throughout the program as well to harvest outcomes and collect evidences of the change process.[7]

The implementation of the approach was developed after consultation with the key-stakeholders involved. Interviews and questionnaires captured the key-challenges and constraints to be tackled regarding drought management, as well as their vision and commitments to the change process.

Following the structure of the Drought NLTA described in its concept note, the change process for each group of partners was grouped under two key-areas (the National and Regional tracks, described above) and for each of these two broader areas, one ambitious long-term outcome was designed (Institutional Change Outcomes), describing the deepest possible transformations the Drought NLTA could influence, without losing sight of the reality of the context and what can be realistically achieved.

Although outputs (as well as inputs and activities) were described in the RF and monitored along the process, the key-focus of the approach has been the design, monitoring, and evaluation of outcomes.

To monitor the program, outcomes were 'harvested' through individual and group interviews, and examining of project documents and related materials to capture the (intended and unintended) relevant political and institutional changes generated throughout the process; allowing the WB team to understand its influence beyond the scope of the program and beyond its outputs. The harvesting of outcomes is not guided by the RF, but compared to it after the harvesting, allowing the identification and acknowledgment of unpredicted/unplanned direct and indirect outcomes (promoted by partners and by partners' partners). Similarly, designed indicators were monitored and provided support to the harvested outcomes. The findings were validated with the management team.

The final steps of the process are the selection of more relevant outcomes and substantiation. The substantiation requires that knowledgeable, independent third parties review the description and confirm the outcomes and the contributions of the program. In the Drought NLTA, the substantiation will be performed after the final harvesting of outcomes (fall/winter of 2015/2016). Substantiation represents an additional source of evidence that helps confirm the harvested outcomes – reinforcing the triangulation process.

---

[7]The harvesting and analysis of outcomes can benefit significantly from opportunities to gather partners and the program team around the table to a shared reflection process, adding another layer of credibility to overcome the risk of reporting outcomes without evidence. It is possible to implement the approach by replacing this step of the process with a group or individual interviews, as it was adopted in the Drought NLTA, but the process may lose some of its richness by placing the collection and analysis in the hands of a single person.

17.5 Monitoring with Outcome Harvesting: Key-Findings and Outcomes[8]

The findings are here presented around the two main Drought NLTA component tracks. Most of the time, the outcomes are strongly influencing each other and are contributing to both of the higher level goals designed for each of the program's tracks. The words in italic are the *milestones* designed in the RF for each group of partners. The numbers (1–11) at the end of each outcome description relate to the mapping of the outcomes (Fig. 17.4, item 17.5.3., below).

This approach does not pay particular attention to outputs and activities, considered as means to achieve the sustainable changes, or outcomes. Rather, the outputs and activities are mentioned as evidence that the WB or the partners have contributed to these outcomes.

17.5.1 National Drought Policy Track: Key Findings

- *Advances in the dialogue towards a drought policy at the national level* happened through the promotion, by the MI, of a series of regional seminars in the Northeast between April and May, 2014, to discuss policies for living/coping within the semi-arid region (1). These discussions included the endorsement of the DM and DPPs in the final recommendations of the discussion process, included in a document released in September, 1st, 2014, and delivered directly to MI. (5)
- A concrete measure to *supporting and leveraging regionally-led drought preparedness* initiatives and a step towards the *institutionalization of a paradigm shift* was a Technical Cooperation Agreement (MoU) signed in Brasília in

---

[8]Throughout the design and implementation of the PM&E approach, more than 43 interviews (formal and informal) and questionnaires were applied with key-stakeholder representatives and Drought NLTA Team members between January 2014 and January 2015.

Other data collection methods used were review of documents and notes; observation of internal and external meetings and workshops; and on-line and press media information collected from March 2014 to January 2015, using Google Alert tool, and as sent to the Consultant by stakeholders and team members.

The sampling criteria adopted on the three phases of the M&E work developed so far were purposeful sampling that could cover a wide range of institutions in different states of the Northeast, at the same time that it considered stakeholders involved in the program in different levels and in both National (drought policy) and Regional tracks.

While in the RF design and the mid-term monitoring report there was no specific concern about gender balance in the sampling, a more equal approach was adopted for the third cycle of interviews, when it was specifically requested that at least one woman should be represented in the groups of stakeholders to be interviewed (along with the other mentioned selection criteria).

Data were organized and analyzed through content analysis, identifying emerging patterns and triangulating to probe findings.

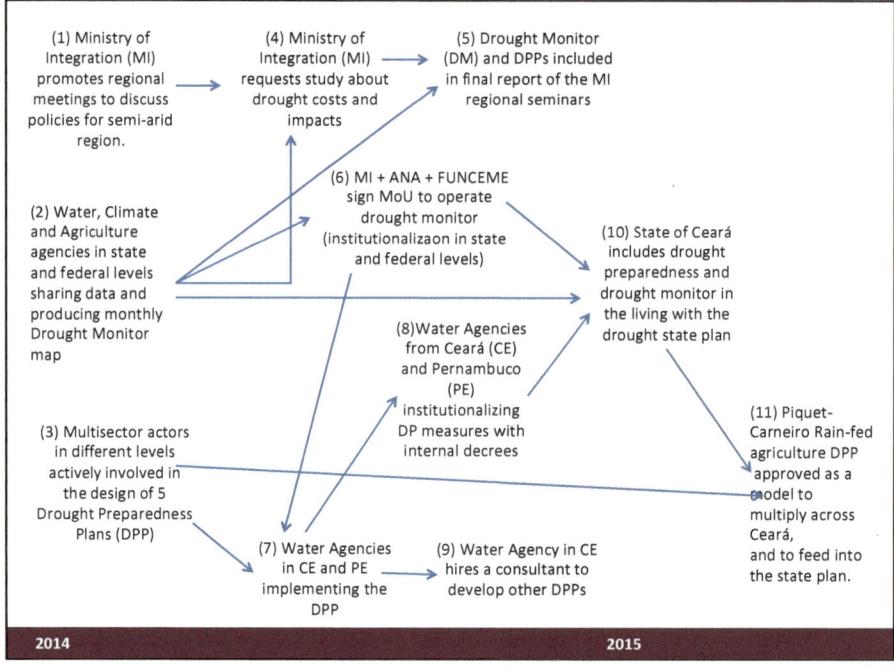

Fig. 17.4 Mapping of Drought NLTA key-outcomes harvested from May 2014 to January 2015

September, 2014, between federal and regional partners (i.e., specifically MI, ANA and FUNCEME), valid until December, 2015. (6) The MoU defines an institutional arrangement for the DM, an operational structure, and transition rules, with MI and ANA assuming key-roles in the governance (roles that are currently being supported by the strongest regional leader of this partnership, FUNCEME). The DM, as evidenced by the MoU, is also evidence of *collaboration through established networks with defined governance rules*, a milestone designed for the Northeast Regional Drought Preparedness track. The DM has been considered a concrete and tangible technological and institutional upgrade, and to some extent, it is buffered from strong political influence and politicization. The GoB considers the DM the foundation upon which any future NDP might be built.

- In mid-2014, halfway through the program, the MI requested additional assistance to the WB to evaluate the impacts and costs of the current drought across the Northeast to support improvements in vulnerability/resilience and impact assessment (progressing towards pillar 2 of the 3 drought preparedness pillars). This represents another concrete step towards *institutionalization of a paradigm shift on drought management*. (*4*)

17.5.2 Northeast Regional Drought Preparedness Pilot Track: Key-Findings

- *Increasing knowledge, data sharing and cooperation* are evidenced by the experimental monthly DM maps that have been voluntarily and cooperatively produced by multi-sector professionals representing the nine northeast semi-arid states and many other stakeholders across the federal, regional, and local levels, since August, 2014. (2)
- Stakeholders involved in the development of each of the five DPPs have *engaged in and promoted capacity development activities*, such as the Drought NLTA regional workshops in January, May, and November, 2014. These workshops took place in Fortaleza (CE), Recife (PE) and Salvador (BA). (*3*)
- The participation and implementation of the DPPs has promoted *increased know-how to plan drought mitigation and response actions*. Water and sanitation companies from PE and CE have *improved permanent and sustainable management capacity* in what was declared to be a *paradigm shift* and a milestone in the history of the partners involved. (7) In November, 2014, and January, 2015, they reported the absence of a water-volume management culture prior to the DPP, with management criteria previously being defined ad hoc by the current manager and no specific operational protocols. They reported that internal decrees were on the way to institutionalizing drought preparedness measures. (8) The DPPs are, thus, have helped the culture of these institutions shift away from crisis to risk management. The water agencies that started the successful implementation of the DPP decided to hire a consultant to develop other DPPs to improve management of other reservoirs in Ceará. (9)
- In February 2015, the Government of Ceará State included drought preparedness measures and the DM in the "Living with Drought" State Plan and presented these measures in high level meetings led by the Governor of Ceará with the presence of the President of the Republic, Ministries, and several Governors of the Northeast. (10)
- The rain-fed agriculture DPP developed in Piquet-Carneiro, a small Municipality in Ceará, has been approved at the State level as a model to follow by other Municipalities across the state and the region, and to feed into the State Plan, as announced by the Governor of Ceará in Piquet-Carneiro on July 31st, 2015. (11)

17.5.3 Mapping Key-Outcomes

The WBI pilots developed a useful dissemination resource tool to highlight key-outcomes through organizing and displaying the outcomes in maps that can be presented in a timeline format, such as presented in the Fig. 17.4.[9]

---

[9]The outcomes can also be organized in a map that presents these and other theory of change elements, such as activities and inputs, as well as in other visual arrangements. Please see World Bank. 2014 c. *Cases in Outcome Harvesting*. Available at http://www.outcomemapping.ca/download/en_Cases%20in%20Outcome%20Harvesting.pdf, pages 15, 26.

When looking at the map below, and reading the above findings, it is important to take into account that the process facilitated by the Drought NLTA is still recent. The activities started in mid-2013, and the first convening workshops to discuss the DM and the DPP started in January, 2014.

The process was fostered by the WB with strong voluntary adhesion of the partners. The relatively young nature of the program means that longer-term outcomes influenced directly or indirectly by the program are yet to crystalize, which explains that many of the milestones (outcomes) above reported are concentrated in the early and mid-term stages of the process.

The outcome map below (Fig. 17.4) shows the key-outcomes, collected along the years of 2014 and 2015 through review of documents and interviews conducted in November 2014 and January 2015 with the Drought NLTA team and partners. Outcomes will be again harvested and then substantiated (i.e., confirmed with key, knowledgeable informants) in early 2016.

17.6 Drought NLTA Implementation Lessons: Program Design and PM&E Approach

17.6.1 Drought NLTA Methodology Strengths and Challenges Assessment

Along with the harvesting of outcomes, a strengths and challenges assessment of the Drought NLTA methodology was conducted between November 2014 and January 2015, seeking to inform future similar collaborations around related topics (e.g., climate change resilience, drought preparedness or other complex issues regarding the influence of policy/institutional changes).

Individual or group interviews (up to three people), in person or via Skype, were conducted with key-partners involved in the DPPs and the DM. Essentially, they were asked to inform what had changed (in their practices, policies, behavior, knowledge, and institutions) since they started participating on the Drought NLTA activities and what were the key-strengths and challenges of the processes.

The results are summarized below.

17.6.2 Challenges in the Drought NLTA Process

Resistance was reported as challenge in the beginning of the program. The past efforts on drought management in Brazil have proven largely unsuccessful, so new initiatives are always looked upon with incredulity. Some respondents also mentioned that potential users of the DM have the perception that it is *just another* indicator or monitoring product. Although this resistance has faded with time, it persists among some stakeholders.

Institutional Fragilities were revealed along the process, ranging from the lack of personnel, high turnover or experience of partners' staffs, and the lack of planning-based management (which includes the lack of integration and knowledge of monitoring and other data).

The perception of these *institutional fragilities* as challenging is well documented in the notes of the third technical workshop (Salvador, INEMA (BA), November 19–20), and more specifically in the discussions about the monitoring network gaps and bottle necks that were identified through an institutional and IT analyses performed to support the DM process.

17.6.3 Strengths in the Drought NLTA Process

Commitment and Participation has been one of the keys to overcome the above-mentioned resistance, resulting from a sum of factors:

- A regionally-led initiative (as opposed to one that is purely top-down);
- A technical-scientific process that is less susceptible to political interference;
- The immediate applicability of the concepts and studies to the ongoing drought in the region;
- The differentiation of the DM from other monitoring products, and;
- The highly participatory methodology that acknowledged the importance of the contributions from technicians as well as of the upper-level managers and policy makers.

Capacity Development and Institutional Strengthening resulted from the participation in the DPPs and the DM. Respondents agreed that there was an unequivocal gain in knowledge and improved capacity to deliver daily duties, with a broader and more complex vision than they had previously possessed before the initiation of the process, as the quote below illustrates:

> *"What can one say, what to argue, to a mayor, a governor?* (. . .) *Now I can say we know the limitations of the dam.* (. . .) *Today I am sure I am adopting the correct measures.* (. . .) *"We feel secure to tell the press we are prepared* (. . .)*".* – Manager of a water resources company in Pernambuco state, participant of a DPP. Interview and workshop notes

The institutional strengthening is directly linked with the capacity development. The interviewees reported that their institutions are stronger and more capable to deliver their services. This strengthening also derives from the methodologies used to develop the plan. The participation in the process and the implementation of the DPPs and DM helped in raising awareness, identifying the gaps and building a strong foundation for stronger institutions. It required partners to look for and organize information that, until then, had never been assembled and interpreted together.

The internalization of the new knowledge and the incorporation of new routines started immediately as the process commenced. These new routines and tools, information, and plans are intended to be used by the organizations beyond the scope of the DPPs/DM.

The contradiction between institutional fragility (pointed to by interviewees as a challenge, as seen above) and the institutional strengthening as a benefit (as indicated by the outcomes) is in fact, two sides of the same coin. Institutional fragility has been identified as a general constraint to development in Brazil, particularly in the same regions of the country as this program (i.e., the Northeast). In the case of the DM, while the key-partners of the process, FUNCEME, INEMA, APAC, ANA, MI, and INMET are more developed and capable of acquiring knowledge from international processes, training professionals in their institutions, adapting the technology and processes to the Brazilian reality, and even advancing it much more than expected and planned, many partners in the Northeast remain in much earlier stages of development. For example, some did not have permanent personnel or appropriate equipment. Institutional fragilities are, therefore, a challenge in the process of developing a shared and voluntary permanent cooperation process that needs periodic and reliable feeding of data.

Thus, the institutional strengthening appears as an absolutely critical benefit of the Drought NLTA process. Stakeholders report, in these early stages, awareness about their institutional fragilities and also a gradual shift in their perspectives, followed by changes in their practices and the institution with respect to new rules and procedures. The Drought NLTA process has also provided these professionals with concrete evidence for justifying requests to their superiors for technical and informational improvements to support the improved management of drought.

Integration of Sectors States and Institutions Institutions that did not initially have much dialogue with one another were brought together, or have further tightened already existing institutional relationships through the participation in the Drought NLTA process. This integration is happening across-sectors (meteorological, agricultural and water sectors), within sectors, and between institutions (hydro-meteorological institutes, water and sanitation companies, etc.), across states, and finally, across state and federal institutions.

The integrated vision of the drought and its management, along with the associated improvement of institutional capacities to deliver services were reported, as highlighted below:

Before, we only monitored our state, now we are looking at the Northeast as a whole and beyond. (...) We expanded not only the knowledge, but our vision of what happens, because nature has no barriers, no limits. Technician at a climate institution, Ceará state, participant of the DM, interview

17.6.4   Limits and Possibilities of the WBI Approach Implementation in the Drought NLTA: Lessons to Be Remembered

The implementation of the WBI pilot approach in the Drought NLTA program has raised some important points of discussion, in terms of methodological conclusions and contributions.

• Although relevant outcomes can happen in early implementation stages, programs framed as multi-stakeholder/multi-sector partnerships and strongly based on voluntary collaboration, such as the Drought NTLA, tend to take time to develop. The harvesting of significant outcomes will likely benefit from more implementation time. When the first harvesting was done (i.e., November 2014– January 2015), the program was in mid-term implementation phase. Results influenced by the program were starting to develop but were not yet ready to be reported as outcomes.

• The risk of having partners over-reporting positive outcomes to which the program has not truly contributed (e.g., to please the donor) can be overcome with rigorous methods. Triangulation of sources (combining document reviews, interviews, and other sources of information) and probing are extremely necessary. The framework provokes the analyst to do just that, by asking for evidence of the reported outcomes.

• Outcomes need to be interpreted taking the context into account (political environment, staff turnover, local and organizational culture, necessary support, etc.). Fostering partnerships needs respect for the various partners' capacities and their specific contexts. This principle allows the collaboration to generate outcomes that sometimes may be more realistic and more likely to be sustainable in the longer term than the planned, non-achieved outcomes. It is the case of the Piquet Carneiro DPP. While this plan did not define policy and management actions triggered as the drought progresses to higher stages, the plan was built through a broad consultation process, including discussions and the development of the plan proposal and intermediate validations with different stakeholders. As a result, it includes coherent and consistent management activities related to the preparation and risk reduction, and touches on response and disaster recovery for extreme drought effects in the municipality. It also provides a series of recommendations for institutional strengthening, adoption of management tools, training and capacity building, and infrastructure investments to provide effective risk management inherent to drought in the municipality of Piquet Carneiro. This rain-fed agriculture DPP has been approved at the State level as a model to follow by other Municipalities across the state and the region, and to feed into the State Plan.

- In such participatory approaches, it is key to involve all partners' representatives in the design of the Results Framework as much as possible – it will be more realistic and promote greater commitment. In the Drought NLTA implementation, because of the different paces of its pieces, not all of partners were already onboard when the RF was designed and reviewed. This resulted in the design of some milestones that only partially happened as the program developed.
- Perhaps the greatest limitation that this approach presents is the difficulty to link long-term, impact evaluation outcomes and, more specifically, indicators, to the program.

Requested by the WB Team, Drought NLTA partners suggested some indicators, but most of the suggested do not capture impact (e.g., # *of downloads of Drought Monitor information maps and narratives from the Drought Monitor website*).

The difficulties with designing impact indicators for a drought resilience program like the Drought NLTA are that: (i) baselines are very challenging to establish and subsequently compare; (ii) attribution of impact of DM and the DPPs in increasing drought resilience might only be possible by comparing against when the next drought happens; and (iii) isolating the influence of an specific tool, such as the DM, from other influences in building such resilience, is very difficult. For example: the suggested indicator # *and distribution of monitoring network points across nine Northeast states*, suggests that it would be possible to identify the impact of the DM in facilitating the expansion and penetration of climate and agriculture monitoring networks in Brazil. However, this impact is difficult to assess because other factors could be contributing equally or more to the expansion, such as economic and political decisions that have no direct relationship with the Drought Monitor.

Other suggested long-term indicators to be monitored that fall in a similar situation are:

- # *of cities that adopt urgent measures (such as water rationing) during drought declaration periods*
- # *of water tank-trucks to provide water for human use during drought declaration periods*
- # *of people with non-interrupted access to water during drought declaration periods*

Designing and evaluating the impact of drought preparedness and other climate change adaptation and resilience initiatives remains a challenge that this specific approach, to date, could only begin to scratch, precisely because of the complexity of the factors that influence the outcomes. However, if *contribution* to impact can be accepted as a measure of success, then OH is an appropriate tool that can be combined with other techniques to evaluate impact.

17.7   Conclusions

In the early-mid stages of the program's implementation, interviews with some of the 80 + professionals involved in the Drought NLTA, have revealed the key constraints and challenges to improving drought management in Brazil. The findings, reported in items 5 and 6 above, show that the initiative is contributing to address some of the reported challenges that Brazil persistently faces in proactively managing droughts.

The program is contributing significantly to both institutional and technical upgrades for better drought management, and two years after the beginning of its implementation, there is evidence of its influence. This evidence has been obtained by implementing a combined PM&E approach, originally piloted by the WBI, which was designed to capture the complexity of institutional and behavioral changes evident in the Drought NLTA. While the methodology has some challenges and limitations, it has proven itself as an effective tool for understanding drought and climate change resilience and adaptation.

References

Earl, S., Carden, F., & Smutylo, T. (2001). *Outcome mapping: Building learning and reflection into development programs*. Ottawa: IDRC.

Gold, J. (2013). *Real-time learning in projects: Outcome management from design to completion*. The World Bank Institute. Slides.

Gold, J. (2014). *Guide for analyzing and learning from outcomes: Making sense of collected outcome information to improve development results*. Washington, DC: The World Bank Institute.

Ministério da Integração Nacional (MI) e Instituto Interamericano de Cooperação para a Agricultura (IICA). (2013). Estudos Referentes ao Diagnóstico da Política nacional de Secas no Brasil: Relatório Contendo Diagnóstico e Embasamento para a Formulação de uma Política Nacional de Secas no Brasil. Consultor, Otamar de Carvalho.

Otoo, S., Agapitova, N., & Behrens, J. (2009). *The capacity development results framework: A strategic and results-oriented approach to learning for capacity development*. Washignton, DC: The World Bank.

Wilhite, D. A. (2011, July 14–15). National drought policies: Addressing impacts and societal vulnerability. In *Towards a compendium on national drought policy: Proceedings of an Expert Meeting*, Washington, DC.

Wilson-Grau, R., & Britt, H. (2012). *Outcome harvesting*. MENA Office, Ford Foundation. Available at http://www.outcomemapping.ca/resource/outcome-harvesting

World Bank. (2014a). *Turn down the heat: Confronting the new climate normal*. Washington, DC: World Bank.

World Bank. (2014b). *Outcome based learning field guide: Tools to harvest and monitor outcomes and systematically learn from outcomes.* Available at https://wbi.worldbank.org/wbi/Data/wbi/wbicms/files/drupal-acquia/wbi/Outcome-Based%20Learning%20Field%20Guide.pdf

World Bank. (2014c). *Cases in outcome harvesting.* Available at http://www.outcomemapping.ca/download/en_Cases%20in%20Outcome%20Harvesting.pdf

World Bank. (2012). *Turn down the heat: Why a 4°C warmer world must be avoided.* Washington, DC: World Bank Group.

World Bank Institute (WBI). (2013a). *Guide for real-time outcome monitoring.*

World Bank Institute (WBI). (2013b). *Designing a multi – stakeholder results framework: A toolkit to guide participatory diagnostics and planning for stronger results and effectiveness.*

World Bank Institute (WBI). (2014). *Steps to frame capacity development results.* Paper.

Chapter 18
The Adaptation M&E Navigator: A Decision Support Tool for the Selection of Suitable Approaches to Monitor and Evaluate Adaptation to Climate Change

Timo Leiter

Abstract With increasing implementation of climate change adaptation policies and projects as well as continued integration of adaptation into planning processes, there is an increasing need to understand the results of these adaptation interventions. Are they achieving their objectives? Are they actually leading to a reduction in vulnerability to climate change?

Monitoring and evaluation (M&E) can help answer these questions. However, due to the context specific and cross-sectoral nature of adaptation there is no one-size fits all approach to M&E. The Adaptation M&E Navigator helps to select a suitable M&E approach by providing a list of specific M&E purposes and matching them to relevant approaches. Key characteristics of each approach are highlighted to enable informed decision making. The Adaptation M&E Navigator also provides links to further guidance and examples from practice. The chapter outlines the rational and structure of the Adaptation M&E Navigator and how it can be used in practice.

Keywords Adaptation • Monitoring and evaluation • M&E approach • Adaptation outcomes • Adaptation process

18.1 Introduction

Preparing for and adjusting to the impacts of climate change through planning, capacity building and adaptation actions is taking place at all levels, on all continents and to an increasing extent (Mimura et al. 2014). According to the 2015 Global Climate Legislation Study, more than 60 countries have frameworks in place for adapting to the impacts of climate change (Nachmany et al. 2015).

T. Leiter (✉)
Climate Change and Climate Policy Group, Deutsche Gesellschaft für Internationale
Zusammenarbeit (GIZ) GmbH, Eschborn, Germany
e-mail: Timo.Leiter@giz.de

© The Author(s) 2017
J.I. Uitto et al. (eds.), *Evaluating Climate Change Action for Sustainable
Development*, DOI 10.1007/978-3-319-43702-6_18

Bilateral and multilateral climate-related finance to developing countries explicitly targeting adaptation to climate change reached USD 10 billion in 2013 (OECD-DAC 2015). The continuous integration of adaptation into planning processes and the technical and financial support to developing countries have resulted in hundreds of adaptation projects around the globe. This leaves decision makers, fund managers and project implementers with the question of what is being achieved. What are the results of all these adaptation interventions? Do they lead to a reduction in vulnerability? How can the outcomes of adaptation be assessed?

Addressing this need, several frameworks and guidebooks for Monitoring and Evaluation (M&E) of adaptation have been developed covering the project or community level (CARE 2012; Olivier et al. 2013; Pringle 2011), the national level (Ford et al. 2013; Price-Kelly et al. 2015; Hammil et al. 2014a) or multiple levels (Brooks et al. 2011; Leiter 2015). An overview of 22 publications and guidebooks for adaptation M&E has been compiled by Bours et al. (2014a).

The increasing number of frameworks and tools for adaptation M&E makes it difficult for decision makers and their advisors to quickly identify an appropriate one that matches their needs. In the field of climate change vulnerability and impact assessment, which is faced with an even greater proliferation of methods and tools, the PROVIA guidance has made an attempt to structure the selection process through decision trees (PROVIA 2013). Whilst the PROVIA guidance provides a useful overview of adaptation M&E literature, the proposed decision tree for M&E focuses on the project level only and consists of rather general questions (e.g. "Have you considered who else needs to be involved in the evaluation?") (PROVIA 2013, p. 52). It is also focusing more on evaluation than on ongoing monitoring and prescribes the use of indicators, which excludes other relevant M&E approaches from the start, including those based on qualitative information. Overall, the PROVIA guidance does not comprehensively identify the breath of specific reasons to engage in M&E of adaptation and does not directly indicate applicable M&E approaches for each of them. Fisher et al. (2015) provide an extensive list of methodologies of potential use for adaptation M&E. Yet, apart from assessing their applicability to simple, complicated or complex interventions they do not link them to initial reasons for undertaking monitoring and evaluation of adaptation.

In fact, decision makers typically encounter M&E in regard to a specific reason or information need such as finding out whether the implementation of an adaptation plan is advancing, or whether a community is better equipped to dealing with climate change impacts as result of an adaptation intervention. Such specific purposes for M&E therefore provide a logical starting point to guide the selection of M&E approaches. Hence, the Adaptation M&E Navigator is structured along specific purposes for undertaking adaptation M&E and matches them to relevant M&E approaches. A short description including benefits and limitations, resources needed for implementation, practical examples and links to further guidance is provided for each approach to facilitate decision-making. The sequence of steps in selecting a suitable M&E approach and the scope of the Adaptation M&E Navigator are shown in Fig. 18.1. The following part of this chapter outlines the content

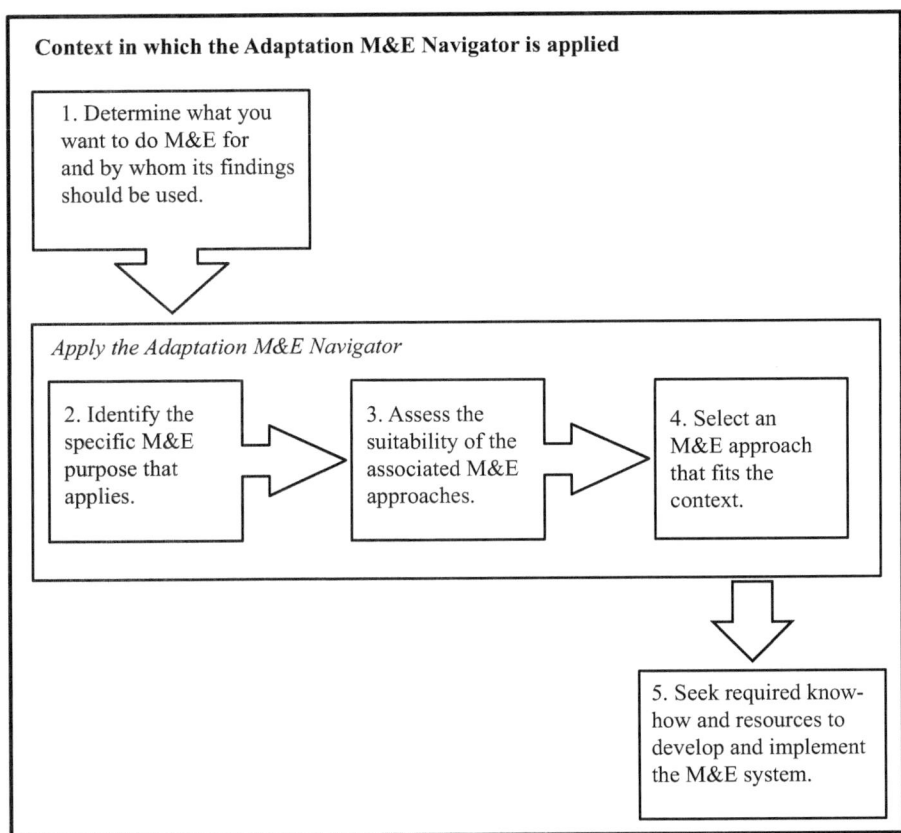

Fig. 18.1 Steps in selecting a suitable M&E approach for adaptation

and structure of the Adaptation M&E Navigator, its limitations and how it can be used in practice.

18.2   Specific Purposes for M&E of Adaptation to Climate Change

The literature identifies a number of general purposes for monitoring and evaluating adaptation interventions, including: assessing whether they are achieving their objectives; supporting management under uncertainty; facilitating learning; and providing accountability (e.g. Pringle 2011; PROVIA 2013; Spearman and McGray 2011). However, decision makers typically encounter the need for M&E of adaptation in light of more particular reasons. Based on a literature review (including amongst others the resources listed in Bours et al. 2014a) and the author's experience in supporting the development of national and sub-national adaptation M&E

systems an initial list of **specific purposes** for adaptation M&E was drafted. These specific purposes are universally formulated to ensure broad applicability and to avoid an unmanageable number of individual cases. The initial list was sent for comments to adaptation and M&E experts, including selected participants of the *2nd International Conference on Evaluating Climate Change and Development*. The resulting list includes **nine specific purposes for adaptation M&E** which are categorised into process or outcome-oriented assessments:

- **Assessing adaptation processes**

  - Monitoring the integration of adaptation into planning processes
  - Monitoring the implementation of adaptation programmes, projects or actions
  - Monitoring the implementation of the National Adaptation Plan (NAP) process
  - Tracking which adaptation activities are taking place at national or sub-national level

- **Assessing adaptation outcomes**

  - Assessing the results of adaptation projects or actions
  - Assessing the results of a programme or portfolio of adaptation projects
  - Assessing whether vulnerability has been reduced as a result of adaptation programmes, projects or actions
  - Assessing progress towards adaptation goals, targets or intended outcomes at national level
  - Assessing whether resilience to climate change has been improved at national level

These nine specific purposes are examples of common reasons for undertaking adaptation M&E – either during or after the implementation of an intervention. The Adaptation M&E Navigator does not, however, include consideration of assessments that typically take place before implementation starts such as identifying climate change impacts and appraising adaptation options (e.g. PROVIA 2013). An exception is the assessment of vulnerability at the start of an intervention if its purpose is to measure adaptation progress over time (e.g. Fritzsche et al. 2014, pp. 155–163). The Adaptation M&E Navigator does not cover tracking financial flows for adaptation (see for example Terpstra and Peterson-Carvalho 2015).

18.3   Connecting the Specific Purposes to Suitable M&E Approaches

The specific purposes for adaptation M&E outlined above differ in regard to what is being monitored or evaluated, at what level, over shorter or longer time periods and whether the focus is on processes or outcomes. Accordingly, each of the specific purposes has different requirements for M&E which makes it possible to preselect M&E approaches that meet those requirements. For example, monitoring the integration of adaptation into planning processes does not require complex statistical analysis. Rather, a qualitative or quantitative approach focusing on the planning processes and involving stakeholders seems more suitable. It could take the form of interviews with key informants or of a set of indicators illustrating progress. This example demonstrates that there is still a variety of possible M&E approaches even for the same specific M&E purpose. Therefore, the Adaptation M&E Navigator does not lead users to the one and only M&E approach, but rather directs them to a short list of relevant M&E approaches. Indeed, the Adaptation M&E Navigator helps to filter among the many existing M&E approaches those that seem most relevant for a given purpose.

The M&E approaches which are associated with the same specific purpose each constitute a very distinctive way of assessment, e.g. assessing improvements in resilience through either a set of high level national indicators or through household level questions as part of a census (Welle et al. 2014). Every M&E approach can in turn be implemented in a variety of ways. For example, the exact interview procedures, number and composition of interviewees of the M&E approach "Qualitative assessment based on interviews" can vary greatly. In fact, the final M&E design is typically tailored to the specific context, as demonstrated by the M&E methodology of the United Kingdom's capacity building support to help implement Ethiopia's Climate-Resilient Green Economy Strategy described in Adler et al. (2015). The Adaptation M&E Navigator is supporting decision makers and their technical advisors to identify a suitable M&E approach which can then be tailored to the particular context.

18.4   Supporting the Selection of a Suitable M&E Approach

The suitability or appropriateness of a particular M&E approach can only be determined in light of the specific context of application. The Adaptation M&E Navigator includes five criteria which are useful to consider:

1. The main intention or general purpose the M&E approach is mainly catering to, i.e. learning, management or accountability
2. A focus on process or outcome-orientation
3. The degree of complexity of implementing the M&E approach
4. The degree of subjectivity of the M&E findings
5. The level of available experience in applying the M&E approach

A detailed description of each criterion and its relevance is provided in Table 18.1. Criteria 3–5 are rated on a 5 point scale (low, low to medium, medium, medium to high, high). The M&E approaches are rated relative to each other, i.e. if one is relatively more complex to carry out or leads to more subjective findings than another M&E approach.

Table 18.1 Decision support criteria to select M&E approaches

Criterion	Description	Relevance
Main intention/ general purpose	Which of the three general purposes (i) management, (ii) learning or (iii) accountability an M&E approach is likely to predominantly address	It is important to consider the **intended use** of an M&E system at the outset. This criterion helps users determine whether a particular M&E approach has the potential to actually meet the intended general purpose
	Management refers to supporting the ongoing management of adaptation actions and processes (in the sense of adaptive management)	
	Learning refers to acquiring a detailed understanding of how and why adaptation interventions have led to certain results or why they did not achieve their objectives	
	Accountability refers to demonstrating that processes and/or actions have taken place and have led to results	
	The nature of an approach, i.e. what data and procedures it uses and what information it provides, determines which of the three general purposes it can best support. For example, if a small number of standard indicators like "Number of beneficiaries" are aggregated for a portfolio of adaptation projects, the resulting information is not adequate to infer how and why adaptation has worked (Chen and Uitto 2014). Thus, this approach is most suitable for accountability purposes, but not for learning	
	M&E approaches can cater to more than one general purpose depending on how exactly they are designed in practice	
Process or outcome orientation	Whether the M&E approach is focusing more on the **process** of implementation or on the **outcomes** of adaptation	The decision to monitor either processes or outcomes, or both, influences the selection of suitable approaches, because it entails different requirements for M&E
	This distinction is common in the literature on adaptation M&E since assessing adaptation outcomes is faced with various challenges (Bours et al. 2014b). As a result, it was suggested to initially focus on process-based indicators and gradually move to outcome-based ones (Harley et al. 2008)	In the context of increasing levels of climate finance it is particularly important to outline which M&E approaches are actually capable of assessing adaptation outcomes, and which only focus on processes

(continued)

Table 18.1 (continued)

Criterion	Description	Relevance
Complexity of implementation	The relative complexity of an M&E approach compared to others. Low complexity indicates that an M&E approach is relatively straight forward to understand. For example, asking beneficiaries about their perceptions is an intuitively understood M&E procedure whereas the details of assessing avoided economic costs are more complex	The degree of complexity provides a rough indication of the ease of applying an M&E approach and of the resources needed (know-how and financial resources)
	Low complexity does not mean that approaches can be easily implemented. Qualitative assessments also require relevant expertise to be carried out in a rigorous manner	
Subjectivity of resulting information	The relative subjectivity of the resulting M&E findings, i.e. the extent to which they can be influenced by those involved in the M&E process. For example, M&E approaches based on surveying beneficiaries will be more subjective than impact evaluations based on quasi experimental designs	It is important to reflect how the M&E process can influence the M&E findings and how this resonates with the purpose and intended use of the M&E results
	Subjectivity does not mean less valuable information. In fact, the views of beneficiaries or key informants may be exactly the type of information needed. Moreover, quantitative approaches cannot be equated with objectivity. Whilst indicator values may be objective, the choice of which indictors are included and how they are defined may not be entirely objective	
Application experience to date	Available experience to date in applying a particular M&E approach to climate change adaptation. Some approaches like theory of change or those used for impact evaluations have been widely used in other fields, but this criterion focuses on the available experience in applying them specifically to adaptation to climate change	Available experience influences the cost and uncertainty of applying a particular M&E approach

In addition to the five criteria, the Adaptation M&E Navigator provides further details for every M&E approach according to a template illustrated in Tables 18.3 and 18.4. The template provides information on the required human and financial resources to implement an M&E approach as well as on benefits and limitations.

Since each M&E approach can be implemented in a variety of ways, the descriptions are based on a general application and cannot take every possible variation into account. Corresponding to its nature as decision support tool, the Adaptation M&E Navigator has to maintain a balance between level of detail and concise, easy to grasp information. Hence, it cannot provide comprehensive detail on how to carry out any of the listed M&E approaches. For the latter purpose the template includes links to practical examples, guidance and further resources that users can refer to. Thus, the Adaptation M&E Navigator equips decision makers and their technical advisors with an overview of relevant approaches and information to support the selection of an M&E approach.

The core of the Adaptation M&E Navigator is provided in Table 18.2 which connects specific purposes to relevant M&E approaches and shows their rating on the five criteria. In the online version of the Adaptation M&E Navigator, colour codes are applied to facilitate a quick interpretation of the ratings. Tables 18.3 and 18.4

Table 18.2 The adaptation M&E navigator: matching specific M&E purposes to relevant M&E approaches

#	Specific purpose	M&E approach	General purpose	Focus on processes or outcomes	Com-plexity	Subjec-tivity	Ex-perience
1	Monitoring the integration of adaptation into planning (mainstreaming)	Qualitative assessment based on interviews	Learning	P	L-M	H	M
		Quantitative or qualitative indicators	Management, Accountability	P	L-M	L-M	M
2	Monitoring the implementation of adaptation programmes, projects or actions	Defining and monitoring activities and outputs	Management, Accountability	P	L	L	H
3	Monitoring the implementation of the National Adaptation Plan process	Defining and monitoring milestones in the NAP process	Management, Accountability	P	L	L-M	L
4	Tracking adaptation activities at national or sub-national level	Database of adaptation activities	Management, Knowledge sharing	P	L-M	M[a]	L-M
5	Assessing the results of adaptation projects or actions — On an ongoing or repeated basis	Qualitative assessment involving beneficiaries	Learning, Management	P/O	L-M	H	M
		Theory of change with adaptation-specific indicators	Management, Accountability	P/O	M	L-M	M
		Repeated vulnerability assessments	See specific purpose #7				
	Assessing the results of adaptation projects or actions — At a certain point in time, typically after completion	Impact evaluation	Learning, Accountability	O	H	L	L
		Assessing avoided economic losses and health benefits	Accountability	O	H	L	L
6	Assessing the results of a programme or portfolio of adaptation projects	Project-specific indicators informing a synthesis of portfolio results	Accountability	P/O	M	M	L
		Common (core) indicators for every project to enable aggregation	Accountability	P/O	M	L-M	M
7	Assessing whether vulnerability has been reduced as a result of adaptation programmes, projects or actions	Measuring vulnerability with indicators as part of a results-based monitoring system	Management, Accountability	O	M	L-M	M
		Repeated vulnerability assessments — Simple	Accountability	O	L	H	M-H
		Repeated vulnerability assessments — Data intensive	Learning, Accountability	O	M-H	L-M	L
8	Assessing progress towards adaptation goals, targets or intended outcomes at national level	Qualitative assessment without indicators	Learning, Management, Knowledge-sharing	P/O	L-M	M-H	L
		Indicator-based assessment — Trend indicators	Management	P/O	M	L	L-M
		Indicator-based assessment — Based on assumptions about how activities lead to outcomes	Management, Accountability	P/O	M-H	L-M	L-M
9	Assessing resilience to climate change at national level	Indicator-based assessments	Management	O	M	L-M	L-M
		Household level questions as part of national census surveys	Management	O	M-H	H	L

Explanation: L = Low, L-M = Low to Medium, M = Medium, M-H = Medium to High, H = High; P = Process, O = Outcome, P/O = Process and/or Outcome

[a]The subjectivity lies in the decision what to count as "adaptation", i.e. what to include in the database of adaptation projects

Table 18.3 Specific purpose: monitoring the integration of adaptation into planning (mainstreaming)

Approach: qualitative assessment based on interviews	
Prospect	To provide in-depth understanding (learning) of the achievements and shortcomings of the mainstreaming process
Potential use of M&E findings	Results of the assessment could be used to improve the mainstreaming process. The target audience includes those who carry out the mainstreaming process and those who can influence it
Description	A qualitative assessment of the mechanism and degree of integration of adaptation into planning processes (mainstreaming) and its effectiveness based on interviews with key informants involved in and/or affected by the implementation of the mainstreaming. Effectiveness can be assessed by the extent to which climate change impacts are taken into account in planning and decision making. A set of guiding questions may be used for interviews
Benefits and limitations	Qualitative assessments can offer a more in-depth understanding than quantitative indicators, particularly in regard to HOW and WHY things work or do not work. Depending on the perspective, number and composition of involved interviewees and on the exact assessment procedures the results may differ in their comprehensiveness and degree of subjectivity. Interviewees involved in the mainstreaming may be hesitant to discuss shortcomings of the process
Resources needed	Qualified interviewers. Know-how to develop the assessment details. Time and financial means to conduct a series of interviews
Example from practice	A study by GIZ (2016) examined the in-country coordination processes for national adaptation planning in Jamaica, Togo and Kenya through qualitative interviews. The results are meant to inform effective coordination mechanisms which facilitate the integration of adaptation into national planning and budgeting processes
Links	Preview of the study by GIZ (2016): http://www.napglobalnetwork.org/wp-content/uploads/2016/03/sNAPshot-Jamaica-1.pdf
Approach: quantitative indicators	
Prospect	To get quantitative expressions of the progress of integrating adaptation into development planning
Potential use of M&E findings	To track implementation and assess results for management and accountability purposes
Description	An indicator-based assessment of selected aspects of the mainstreaming process based on quantitative and/or qualitative information. The criteria for scoring, i.e. what needs to be achieved to get a certain indicator value, need to be clearly defined. This way, qualitative information can be converted into quantitative scores
Benefits and limitations	Quantitative indicators can provide a snapshot of the status quo of the mainstreaming process, albeit being limited to aspects which can be more easily quantified. Quantitative indicators are not well suited to get an in-depth understanding of how and why the mainstreaming process works and where the shortcomings are
Resources needed	Resource requirements largely depend on the efforts needed to gather the respective data and on the number of indicators. If the data can be collected with relative ease then resource needs can be lower than for qualitative assessments

(continued)

Table 18.3 (continued)

Examples from practice	The Climate Investment Funds' Pilot Program for Climate Resilience (PPCR) has operationalized the indicators "Degree of integration of climate change in national, including sector, planning" and the "Evidence of strengthened government capacity and coordination mechanism to mainstream climate resilience" through scorecards (Röhrer and Kouadio 2015). The indicators are specified through five sub-questions which are measured at national level against criteria to be defined by the national stakeholders
	To assess the development of mainstreaming capacity of line ministries executing the Government of Ethiopia's Climate-Resilient Green Economy (CRGE) strategy a participatory self-assessment approach was designed (Adler et al. 2015). An assessment matrix covering three aspects of mainstreaming (planning, staff awareness and skills as well as safeguards and equity) provides the scoring criteria. A qualified assessor and the interviewees jointly agree on the score for each component based on the assessment matrix
	IIED's Tracking Adaptation and Measuring Development (TAMD) framework suggests indicators for climate risk management (track 1) and for adaptation and development performance (track 2) based on a theory of change. A number of generic indicators for track 1 have been defined and can be assessed through scorecards (Brooks et al. 2014)
Links	The Climate Investment Fund's website on measuring results: http://www.climateinvestmentfunds.org/cif/measuring-results
	IIED's website on the Tracking Adaptation and Measuring Development (TAMD) framework: http://www.iied.org/tracking-adaptation-measuring-development-tamd
	Repository of adaptation indicators: examples from national monitoring and evaluation systems (Hammil et al. 2014b)

Table 18.4 Specific purpose: monitoring the implementation of National Adaptation Plan process (NAP process)

Approach: defining and monitoring milestones in the NAP process	
Background: The National Adaptation Plan (NAP) process was established by the parties to the UNFCCC to reduce vulnerability and integrate adaptation into policies and planning processes at all levels (UNFCCC 2011). The initial guidelines for the formulation of NAPs state that least developed country parties should "provide information in their national communications on the progress made and the effectiveness of the national adaptation plan process." (UNFCCC 2011, p. 86)	
Prospect	Knowing whether the NAP process in a particular country is advancing in accordance to predefined milestones or targets
Potential use of M&E findings	To track the implementation of the NAP process for management and accountability purposes
Description	Milestones or targets for the NAP process in a particular country are defined and their achievement monitored at agreed points in time. The milestones or targets need to be specific enough to enable an unambiguous assessment based on document analysis or interviews

(continued)

Table 18.4 (continued)

Approach: defining and monitoring milestones in the NAP process	
Benefits and limitations	Agreeing on milestones or targets for the NAP process can provide orientation for its implementation. Comparing actual progress with milestones does not directly provide an understanding of how and why the mainstreaming process works or not, but it can indicate the need for adjustments or further analysis
Resources needed	In general, resource requirements are low compared to other M&E approaches since some of the data is expected to be readily available from document analysis
Examples from practice	The Least Developed Countries Expert Group (LEG) has defined ten "Essential functions" that the NAP process should deliver to countries (UNFCCC 2013). The NAP process can subsequently be monitored on whether these functions are fulfilled in a given country. The LEG has developed a tool for this purpose ("PEG tool") which defines expected outcomes and a list of specific questions for each essential function
	The Stocktaking for National Adaptation Planning (SNAP) tool by GIZ (2014) defines seven success factors for the NAP process. Countries can assess their current and intended future level on these success factors. Progress over time can be illustrated in a radar chart (see GIZ 2014)
Links	Guidebook on the development of national adaptation M&E systems (Price-Kelly et al. 2015)
	Website of the Least Developed Countries Expert Group (LEG) where information on the PEG tool will be posted: http://unfccc.int/coopera tion_support/least_developed_countries_portal/ldc_expert_group/items/6110.php
	Information on the NAP process including the SNAP tool: https://gc21.giz.de/ibt/var/app/wp342deP/1443/index.php/knowledge/mainstreaming/

showcase detailed descriptions of selected M&E approaches. The complete version of the Adaptation M&E Navigator including descriptions of all M&E approaches is available on www.AdaptationCommunity.net under "Monitoring & Evaluation" (see below).

18.5   Using the Adaptation M&E Navigator

The Adaptation M&E Navigator is available as online tool on www. AdaptationCommunity.net under "Monitoring & Evaluation". Since early 2013, the knowledge portal AdaptationCommunity.net provides introductions to key topics, examples from practice, webinar recordings and publications on four focal topics including climate information, vulnerability assessment, mainstreaming and National Adaptation Planning as well as monitoring and evaluation. It is operated by GIZ (Deutsche Gesellschaft für Internationale Zusammenarbeit GmbH), the

German technical development cooperation agency on behalf of the Federal Ministry for the Environment, Nature Conservation, Building and Nuclear Safety (BMUB) as well as the Federal Ministry for Economic Cooperation and Development (BMZ). Recently the topics of ecosystem-based adaptation and private sector adaptation have been added to the site. The website has so far reached the highest amount of users during the UNFCCC Conferences of the Parties and currently peaks at more than 2,000 accesses per day. Hosting the Adaptation M&E Navigator on AdaptationCommunity.net not only ensures high accessibility and a relevant audience, but also enables updates of the tool as new experiences and publications become available.

18.6   Limitations

As a decision support tool, the Adaptation M&E Navigator must be concise, easy to navigate, understandable to non-experts and applicable to a broad variety of contexts. It is therefore facing a number of tradeoffs. First, it has to strike a careful balance between being concise and providing sufficient degree of detail. As shown in Fig. 18.1, the scope of the Adaptation M&E Navigator is limited to providing an overview of relevant approaches in form of a brief description. Additional guidance may be needed to design and implement a particular approach. Second, in order to keep the approaches to a manageable number they have to be applicable to a relatively broad context and cannot account for every possible variation. As a result, the ratings provided for the three criteria of complexity, subjectivity and available experience are indicative only and could deviate in practice depending on the details of implementation. Third, some of the specific M&E purposes are more suited to standardized M&E approaches than others. Practice has shown that national adaptation M&E systems developed to date are diverse and very context dependent (EEA 2015; Hammil et al. 2014a; Leiter 2013). Thus, whilst the Adaptation M&E Navigator can point to a direction in regard to a suitable M&E approach, the development of the actual M&E system may require a more complex process (considerations for developing national adaptation M&E systems are outlined in Leiter (2013) and Price-Kelly et al. (2015)).

Furthermore, whilst there was general agreement on the common M&E purposes featured in the Adaptation M&E Navigator, feedback by colleagues who commented on a draft version suggests that the purposes could be arranged in slightly different ways. For instance, if monitoring the implementation of projects (purpose #2) was broadened to include monitoring of adaptation plans, then monitoring the National Adaptation Plan process (purpose #3) could be grouped as a special case under it. Nevertheless, it was maintained as a separate item due to its importance for countries under the UNFCCC negotiations. Finally, as pointed out by Fisher et al. (2015, p. 30): "What makes a method most appropriate to climate change adaptation is not necessarily its intrinsic qualities, (. . .), but instead how the method is applied." Thus, the decision support provided by the Adaptation M&E Navigator is only part of the total process that leads to an effective application of M&E for adaptation to climate change (compare Fig. 18.1).

18.7   Conclusion

The Adaptation M&E Navigator is closing a gap in the existing landscape of guidebooks and tools for adaptation M&E. First, it provides a list of specific purposes for undertaking adaptation M&E in practice. In doing so it goes beyond the frequently stated general purposes like accountability and learning which, taken on their own, are not sufficient to decide upon particular M&E approaches. Secondly, the Adaptation M&E Navigator illustrates to decision makers the range of available options and equips them with the necessary information to select among those the most suitable one for their particular purpose. The Adaptation M&E Navigator is hosted on an established online platform (www.AdaptationCommunity.net) in the form of an easy to use web interface. By drawing upon adaptation M&E approaches and examples available to date, the Adaptation M&E Navigator also demonstrates the progress which has been made in this subject area since the first International Conference on Evaluating Climate Change and Development took place in 2009.

References

Adler, R., Wilson, K., Abbot, P., & Blackshaw, U. (2015). Approach to monitoring and evaluation of institutional capacity for adaptation to climate change: The case of the United Kingdom's investment to Ethiopia's climate-resilient green economy. In D. Bours, C. McGinn, & P. Pringle (Eds.), *Monitoring and evaluation of climate change adaptation: A review of the landscape. New Directions for Evaluation, 147,* 61–74.

Bours, D., McGinn, C., & Pringle, P. (2014a). *Monitoring & evaluation for climate change adaptation and resilience: A synthesis of tools, frameworks and approaches* (2nd ed.). Phnom Penh: SEA Change Community of Practice, and Oxford UKCIP. Retrieved from: http://www.seachangecop.org/node/3258

Bours, D., McGinn, C., & Pringle, P. (2014b). *Guidance note 1: Twelve reasons why climate change adaptation M&E is challenging.* Phnom Penh: SEA Change CoP, and Oxford, United Kingdom: UKCIP. Retrieved from http://www.ukcip.org.uk/wp-content/PDFs/MandE-Guidance-Note1.pdf.

Brooks, N., Anderson, S., Ayers, J., Burton, I., & Tellam, I. (2011). *Tracking adaptation and measuring development (TAMD) (Climate ChangeWorking Paper No. 1).* London: International Institute for Environmental Development (IIED). Retrieved from http://pubs.iied.org/pdfs/10031IIED.pdf

Brooks, N., Fisher, S., Rai, N., Anderson, S., Karani, I., Levine, T., & Steinbach, D. (2014). *Tracking adaptation and measuring development. A step-by-step guide.* London: International Institute for Environmental Development (IIED). Retrieved from http://pubs.iied.org/10100IIED.html

CARE. (2012). *Participatory monitoring, evaluation, reflection and learning for community-based adaptation: A manual for local practitioners.* CARE International and IIED. Retrieved from: http://www.care.org/sites/default/files/documents/CC-2012-CARE_PMERL_Manual_2012.pdf

Chen, S., & Uitto, J. I. (2014). Small grants, big impacts. Aggregation challenges. In J. I. Uitto (Ed.), *Evaluating environment in international development* (pp. 105–122). New York: Routledge.

European Environment Agency (EEA). (2015). *National monitoring, reporting and evaluation of climate change adaptation in Europe.* Copenhagen: European Environment Agency (EEA).

Fisher, S., Dinshaw, A., McGray, H., Rai, N., & Schaar, J. (2015). Evaluating climate change adaptation: Learning from methods in international development. In D. Bours, C. McGinn, & P. Pringle (Eds.), *Monitoring and evaluation of climate change adaptation: A review of the landscape. New Directions for Evaluation, 147*, 13–35.

Ford, J. D., Berrang-Ford, L., Lesnikowski, A., Barrera, M., & Heymann, S. J. (2013). How to track adaptation to climate change: A typology of approaches for national-level application. *Ecology and Society, 18*(3), 40.

Fritzsche, K., Schneiderbauer, S., Bubeck, P., Kienberger, S., Buth, M., Zebisch, M., & Kahlenborn, W. (2014). *The vulnerability sourcebook. Concept and guidelines for standardised vulnerability assessments*. Eschborn: Deutsche Gesellschaft für Internationale Zusammenarbeit (GIZ) GmbH. Retrieved from http://www.adaptationcommunity.net/knowl edge/vulnerability-assessment/vulnerability-sourcebook/

GIZ. (2014). *The stocktaking for National Adaptation Planning (SNAP) tool*. Eschborn: Deutsche Gesellschaft für Internationale Zusammenarbeit (GIZ) GmbH. Retrieved from http://www. adaptationcommunity.net/?wpfb_dl=148

GIZ. (2016). *Institutional context analyses for national climate adaptation planning processes: Lessons from Jamaica, Kenya and Togo*. Eschborn: Deutsche Gesellschaft für Internationale Zusammenarbeit (GIZ) GmbH.

Hammil, A., Dekens, J., Olivier, J., Leiter, T., & Klockemann, L. (2014a). *Monitoring and evaluating adaptation at aggregated levels: A comparative analysis of ten systems*. Eschborn: Deutsche Gesellschaft für Internationale Zusammenarbeit (GIZ) GmbH. Retrieved from http:// www.adaptationcommunity.net/knowledge/monitoring-evaluation-2/national-level-adaptation- me/overview-and-analysis-of-10-nationaladaptation-systems/

Hammil, A., Dekens, J., Leiter, T., Olivier, J., & Klockemann, L. (2014b). *Repository of adaptation indicators. Real case examples from national monitoring and evaluation systems*. Eschborn: Deutsche Gesellschaft für Internationale Zusammenarbeit (GIZ) GmbH. Retrieved from   http://www.adaptationcommunity.net/knowledge/monitoring-evaluation-2/national- level-adaptation-me/repository-of-adaptation-ndicators/

Harley, M., Horrocks, L., Hodgson, N., & Van Minnen, J. (2008). *Climate change vulnerability and adaptation indicators* (ETC/ACC Technical Paper 2008/9). Bilthoven: European Topic Centre on Air and Climate Change, European Environmental Agency.

Leiter, T. (2013). *Recommendations for adaptation M&E in practice*. Discussion paper. Eschborn: Deutsche Gesellschaft für Internationale Zusammenarbeit (GIZ) GmbH. Retrieved from: http://www.adaptationcommunity.net/?wpfb_dl=132

Leiter, T. (2015). Linking monitoring and evaluation of adaptation to climate change across scales: Avenues and practical approaches. In D. Bours, C. McGinn, & P. Pringle (Eds.), *Monitoring and evaluation of climate change adaptation: A review of the landscape. New Directions for Evaluation, 147*, 117–127.

Mimura, N., Pulwarty, R. S., Duc, D. M., Elshinnawy, I., Redsteer, M. H., Huang, H. Q., Nkem, J. N., & Sanchez Rodriguez, R. A. (2014). Adaptation planning and implementation. In C. B. Field, V. R. Barros, D. J. Dokken, K. J. Mach, M. D. Mastrandrea, T. E. Bilir, M. Chatterjee, K. L. Ebi, Y. O. Estrada, R. C. Genova, B. Girma, E. S. Kissel, A. N. Levy, S. MacCracken, P. R. Mastrandrea, & L. L. White (Eds.), *Climate change 2014: Impacts, adaptation, and vulnerability. Part A: Global and sectoral aspects. Contribution of Working Group II to the Fifth Assessment Report of the Intergovernmental Panel on Climate Change*. Cambridge: Cambridge University Press.

Nachmany, M., Frankhauser, S., Davidova, J., Kingsmill, N., Landesman, T., Roppongi, H., ...Townshend, T. (2015). *The 2015 global climate legislation study. A review of climate change legislation in 99 countries*. London: Grantham Research Institute on Climate Change and the Environment.

OECD Development Assistance Committee (OECD-DAC). (2015). *Climate-related development finance in 2013*. Retrieved from: http://www.oecd.org/dac/environment-development/Climate- related%20development%20finance_June%202015.pdf

Olivier, J., Leiter, T., & Linke, J. (2013). *Adaptation made to measure: A guidebook to the design and results-based monitoring of climate change adaptation projects* (2nd ed.). Eschborn:

Deutsche Gesellschaft für Internationale Zusammenarbeit (GIZ) GmbH. Retrieved from http://www.adaptationcommunity.net/knowledge/monitoring-evaluation-2/project-level-adaptation-me-2/adaptation-made-to-measure/

Price-Kelly, H., Leiter, T., Olivier, J., & Hammil, A. (2015). *Developing national adaptation monitoring and evaluation systems: A guidebook.* Eschborn: Deutsche Gesellschaft für Internationale Zusammenarbeit (GIZ) GmbH. Retrieved from http://www.adaptationcommunity.net/knowledge/monitoring-evaluation-2/national-level-adaptation-me/developing-national-adaptation-me-systems/

Pringle, P. (2011). *AdaptME toolkit: Adaptation monitoring and evaluation.* Oxford, UK: UKCIP. Retrieved from: http://www.ukcip.org.uk/wp-content/PDFs/UKCIP-AdaptME.pdf

PROVIA. (2013). *PROVIA guidance on assessing vulnerability, impacts and adaptation to climate change.* Published by the United Nations Environment Programme (UNEP). Retrieved from: http://www.unep.org/provia/RESOURCES/Publications/PROVIAGuidancereport/tabid/130752/Default.aspx

Röhrer, C., & Kouadio, K. E. (2015). Monitoring, reporting, and evidence-based learning in the climate investment fund's pilot program for climate resilience. In D. Bours, C. McGinn, & P. Pringle (Eds.), *Monitoring and evaluation of climate change adaptation: A review of the landscape. New Directions for Evaluation,* 147, 129–145.

Spearman, M., & McGray, H. (2011*). Making adaptation count: Concepts and options for monitoring and evaluation of climate change adaptation.* Eschborn: Gesellschaft für Internationale Zusammenarbeit (GIZ) GmbH. Retrieved from http://pdf.wri.org/making_adaptation_count.pdf

Terpstra, P., & Peterson-Carvalho, A. (2015). *Tracking adaptation finance. An approach for civil society organizations to improve accountability for climate change.* Washington, DC: World Resources Institute, and Oxfam America. Retrieved from: http://www.wri.org/publication/tracking-adaptation-finance

United Nations Framework Convention on Climate Change (UNFCCC). (2011). *Report of the conference of the parties on its seventeenth session* (Addendum Part Two: Action taken by the Conference of the Parties at its seventeenth session. FCCC/CP/2011/9/Add.1). Bonn: United Nations Framework Convention on Climate Change.

United Nations Framework Convention on Climate Change (UNFCCC). (2013). *Report on the 24th meeting of the Least Developed Countries Expert Group (FCCC/SBI/2013/15).* Bonn: United Nations Framework Convention on Climate Change.

Welle, T., Witting, M., Birkmann, J., & Brossmann, M. (2014). *Assessing and monitoring climate resilience. From theoretical considerations to practically applicable tools – a discussion paper.* Eschborn: Deutsche Gesellschaft für Internationale Zusammenarbeit (GIZ) GmbH. Retrieved from http://www.adaptationcommunity.net/knowledge/monitoring-evaluation-2/national-level-adaptation-me/ssessing-and-monitoring-climate-resilience/

About the Authors

Aryanie Amellina is a climate change and development professional with degrees in environmental engineering and development studies. She worked with the National Council of Climate Change and Indonesia Joint Crediting Mechanism (JCM) Secretariat on the JCM implementation focusing on MRV structure and process including methodology development, validation and verification, project assessment and evaluation, as well as stakeholders' engagement. She currently focuses on the JCM implementation and market mechanisms in Asian countries, including MRV systems and capacity development, as a Climate and Energy Policy Researcher at the Institute for Global Environmental Studies.

Babou André Bationo is a Senior Forestry Biology and Ecology Scientist. He holds a PhD from University of Ouagadougou (Burkina Faso). He is working for Agriculture and Environmental research Institute (INERA) of Burkina Faso. He is an associate researcher and the Focal point of ICRAF in Burkina Faso. His researches include participatory regeneration and ecology of agroforestry tree species in the agro-systems.

Emilia Bretan is a Brazilian independent consultant in planning, management, monitoring and evaluation of human rights and local and international development initiatives. She holds a BA, MS and a Phd in law from the University of São Paulo, Brazil, and has provided services for The World Bank, UNDP, Canadian DFAT-D and Plan International Brazil, among others. She is a trainer in Outcome Mapping and Outcome Harvesting approaches, an active member of IDEAS (leader of the Climate ITIG) and of the Outcome Mapping Learning Community. IPDET 2014 graduate.

Saaka Buah is an agronomist and soil scientist with a PhD in Soil Fertility and Plant Nutrition from Iowa State University, Ames, Iowa, USA. He is currently working with CSIR-SARI, Ghana where his research activities focus on solving agricultural production problems in the savanna zone of Ghana. He also provides

© The Author(s) 2017
J.I. Uitto et al. (eds.), *Evaluating Climate Change Action for Sustainable Development*, DOI 10.1007/978-3-319-43702-6

technical assistance to increase the availability of appropriate and affordable Integrated Soil Fertility Management technologies to sustainably improve agricultural productivity in northern Ghana. Also he is currently the head of the farming system research team based in the Upper West region of Ghana.

Lee Cando-Noordhuizen joined the GEF Independent Evaluation Office as a consultant in March 2015. She has worked for various UN agencies including the Executive Office of the UN Secretary-General, UNDP, UNICEF and UNFCCC in the fields of climate change (mitigation, adaptation, finance, development impacts) and sustainable development. Recently, Lee was posted in Indonesia where she worked as a Climate Change Communications Coordinator for the Center for International Forestry Research (CIFOR), and as a Senior Advisor to the German International Cooperation's (GIZ) ASEAN-German Programme on Response to Climate Change.

Michael Carbon has joined the Evaluation Office of UNEP in Nairobi in April 2010, after working as an Evaluation Officer for IFAD in Rome since 2005. Michael led and contributed to a great variety of ex-post evaluations, including evaluations of rural development projects, environmental projects, country programmes and cross-cutting themes. Before he became an evaluation professional, Michael worked in northern Vietnam, as a technical assistant and later as a project coordinator in the field of rural extension and farmer organization, and conducted agri-systems research in Guinée-Conarky, El Salvador and Cuba. On 01 February 2016, Michael returned to the Independent Office of Evaluation of IFAD as a Senior Evaluation Officer. Michael holds a MSc in Bio-engineering, specializing in Forestry, Nature Conservation and Tropical Agriculture from KUL (Leuven, Belgium), an Engineering Diploma in Tropical Agronomy from CNEARC (Montpellier, France) and a PhD Preparatory Studies Degree in Geography and Development Practice from INA-PG (Paris).

Joanne Chong is a Research Director at the Institute for Sustainable Futures, University of Technology Sydney. Joanne leads collaborative, transdisciplinary research projects that focus on improving people's resilience to drought, extreme events and other impacts of climate change. Joanne's research addresses policy, institutional arrangements, planning and incentives in the water resources sector and for cross-sectoral local and regional development. As a monitoring and evaluation specialist, Joanne works in partnership with communities, businesses, governments, civil society and donors to investigate and apply practice learnings at local to global scales. Joanne holds qualifications in economics, engineering and environmental law.

Monika Egger Kissling was educated as an economist at the University of St. Gallen, Switzerland. From 1984–1996 researcher at the Institute for Development Studies at the University of Geneva, Switzerland. More than 15 years of experience in evaluation and consultancy in development cooperation as an independent evaluator. Since 2010 employee in the Federal Department of Foreign Affairs (FDFA) as Program Officer Evaluation at the Swiss Agency for Development and Cooperation (SDC) in charge for the conceptualization, mandating and process management of complex thematic and institutional strategic evaluations. 2013–2014 responsible for the production of the Report on Effectiveness of Swiss International Cooperation in Climate Change 2000–2012.

Nathan L. Engle is Climate Change Specialist for the Climate Policy Team at the World Bank. He is involved with efforts to integrate climate resilience into World Bank operations and is helping to lead projects on adaptation, drought policy and management, and water resources. Nate completed his Ph.D. at the University of Michigan's School of Natural Resources & Environment, and also earned masters' degrees at the University of Michigan in public policy (MPP) and natural resources and environment (MSc).

Wiebke Förch works with the CGIAR Research Program on Climate Change, Agriculture and Food Security (CCAFS) as Science Officer, based at the International Livestock Research Institute (ILRI) in Nairobi, Kenya. She helped coordinating the shift towards a learning-based impact pathway and results-based management approach for CCAFS, and is supporting the overall ME&L of the program. She holds a PhD from the Arid Lands Resource Sciences Interdisciplinary Program at the University of Arizona and a BSc(Hons) in Agriculture from the University of Reading. Her professional experience and research interests revolve around food security, resilience and vulnerability, global change impacts on society and the enabling environments, governance and institutions to support communities in adapting to climate change. She is also interested in research for development processes and transformative learning, how to work with theories of change and impact pathways to increase the effectiveness of research for development.

Yann François is an Engineer specialized in natural resources management, climate change adaptation and mitigation projects and climate finance mechanisms. He has occupied various management and technical positions in the international NGO GERES which allowed him to develop skills in project development, implementation and evaluation in the fields of sustainable forest management, climate-smart agriculture and energy in West Africa, Central Asia and Southeast Asia. As a Climate and Development Technical Advisor is supporting evidence-based projects and policy development, collaborating with political decision-makers on Nationally Appropriate Mitigation Actions (NAMA) and REDD+ strategies as well as supporting the long-term financing of projects through carbon financing under the UNFCCC CDM and Gold Standard mechanisms.

Marina Gavaldão is a Forestry Engineer specialist on ecosystems services and climate finance. Since 2003, Marina has gathered experience in the design, implementation and the coordination of climate change mitigation projects and environmental economics-related research. As Technical Director of the Climate Change

Unit of the NGO GERES she has been responsible for the technical coordination of climate-related projects including designing and coordinating adaptation and mitigation projects, notably carbon projects in West Africa, Cambodia and Afghanistan. Additionally, she coordinated researches on climate finance mechanisms, such as the suppressed demand working-group and on standardized baselines for the CDM. Marina also contributes as technical advisor to the development and ground implementation of LEDS and for the improvement of carbon standards.

Anna Gero is a Senior Research Consultant at the Institute for Sustainable Futures, University of Technology Sydney. Anna's research focuses on the intersection between climate change adaptation, disaster risk management and development in the Asia-Pacific region. Anna's work includes evaluating and developing national climate change adaptation policy and strategies and conducting participatory climate change vulnerability and capacity assessments. Anna holds degrees in atmospheric science and environmental management.

Jasmine Hyman is an Economic Affairs Officer at the United Nations Department of Economic and Social Affairs. She is also currently completing a doctorate at the Yale School of Forestry and the Environment, where she seeks to identify design principles for global climate finance schemes that promote equitable development and social justice. Prior to her research at Yale, Jasmine helped build the Gold Standard Foundation, a certification scheme for greenhouse gas emission reduction projects under the Kyoto Protocol's offset scheme and for the voluntary carbon offset markets in the US and Australia.

Irene Karani is a Director of LTS Africa. She has over 20 years in climate change related issues, natural resource management and rural livelihoods with communities in Africa. is highly experienced in the design, management and implementation and evaluation of projects/programmes related to climate change adaptation, conservation and general environmental management, food security, water and sanitation, pastoralism, gender mainstreaming, capacity building, drought cycle management, eco-tourism and poverty alleviation. Irene has very strong monitoring and evaluation skills having worked to design project and programme M&E systems at community, sub-national, national and regional levels.

Nyachomba Kariuki is a consultant with LTS Africa. She has over 4 years, experience, in testing the TAMD framework in Isiolo and is highly experienced in participatory approaches to M&E through training, analysis and documentation of theories of change, development of indicators and assumptions at community level.

Timo Leiter is an Advisor for adaptation to climate change at German Development Cooperation's (GIZ) Competence Centre for Climate Change. Working with government and non-government partners he is developing methods and approaches to plan and implement adaptation from national to local level. His experience covers the whole adaptation cycle from climate information over vulnerability assessments to mainstreaming and monitoring and evaluation. Mr. Leiter is specialized in monitoring and evaluation of adaptation and has authored several publications on the topic in academic journals such as *New Directions for Evaluation* as well as guidebooks for practitioners including GIZ's "Adaptation made to measure".

Debora Ley has a PhD in Geography and the Environment from the University of Oxford. Her dissertation topic analyzed if rural renewable energy projects can simultaneously meet the multiple goals of sustainable development, and climate change mitigation and adaptation, and if so, under what conditions. She has a MSc in Civil Engineering from the University of Colorado at Boulder and a BSc in Electromechanical Engineering. She has worked at Mexico's National Energy Savings Commission, Sandia National Laboratories, the UNECLAC, and most recently with TetraTech as a Renewable Energy Expert under Central America's Regional Clean Energy Initiative. She is a volunteer with Engineers Without Borders.

Takaaki Miyaguchi, PhD, is Associate Professor at Ritsumeikan University's College of International Relations (Kyoto, Japan). His research topics include the role of the private sector in climate change, and evaluation of donor-funded climate change mitigation and adaptation projects. Prior to joining the university in 2012, he has worked in UNDP and other UN agencies for 8 years, developing and managing both climate change mitigation and adaptation projects in more than 20 countries.

Neeraj Kumar Negi is a public affairs professional by training. He has been working at the GEF Independent Evaluation Office since 2005 in different capacities. Presently, he works at the Office as a Senior Evaluation Officer. He has conducted evaluations on topics such as the GEF system for resource allocation; impact of climate change mitigation activities in emerging economies; accreditation process for the expansion of GEF partnership; and, the GEF India country portfolio. He has also been involved in evaluations on incremental costs, co-financing, project cycle, and the GEF Small Grants Program. He supervises preparation of the annual performance report of the GEF. Neeraj also worked at Seva Mandir, an NGO based in Western India, from 1998 to 2003. During this engagement he planned and implemented projects that focused on community-based natural resource management, agroforestry, rural livelihoods, village institutions, and drought relief. During 2002–03, he also had a yearlong co-terminus appointment at the Poverty Action Lab and coordinated field implementation of a survey on health status and health care delivery. Neeraj holds a Master's Degree in Public Affairs from the Woodrow Wilson School of Public and International Affairs, Princeton University, with specialization in Economics (2003–05). He also holds a Post Graduate Diploma in Forestry Management (PGDFM) from the Indian Institute of Forest Management (IIFM) (1996–98).

Jyotsna (Jo) Puri is currently Deputy Executive Director and Head of Evaluation at the International Initiative of Impact Evaluation (3ie). Jo is also adjunct faculty at the School of International and Public Affairs (SIPA), Columbia University, New York, where she teaches development evaluation. She is currently advising IFAD, WFP and AGRA on impact evaluations of programmes related to food security and nutrition. Jo leads 3ie sector based work on themes such as social norms, agriculture, nutrition, community engagement and environment. Jo has more than 18 years of experience in policy research and development evaluation. She has undertaken and led evaluation related work for UNDP, UNICEF, GEF and the MacArthur Foundation. Her research has focused on analyzing poverty impacts of policy and infrastructure investments in Asia and Latin America. Her other areas of work include examining impacts of policies in the areas of environment, agriculture, health and climate change. As policy adviser at UNEP she has provided thematic and strategic advice on program development and engaging governments at various levels for effective delivery of outcomes for equitable, growth transitions. She is the lead author of a book on measuring and interpreting monitoring and evaluation indicators prepared for the Human Development Report Office and published by UNDP; Co-author of a book examining implications of Joint Implementation of Climate Change commitments for developing countries and led the publication of a synthesis report on Forests in a Green Economy published by UNEP. She sits on the board of Community of Evaluators, South Asia and the Humanitarian Certification Initiative. Jo's academic qualifications include a Ph.D. and M.Sc. in Resource Economics and a Masters in Development Economics.

Issa Sawadogo is animal and pasture scientist with a doctoral degree in agropastorlism for the Museum National d'Histoire Naturelle, France. He is a project officer at the International union for conservation of nature, program of Burkina. His research work focuses on adaptation to climate change monitoring and evaluation, pasture management and breeders' practices.

Tonya Schuetz has 15 years of experience in change and results-based management, monitoring and evaluation, personnel and capacity development: five years in the private sector focusing on process analysis and resource optimization, and ten years in research for development projects and programs in over 20 countries in Sub-Sahara Africa and Asia with substantial field experience on the project and policy influencing level. She has worked across a range of sectors, including agriculture, water, health and education for organizations like IFAD, CGIAR Centers and Research Programs. Her cross-cutting experience includes institutional analysis and assessment of their capacity, quality management, knowledge and innovation management with hands-on skills in project/ program design, monitoring and evaluation of research projects for outcomes and impact, institutional and organizational learning, conceptualization of adult learning. Currently, she works as an independent consultant with different CGIAR Research Programs, like Climate Change Agriculture Food Security, on results-based management, monitoring, evaluation, learning and process facilitation, and with private sector companies to strengthen their sustainability development.

Jacques Somda is a senior regional program officer with the International Union for conservation of nature, Central and West Africa program. He holds a doctoral degree in rural economics for the University of Cocody, Côte d'Ivoire. His research work focuses on project/program monitoring and evaluation, climate change M&E, environmental economics, technology adoption and policy analysis.

Philip Thornton leads the flagship on "Policies and institutions for climate-resilient food systems" of the CGIAR Research Programme on Climate Change, Agriculture and Food Security (CCAFS). He is hosted at the International Livestock Research Institute (ILRI) in Nairobi, Kenya, and based in Edinburgh. He is an Honorary Research Fellow in the School of Geosciences at the University of Edinburgh. He has a PhD in Farm Management and Agricultural Economics from Lincoln College, New Zealand. His research interests include integrated modelling at different scales and evaluating climate change impacts and adaptation options in smallholder farming systems in the tropics and subtropics. He has contributed to several global assessments including the IPCC's Fourth and Fifth Assessments.

Abasse Tougiani is a Senior Scientist at National Agricultural Research Institute of Niger (INRAN) and he is a ICRAF focal point in Niamey, Niger. He holds a PhD in Biology and Silviculture from the University of Ibadan, Ibadan, Nigeria. His research interests include Agro forestry, Climate Change, Iintegration of Agriculture Livestock and Tree species production system and cultivation of high-value indigenous tree species.

Pia Treichel is the Program Manager for Climate Change Adaptation at Plan International Australia. With over 9 years' experience working on climate change with a focus ranging from local government to the United Nations, Pia currently works on child- and youth-centred climate change with Plan International. Pia has a Master of Environment degree from Melbourne University.

Juha I. Uitto is Director of the GEF Independent Evaluation Office. He has previously held positions as Deputy Director and Evaluation Adviser in the Independent Evaluation Office of the United Nations Development Programme (UNDP) and as Monitoring and Evaluation Coordinator/Specialist with the GEF. Since the late-1990s, he has conducted and managed a large number of programmatic and thematic evaluations of international cooperation at the global, regional and country levels, in particular related to environmental management and poverty-environment linkages. Dr. Uitto spent the 1990s with the United Nations University coordinating the university's environment and sustainable development research and training programs. His earlier work included positions in the Nordic Africa Institute, the International Fund for Agricultural Development and as consultant in development cooperation. He has had visiting positions in the Graduate School of Global

Environmental Studies, Kyoto University in Japan; Division for Global Affairs, Rutgers—The State University of New Jersey; and the International Studies Program of the University of Montana, Missoula. He was educated at the Universities of Helsinki and Lund, and holds a PhD in Social and Economic Geography. He has authored/edited several books and published more than 30 peer reviewed articles and book chapters on topics related to the environment, natural resources management, environmental hazards, and evaluation. His book Evaluating Environment in International Development was published by Routledge-Earthscan in May 2014. In October 2012, the European Evaluation Society awarded his paper on environmental evaluation for a "distinguished contribution to evaluation practice."

Rob D. van den Berg President, International Development Evaluation Association and Visiting Professor, International Development Institute, King's College (London), is a graduate of the State University of Groningen in the Netherlands. He is also a Visiting Fellow at the Centre for Development Impact at the Institute of Social Studies in Brighton, UK. He is a member of the Monitoring Advisory Group of the Global Partnership for Effective Development Cooperation. From 2004 to 2014 he was Director of the Independent Evaluation Office of the Global Environment Facility in Washington, DC and before that Director of Evaluation of the Dutch Ministry of Foreign Affairs from 1999 to 2004. He has been active in the DAC Evaluation Network (Chair from 2002–2004), the UN Evaluation Group and the Evaluation Cooperation Group of the International Financial Institutions. Since 2015 he is providing independent evaluation advice to development agencies.

Ioannis Vasileiou is an Agricultural Specialist, with the Climate-Smart Agriculture (CSA) Upscaling Team, at the World Bank, supported by CCAFS and IFPRI. Ioannis also serves as a Science Officer on Policies & Institutions for the CGIAR Program on Climate Change, Agriculture and Food Security (CCAFS). His current work focuses on issues related to global change and resilience, as well as enabling governance, institutions and financing mechanisms for agriculture. He has previously worked on sustainability issues in the European Commission, the United Nations, and the non-profit sector. He holds a Master of International Affairs from Columbia University, where he specialized in Economic and Political Development, as well as a joint Msc-MBA in Industrial Management, focusing on Energy and Environment, from the National Technical University of Athens and the University of Piraeus.

Roman Windisch is the Deputy Country Director of Swiss State Secretariat for Economic Affairs SECO in Vietnam. He is responsible for Switzerland's economic development cooperation program on infrastructure financing and private sector development. Before, he worked for several years at headquarters as an evaluation officer in SECO's quality assurance division. He holds a Master degree in Geography from the University of Zurich and a Diploma on Sustainable Development from De Montfort University Leicester.

Aaron Zazueta is a social anthropologist and was until recently the Chief Evaluation Officer and the team leader on impact evaluation for the GEF. He joined the GEF in 2002 and was responsible for the GEF Annual Performance Report. He has carried out evaluations on international waters, climate change and biodiversity in East and Southeast Asia, Eastern Europe, Africa and Latin America. Prior to joining the GEF, he was part of the team that evaluated the World Bank's 1991 Forest Policy and the application of the Bank's Indigenous People's Safeguards and conducted project evaluations for the World Bank. He was also the lead advisor for the Inter-American Development Bank Strategy for Citizen Participation. Between 1990 and 1996, he was Regional Director for Latin America at the World Resources Institute, where he worked with governments, NGOs and community groups on the formulation of environmental policies in Latin America. He also served as a strategic and policy advisor to several governments in Latin America. His previous publications are on issues related to environmental policy, citizen participation and sustainable development. He holds a Ph. D in Anthropology from the University of California.

Robert Zougmoré is an agronomist and soil scientist with a PhD in Production Ecology & Resources Conservation, University of Wageningen. He is based at ICRISAT Bamako where he currently leads the CGIAR research program on climate change, agriculture and food security (CCAFS) in West Africa. His work focuses on the development of climate-smart agriculture technologies, practices, institutions and policies for better climate risk management in West Africa. Before joining CCAFS, he was a senior staff within the Sahara & Sahel Observatory (Tunisia) where he was focused on: (1) Desertification, land Degradation and Drought (DLDD) including environmental surveillance, monitoring and evaluation of DLDD, drought early warning, etc.; (2) climate change adaptation in Africa (analyzing adaptation strategies of vulnerable populations in arid and semi-arid zones, etc.). He has published more than 50 papers and book chapters on soil erosion, integrated soil, water and nutrient management options and their economic benefits, and climate change.

Index

© The Author(s) 2017
J.I. Uitto et al. (eds.), *Evaluating Climate Change Action for Sustainable
Development*, DOI 10.1007/978-3-319-43702-6